METHODS IN MOLECULAR BIOLOGY

Series Editor
John M. Walker
School of Life and Medical Sciences
University of Hertfordshire
Hatfield, Hertfordshire, UK

For further volumes:
http://www.springer.com/series/7651

For over 35 years, biological scientists have come to rely on the research protocols and methodologies in the critically acclaimed *Methods in Molecular Biology* series. The series was the first to introduce the step-by-step protocols approach that has become the standard in all biomedical protocol publishing. Each protocol is provided in readily-reproducible step-by-step fashion, opening with an introductory overview, a list of the materials and reagents needed to complete the experiment, and followed by a detailed procedure that is supported with a helpful notes section offering tips and tricks of the trade as well as troubleshooting advice. These hallmark features were introduced by series editor Dr. John Walker and constitute the key ingredient in each and every volume of the *Methods in Molecular Biology* series. Tested and trusted, comprehensive and reliable, all protocols from the series are indexed in PubMed.

Candida auris

Methods and Protocols

Edited by

Alexander Lorenz

Institute of Medical Sciences (IMS), University of Aberdeen, Aberdeen, UK

Humana Press

Editor
Alexander Lorenz
Institute of Medical Sciences (IMS)
University of Aberdeen
Aberdeen, UK

ISSN 1064-3745 ISSN 1940-6029 (electronic)
Methods in Molecular Biology
ISBN 978-1-0716-2416-6 ISBN 978-1-0716-2417-3 (eBook)
https://doi.org/10.1007/978-1-0716-2417-3

Cover Caption: Transmission electron micrograph of Candida auris B11221. Image Credit: Chloe Pelletier.

This Humana imprint is published by the registered company Springer Science+Business Media, LLC, part of Springer Nature.
The registered company address is: 1 New York Plaza, New York, NY 10004, U.S.A.

Preface

This *Methods in Molecular Biology* volume covers experimental protocols for the recently emerged fungal pathogen of humans, *Candida auris*. This yeast has shown unique features, which make it a particularly worrisome threat in healthcare settings worldwide. It easily acquires resistance to (multiple) antifungal drugs, and has a propensity to stick to human skin and to survive on smooth surfaces of hospital equipment and furniture for a prolonged time. As *C. auris* has only been identified and described in 2009, we are still largely ignorant about many fundamental aspects of its biology. I think that this book highlights the technical challenges ahead in elucidating the traits of this newly emerged pathogen underpinning its antifungal resistance, virulence, and pathogenicity. At the same time, this volume will hopefully motivate and enable clinical and basic researchers to turn their attention to *C. auris* by providing detailed experimental protocols to tackle this important source of hospital-acquired infections. The protocols presented in this volume cover the identification by PCR and mass spectrometry, and molecular typing to identify clade affiliation, determining resistance to antifungal drugs, genetic manipulation, bioinformatic genome analysis, and phenotypic characterization using various biochemical, mass spectrometry, microscopy, and flow cytometry approaches, as well as methods to elucidate host-pathogen interactions. I am very much indebted to the many colleagues who willingly gave their time to write the chapters for this volume which, I hope, provides a useful compendium of research tools to delve into the biology of *C. auris*.

Aberdeen, UK *Alexander Lorenz*

Contents

Part IV Phenotypic and Biochemical Characterization of *Candida auris*

Part V Bioinformatic Methods for *Candida auris*

Part VI Exploring Host-Pathogen Interactions of *Candida auris*

Contributors

ALIREZA ABDOLRASOULI • *Department of Medical Microbiology, King's College Hospital, London, UK; Department of Infectious Diseases, Imperial College London, London, UK*
NATHANIEL D. ALBERT • *Department of Infectious Diseases, Infection Control and Employee Health, The University of Texas MD Anderson Cancer Center, Houston, TX, USA*
LYSANGELA R. ALVES • *Instituto Carlos Chagas, Fundacao Oswaldo Cruz (Fiocruz), Curitiba, Brazil, Curitiba, Brazil*
UJWAL R. BAGAL • *Mycotic Diseases Branch, Centers for Disease Control and Prevention, Atlanta, GA, USA*
ATANU BANERJEE • *Amity Institute of Biotechnology and Amity Institute of Integrative Sciences and Health, Amity University Haryana, Gurugram, India*
ANDREW M. BORMAN • *UK National Mycology Reference Laboratory, Public Health England South-West Regional Laboratory, Southmead Hospital, Bristol, UK; MRC Centre for Medical Mycology, University of Exeter, Exeter, UK*
JEAN-PHILIPPE BOUCHARA • *Univ Angers, Univ Brest, IRF, SFR ICAT, Angers, France*
GUSTAVO BRAVO RUIZ • *Institute of Medical Sciences (IMS), University of Aberdeen, Aberdeen, UK; Wellcome Centre for Anti-Infectives Research, School of Life Sciences, University of Dundee, Dundee, UK*
JASON L. BROWN • *Oral Sciences Research Group, College of Medical, Veterinary and Life Sciences, Glasgow University, Glasgow, UK; Glasgow Biofilm Research Network, Glasgow, UK*
GEMA BRU • *Genetic PCR Solutions™, Orihuela, Alicante, Spain*
J. Z. ALEX CHEONG • *Department of Medicine, University of Wisconsin, Madison, WI, USA; Department of Medical Microbiology and Immunology, University of Wisconsin, Madison, WI, USA*
NANCY A. CHOW • *Mycotic Diseases Branch, Centers for Disease Control and Prevention, Atlanta, GA, USA*
ANURADHA CHOWDHARY • *Medical Mycology Unit, Department of Microbiology, Vallabhbhai Patel Chest Institute, University of Delhi, Delhi, India*
DORA E. CORZO-LEÓN • *Institute of Medical Sciences, School of Medicine, Medical Sciences and Nutrition, University of Aberdeen, Foresterhill, Aberdeen, UK; Unidad de Epidemiología Hospitalaria e Infectologia, Instituto Nacional de Enfermedades Respiratorias, Mexico City, Mexico*
VINCENT COURDAVAULT • *Université de Tours, EA2106 Biomolécules et Biotechnologies Végé tales, Tours, France*
LEAH E. COWEN • *Department of Molecular Genetics, University of Toronto, Toronto, ON, Canada*
CHRISTINA A. CUOMO • *Infectious Disease and Microbiome Program, Broad Institute of MIT and Harvard, Cambridge, MA, USA*
AUKE W. DE JONG • *Westerdijk Fungal Biodiversity Institute, Utrecht, The Netherlands; Department of Medical Microbiology, University Medical Center Utrecht, Utrecht, The Netherlands*

BAHGAT FAYED • *Research Institute for Medical and Health sciences, University of Sharjah, Sharjah, UAE; Chemistry of Natural and Microbial Product Department, National Research Centre, Cairo, Egypt*

MARK A. FRASER • *National Mycology Reference Laboratory, UK Health Security Agency, Bristol, UK*

LALITHA GADE • *Mycotic Diseases Branch, Centers for Disease Control and Prevention, Atlanta, GA, USA*

AHMED GAMAL • *Center for Medical Mycology, University Hospitals Cleveland Medical Center and Case Western Reserve University, Cleveland, OH, USA*

BERT GERRITS VAN DEN ENDE • *Westerdijk Fungal Biodiversity Institute, Utrecht, The Netherlands*

MAHMOUD A. GHANNOUM • *Center for Medical Mycology, University Hospitals Cleveland Medical Center and Case Western Reserve University, Cleveland, OH, USA*

FERRY HAGEN • *Westerdijk Fungal Biodiversity Institute, Utrecht, The Netherlands; Department of Medical Microbiology, University Medical Center Utrecht, Utrecht, The Netherlands; Institute for Biodiversity and Ecosystem Dynamics, University of Amsterdam, Amsterdam, The Netherlands*

MONZER HAMZE • *Laboratoire Microbiologie Santé et Environnement (LMSE), Doctoral School of Sciences and Technology, Faculty of Public Health, Lebanese University, Tripoli, Lebanon*

JANET HERRADA • *Center for Medical Mycology, University Hospitals Cleveland Medical Center and Case Western Reserve University, Cleveland, OH, USA*

KALI R. IYER • *Department of Molecular Genetics, University of Toronto, Toronto, ON, Canada*

CHAD J. JOHNSON • *Department of Medicine, University of Wisconsin, Madison, WI, USA*

ELIZABETH M. JOHNSON • *UK National Mycology Reference Laboratory, Public Health England South-West Regional Laboratory, Southmead Hospital, Bristol, UK; MRC Centre for Medical Mycology, University of Exeter, Exeter, UK*

LINDSAY R. KALAN • *Department of Medicine, University of Wisconsin, Madison, WI, USA; Department of Medical Microbiology and Immunology, University of Wisconsin, Madison, WI, USA*

DALAL KASIR • *Laboratoire Microbiologie Santé et Environnement (LMSE), Doctoral School of Sciences and Technology, Faculty of Public Health, Lebanese University, Tripoli, Lebanon*

RYAN KEAN • *Glasgow Biofilm Research Network, Glasgow, UK; Department of Biological and Biomedical, School of Health and Life Sciences, Glasgow Caledonian University, Glasgow, UK*

NITESH KUMAR KHANDELWAL • *Department of Chemistry and Biochemistry, University of Arizona, Tucson, AZ, USA*

SANG HU KIM • *Department of Molecular Genetics, University of Toronto, Toronto, ON, Canada*

GHALIA KOHDER • *Research Institute for Medical and Health sciences, University of Sharjah, Sharjah, UAE; College of Pharmacy, University of Sharjah, Sharjah, UAE*

DIMITRIOS P. KONTOYIANNIS • *Department of Infectious Diseases, Infection Control and Employee Health, The University of Texas MD Anderson Cancer Center, Houston, TX, USA*

MILENA KORDALEWSKA • *Center for Discovery and Innovation, Hackensack Meridian Health, Nutley, NJ, USA*

ANASTASIA P. LITVINTSEVA • *Mycotic Diseases Branch, Centers for Disease Control and Prevention, Atlanta, GA, USA*
LISA LONG • *Center for Medical Mycology, University Hospitals Cleveland Medical Center and Case Western Reserve University, Cleveland, OH, USA*
ALEXANDER LORENZ • *Institute of Medical Sciences (IMS), University of Aberdeen, Aberdeen, UK*
ANDREW M. LYNN • *School of Computational and Integrative Sciences, Jawaharlal Nehru University, New Delhi, India*
DONNA M. MACCALLUM • *Institute of Medical Sciences, School of Medicine, Medical Sciences and Nutrition, University of Aberdeen, Foresterhill, Aberdeen, UK; Aberdeen Fungal Group, Institute of Medical Sciences, School of Medicine, Medical Sciences and Nutrition, University of Aberdeen, Aberdeen, UK*
CATHERINE MARK • *Institute of Medical Sciences, School of Medicine, Medical Sciences and Nutrition, University of Aberdeen, Foresterhill, Aberdeen, UK*
ANTONIO MARTÍNEZ-MURCIA • *Department of Microbiology, University Miguel Hernández, Orihuela, Alicante, Spain; Genetic PCR Solutions™, Orihuela, Alicante, Spain*
NAVEEN KUMAR MEENA • *School of Computational and Integrative Sciences, Jawaharlal Nehru University, New Delhi, India*
GILLIAN MILNE • *Institute of Medical Sciences (IMS), School of Medicine, Medical Sciences and Nutrition, University of Aberdeen, Aberdeen, UK*
ELIZABETH MISAS • *Mycotic Diseases Branch, Centers for Disease Control and Prevention, Atlanta, GA, USA*
CAROL A. MUNRO • *Institute of Medical Sciences, School of Medicine, Medical Sciences and Nutrition, University of Aberdeen, Foresterhill, Aberdeen, UK*
REMYA NAIR • *Department of Hematology and Medical Oncology, Winship Cancer Institute, School of Medicine, Emory University, Atlanta, GA, USA*
AARON NAVARRO • *Genetic PCR Solutions™, Orihuela, Alicante, Spain*
JENIEL E. NETT • *Department of Medicine, University of Wisconsin, Madison, WI, USA; Department of Medical Microbiology & Immunology, University of Wisconsin, Madison, WI, USA*
LEONARDO NIMRICHTER • *Laboratório de Glicobiologia de Eucariotos, Instituto de Microbiologia Paulo de Góes, Universidade Federal do Rio de Janeiro, Rio de Janeiro, Brazil*
JOSHUA D. NOSANCHUK • *Department of Microbiology and Immunology, Albert Einstein College of Medicine, Bronx, NY, USA; Division of Infectious Diseases, Department of Medicine, Albert Einstein College of Medicine, Bronx, NY, USA*
MARWAN OSMAN • *Laboratoire Microbiologie Santé et Environnement (LMSE), Doctoral School of Sciences and Technology, Faculty of Public Health, Lebanese University, Tripoli, Lebanon; Department of Public and Ecosystem Health, College of Veterinary Medicine, Cornell University, Ithaca, NY, USA*
NICOLAS PAPON • *Univ Angers, Univ Brest, IRF, SFR ICAT, Angers, France*
CHLOE PELLETIER • *Institute of Medical Sciences (IMS), University of Aberdeen, Aberdeen, UK; MRC Centre for Medical Mycology, Department of Biosciences, University of Exeter, Exeter, UK*
DAVID S. PERLIN • *Center for Discovery and Innovation, Hackensack Meridian Health, Nutley, NJ, USA*
JOHN PHAN • *Centers for Disease Control and Prevention, Atlanta, GA, USA*

RAJENDRA PRASAD • *Amity Institute of Biotechnology and Integrative Sciences and Health, Amity University Haryana, Amity Education Valley, Gurgaon, India; Amity Institute of Biotechnology and Amity Institute of Integrative Sciences and Health, Amity University Haryana, Gurugram, India*

GORDON RAMAGE • *Oral Sciences Research Group, College of Medical, Veterinary and Life Sciences, Glasgow University, Glasgow, UK; Glasgow Biofilm Research Network, Glasgow, UK*

NICOLE ROBBINS • *Department of Molecular Genetics, University of Toronto, Toronto, ON, Canada*

KYLE ROBERTS • *Center for Medical Mycology, University Hospitals Cleveland Medical Center and Case Western Reserve University, Cleveland, OH, USA*

MARCIO L. RODRIGUES • *Instituto Carlos Chagas, Fundacao Oswaldo Cruz (Fiocruz), Curitiba, Brazil, Curitiba, Brazil; Instituto de Microbiologia Paulo de Góes, Universidade Federal do Rio de Janeiro, Rio de Janeiro, Brazil*

DHANANSAYAN SHANMUGANAYAGAM • *Department of Animal and Dairy Sciences, University of Wisconsin, Madison, WI, USA; Department of Surgery, University of Wisconsin School of Medicine and Public Health, Madison, WI, USA; Center for Biomedical Swine Research and Innovation, University of Wisconsin School of Medicine and Public Health, Madison, WI, USA*

TERRANCE P. SHEA • *Infectious Disease and Microbiome Program, Broad Institute of MIT and Harvard, Cambridge, MA, USA*

LEIGHANN SHERRY • *Oral Sciences Research Group, College of Medical, Veterinary and Life Sciences, Glasgow University, Glasgow, UK; Glasgow Biofilm Research Network, Glasgow, UK*

BRYN SHORT • *Oral Sciences Research Group, College of Medical, Veterinary and Life Sciences, Glasgow University, Glasgow, UK; Glasgow Biofilm Research Network, Glasgow, UK*

STYLIANOS C. SIMANTIRAKIS • *Aberdeen Fungal Group, Institute of Medical Sciences, School of Medicine, Medical Sciences and Nutrition, University of Aberdeen, Aberdeen, UK*

ASHUTOSH SINGH • *Medical Mycology Unit, Department of Microbiology, Vallabhbhai Patel Chest Institute, University of Delhi, Delhi, India*

SAMEH S. M. SOLIMAN • *Research Institute for Medical and Health sciences, University of Sharjah, Sharjah, UAE; College of Pharmacy, University of Sharjah, Sharjah, UAE*

POONAM VISHWAKARMA • *School of Computational and Integrative Sciences, Jawaharlal Nehru University, New Delhi, India; DSIMB Laboratory, Faculty of Sciences and Technology, University of Reunion Island, Saint Denis, France*

DARLENE WAGNER • *IHRC, Inc., Atlanta, GA, USA*

LOUISE A. WALKER • *Institute of Medical Sciences (IMS), School of Medicine, Medical Sciences and Nutrition, University of Aberdeen, Aberdeen, UK*

ALICIA WARE • *Glasgow Biofilm Research Network, Glasgow, UK; Department of Biological and Biomedical, School of Health and Life Sciences, Glasgow Caledonian University, Glasgow, UK*

MOHD WASI • *School of Life Sciences, Jawaharlal Nehru University, New Delhi, India*

RORY M. WELSH • *Mycotic Diseases Branch, Centers for Disease Control and Prevention, Atlanta, GA, USA*

SEBASTIAN WURSTER • *Department of Infectious Diseases, Infection Control and Employee Health, The University of Texas MD Anderson Cancer Center, Houston, TX, USA*

ANAMIKA YADAV • *Medical Mycology Unit, Department of Microbiology, Vallabhbhai Patel Chest Institute, University of Delhi, Delhi, India; Department of Zoology, Ramjas College, University of Delhi, Delhi, India*
DANIEL ZAMITH-MIRANDA • *Department of Microbiology and Immunology, Albert Einstein College of Medicine, Bronx, NY, USA; Division of Infectious Diseases, Department of Medicine, Albert Einstein College of Medicine, Bronx, NY, USA*

Part I

Isolation and Identification of *Candida auris*

Chapter 1

Isolation of *Candida auris* in Clinical Specimens

Anamika Yadav, Ashutosh Singh, and Anuradha Chowdhary

Abstract

Candida auris is a multidrug-resistant yeast causing healthcare-associated outbreaks of blood stream infections worldwide. Currently, *C. auris* isolation and identification is complicated by issues such as misidentification and long turnaround time associated with application of commonly used diagnostic tools. Based on phenotypic characteristics, differentiation of *C. auris* from related *Candida haemulonii* complex spp. is problematic. *Candida auris* can be misidentified using biochemical-based systems such as VITEK 2 YST, API 20C, BD Phoenix yeast identification system, and MicroScan. *C. auris* growth at 42 °C and in the presence of 10% NaCl helps in presumptive identification of this yeast from related *Candida haemulonii* complex spp. A new CHROMagar™ Candida Plus agar is an excellent alternative to current conventional mycological media for the screening of patients colonized/infected with *Candida auris*. Matrix-assisted laser desorption/ionization time-of-flight (MALDI–TOF) can differentiate *C. auris* from other *Candida* species, but not all the reference databases included in MALDI–TOF devices allow for detection. Currently, accurate identification of *C. auris* can be performed using the updated FDA-approved libraries or "research use-only" libraries. Molecular techniques have greatly enhanced the diagnosis of *C. auris*. Sequencing of rDNA genetic loci, namely, internal transcribed spacer and D1/D2 region of large subunit (LSU), and PCR/qPCR assays has successfully been applied for identification of *C. auris*. Real-time PCR assays bear incomparable potential of being the most efficient tool for high-throughput screening of surveillance samples. If properly validated, they can deliver the diagnostic result within several hours, since the DNA can be isolated directly from the patient specimen without the need of obtaining a colony. In this chapter we detailed the isolation of *Candida auris* from various clinical specimens and its currently available identification methods and hitches.

Key words *Candida auris*, Molecular identification, Misidentification, Phenotypic identification, CHROMagar Candida

1 Introduction

Candida auris is a global nosocomial pathogen that has been reported to cause healthcare-associated infections in more than 40 countries on 5 continents [1]. About two thirds of infections caused by *C. auris* are bloodstream infections that are associated with high mortality in patients admitted to intensive care units. Unlike other yeasts, *C. auris* has the capacity to colonize the skin

Alexander Lorenz (ed.), *Candida auris: Methods and Protocols*, Methods in Molecular Biology, vol. 2517,
https://doi.org/10.1007/978-1-0716-2417-3_1,

surfaces of the patients, and subsequent nosocomial transmission among hospitalized patients occurs via fomites and hands of healthcare personnel [2, 3]. Further, one in five colonized patients can progress to invasive infections. It is, thus, extremely important to rapidly and correctly identify subjects with *C. auris*, in order to contain its spread in healthcare settings. In the past few years *C. auris* colonization in hospitalized patients has been increasingly reported worldwide. The highest prevalence of colonization (23–71%) is reported in patients of ventilator-capable skilled nursing facilities in the United States that probably serve as sources/reservoir of transmission of *C. auris* to other healthcare facilities via transferred patients [4]. The major obstacles impacting the control of *C. auris* spread include common misidentification by commercial biochemical diagnostic platforms such as VITEK 2 (bioMérieux), API 20C AUX (bioMérieux), BD Phoenix (Becton Dickinson), and MicroScan (Beckman Coulter) available in clinical and public health laboratories. However, in the last few years, several of these commercial identification systems have updated their database to include *C. auris* in their systems resulting in accurate identification of this yeast. Further, MALDI–TOF MS including VITEK MS (bioMérieux) and the MALDI Biotyper (OC version 3.1 CA Bruker Daltonics) with updated databases are excellent rapid platforms for identification of *C. auris* in diagnostic microbiology laboratories [5]. Isolation of *C. auris* from clinical samples including currently available diagnostic methods is detailed below.

2 Collection, Transportation, and Processing of Samples for Isolation of *Candida auris* from Clinical Specimens

Appropriate collection of clinical specimens is essential for mycological investigation. Further, timely transportation of specimens within 4 h is required as bacterial overgrowth can inhibit slow-growing fungi or reduce the fungal viability. For transportation of samples for fungal culture, ambient room temperature (25 °C) is recommended. It is important to collect an adequate volume of samples for fungal culture as certain specimens need to be concentrated and pre-treated before plating.

In case of delay, samples are to be stored at 4 °C after proper packaging. Amies medium has been widely used for the transport of swab specimens collected for surveillance purposes. To detect *C. auris* colonization, swabbing of multiple body sites by rayon-tipped swabs or nylon-flocked swabs is recommended by Centers for Disease Control and Prevention (CDC), USA [6].

An overview of processing of clinical specimens for isolation of *Candida auris* is given below.

2.1 Blood Culture

Many automated blood culture systems like BACTEC (Becton Dickinson), BacT/ALERT 3D (bioMérieux), and VersaTREK (Trek diagnostics system) are available with medium modifications to enhance fungal growth; these can recover *C. auris*.

(a) While collecting blood samples, blood volume and blood to broth ratio are important factors to be kept in mind. For adults, 20–30 mL of blood is recommended in two different blood culture bottles. A five- to tenfold dilution of blood in broth (BacT/ALERT® culture media) is important for better recovery of organisms. A thorough decontamination of skin is required before sample collection, and the sample should be transported within 2 h at room temperature. Peripheral blood cultures from venipuncture following skin preparation with 70% isopropyl alcohol or 0.5% chlorhexidine in 70% alcohol are standard procedures for detection of *C. auris* bloodstream infections. In case of patients who have a central line, saline flush the line prior to the withdrawal of the blood sample for culture [7].

(b) Catheter-related bloodstream infections: Blood culture must be obtained from both, one set through the device and one set from another venipuncture. When collecting intravenous catheter tips for culture, aseptically cut a ~4 cm segment from the tip and place in a sterile container; transport rapidly, within 15–20 min, at room temperature to prevent drying out. In case of delay, store at 4 °C [8].

2.2 Sterile Body Fluids (Cerebrospinal, Peritoneal, Synovial, Vitreous Humor, etc.)

C. auris has been isolated from sterile body fluids such as cerebrospinal, peritoneal, and pericardial fluid. About 3–5 mL of body fluid is to be collected in most cases, except for bone marrow aspirate and vitreous fluid where a small volume (0.5 mL) is obtained with a syringe and kept in a vacutainer tube with heparin to prevent blood clotting. Lysis centrifugation tubes can also be used for this purpose. While processing the specimens, in case specimen volume is $\geq$2 mL, then centrifuge at 2000 × g for 10 min, and use sediment to inoculate the culture medium. However, if the volume is <2 mL, then specimens should be placed directly onto agar plates [8].

2.3 Respiratory Specimens (Sputum, Bronchial Aspirate, Bronchoalveolar Lavage)

Isolation of *C. auris* from the respiratory tract suggests colonization.

Sputum Sputum specimens require pre-treatment with mucolytic enzymes such as N-acetyl-L-cysteine, 5% oxalic acid, or dithiothreitol (Sputolysin). A complete homogenization is mandatory for sputum samples before inoculating on agar plates.

Bronchial aspirate/bronchoalveolar lavage (BAL) 5–10 mL of bronchial aspirate/BAL specimens are centrifuged at 1200 × g for 10 min, and 0.5 mL of centrifuged sediment is inoculated on routine mycological media such as Sabouraud Dextrose Agar (SDA) plates [1% (w/v) peptone, 4% (w/v) glucose, and 1.5–2.0% (w/v) agar with a final pH of 5.6] and incubated at 28 °C and 37 °C. Culture of respiratory samples is a slow and insensitive method for the diagnosis of respiratory fungal infections, (primarily caused by molds and dimorphic fungi) although its sensitivity can be improved by increasing the culture volume. In a recent study, positivity rate of conventional cultures of respiratory specimens yielded *Aspergillus* spp. in 15.7%, and that of high-volume culture (HVC) was 54.2% [9].

2.4 Urine

A midstream urine specimen is to be collected in a sterile container. From catheterized patients, collect specimens aseptically by aspirating from the port of the urinary catheter. Urine specimens need to be transported within 2–4 h at room temperature and centrifuged at 2000 × g for maximum recovery of yeast isolates [7].

2.5 Abscess (Exudates/Pus/Wound Material) and Tissue Specimens)

Specimens can be collected from active peripheral edges of open abscesses; if the abscess is closed, then aspirate the specimen with a syringe after decontaminating the lesion, and transfer to a sterile container. Swabs should never be used for closed lesions. Tissue samples should be minced with a scalpel or ground with a mortar and pestle in order to release the intracellular yeast and enable their growth on the culture medium [8].

2.6 Surveillance Cultures from Body Sites for Detection of Colonization by Candida auris

As described above, rayon-tipped or nylon-flocked swabs can be used for swabbing different body sites like the axilla, groin, external nares, and ear. These locations were determined to be high-yield sites for detection of colonization by *C. auris*. Single swabs for each site or composite swabs (axilla + groin) can be collected. To collect swabs from the axillae, target the crease in the skin where the arm meets the body, and for the groin region, target the inguinal crease in the skin where the leg meets the pelvic region. Roll the swab on an area no larger than 26 cm^2 in both horizontal and vertical sweeps for approximately 1 min after dipping in normal saline. Swabs immersed in Amies transport medium are preferred over dry swabs as they promote the viability of the organism. The swabs should be transported within 24 h at room temperature and inoculated onto agar plates after vortex-mixing the swab along with the transport medium. Swabs can be directly streaked over agar plates while rotating in every possible direction. Both the groin and axilla are the most probable site of colonization with a positivity rate of 37% and 35%, respectively [10]. Further, a high positivity rate of 75% from groin samples was observed for *C. auris* colonization in another study [3]. Additional body sites such as the palms, finger tips, and toe webs screened for *C. auris* colonization also yield high positivity rates [11].

3 Phenotypic Characteristics of *Candida auris*

Like other yeasts, *C. auris* can also grow on routine mycological media used for isolation of yeasts. On Sabouraud Dextrose Agar (SDA) plates, *C. auris* forms cream-colored, smooth colonies at 35–37 °C after 24–48 h of incubation. To prevent bacterial growth, chloramphenicol (25 mg/L) and gentamicin (40 mg/L) are added in SDA culture plates [3]. *C. auris* is a budding yeast, which does not form germ tubes and lacks chlamydospore formation on cornmeal agar at 30 °C [12]. On microscopic examination of SDA growth, *C. auris* exhibit oval or elongated yeast cells [13] of approximately 2.0–3.0 × 2.5–5.0 μm in size which are capable of existing as aggregated or nonaggregated cells [14]. However, elongated cells are observed on yeast extract–peptone–dextrose (YPD) medium supplemented with 10% NaCl [15]. A high concentration of NaCl (10%) induces the development of elongated and pseudohyphal cells in *C. auris* South Asian clade (Clade I) [16].

Unlike closely related *Candida* species like *C. haemulonii* and *C. duobushaemulonii*, *C. auris* can grow well at 42 °C and tolerate high salt concentrations (>10% NaCl, wt/vol) [17]. This thermotolerant and osmotolerant characteristics probably contribute to the persistence and survival of *C. auris* on biotic and abiotic surfaces for long periods of time. Some phenotypic interclade variations have been reported. *Candida auris* strains belonging to Clade I (South Asian), Clade III (South African), and Clade IV (South American) are recovered from outbreaks in healthcare facilities worldwide, whereas Clade II (East Asian) strains are predominantly associated with cases of ear infection [18, 19]. Unlike Clade I lineage isolates, Clade II and Clade III isolates are unable to produce even rudimentary pseudohyphae on cornmeal agar plates at 30 °C for 48 h [20]. Also, clinical Clade I strains of *C. auris* exhibit inhibition at ≥0.01% concentration of cycloheximide (actidione), while *C. auris* strains belonging to Clade II and Clade III grow better in the presence of actidione (0.05%) [20]. Further, *C. auris* Clade I strains assimilate N-acetyl glucosamine, whereas Japanese and Korean reference isolates, which belong to Clade II, do not assimilate N-acetyl glucosamine [21].

4 Isolation of *Candida auris* from Clinical Samples Harboring Mixed Yeast Flora

Usage of differential selective media enhances the recognition and isolation of *C. auris* from mixed yeast colonies obtained in the primary culture.

CHROMagar™ is a routinely used mycological medium for screening of mixed yeast colonies in primary culture. CHROMagar™ Candida medium is a selective differential medium that

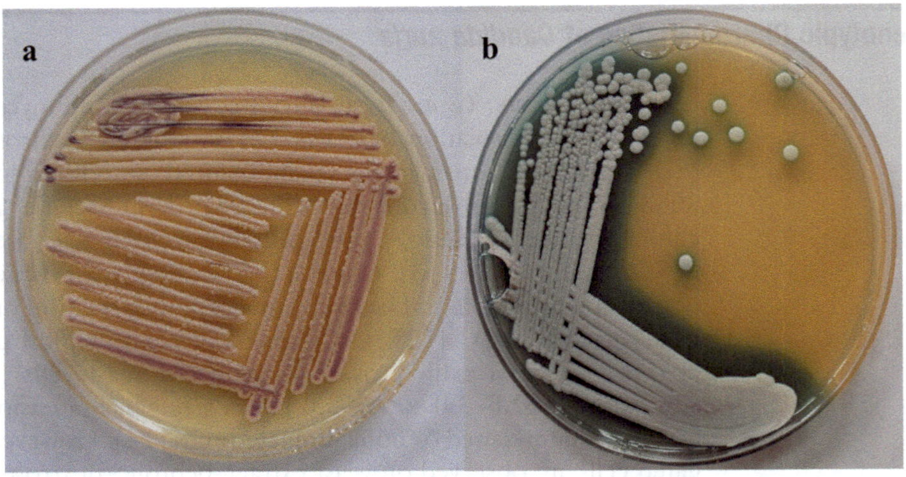

Fig. 1 *Candida auris* colonies showing (**a**) pink color on CHROMagar Candida medium; and (**b**) distinctive pale cream with blue halo on CHROMagar™ Candida Plus medium

imparts different colored colonies to *Candida* spp. *C. auris* shows cream to pink color colonies on CHROMagar™ Candida medium which is common to *Candida krusei*, *Candida guilliermondii*, or the *Candida glabrata* complex, hence does not permit differentiation of *C. auris*. The recently developed CHROMagar™ Candida Plus medium is an invaluable tool for presumptive identification of *C. auris*. Borman et al. [22] tested the media on 10 *C. auris* isolates and observed that *C. auris* exhibit pale cream colonies with a distinctive blue halo after 36 h of incubation at 35–37 °C (Fig. 1). Further, 52 different yeast spp. were screened, and only *Candida diddensiae* showed similar appearance. Also, enhanced rate of growth and 100% sensitivity and specificity of *C. auris* on CHROMagar™ Candida Plus agar suggests that it is an excellent alternative to current conventional mycological media for the isolation of *C. auris* [22].

5 Selective Enrichment Broth for Isolation of *Candida auris* from Surveillance Cultures

To detect colonized subjects, the use of selective enrichment broth media (see below for recipes) results in specific and enhanced recovery of *C. auris* in surveillance cultures collected from various body sites. These selective enrichment broth media have hypersaline conditions and contain dulcitol or mannitol as carbon sources in place of glucose. Like *C. auris*, *Candida glabrata* is one of the species that can sometimes grow at high salt, and therefore for differentiation on *C. auris*, selective enrichment broth dulcitol is used as a carbon source rather than glucose as *C. glabrata* cannot assimilate or ferment dulcitol. Inoculated broth tubes are vortex-mixed and placed in an orbital shaking incubator at 250 rpm at 40 °C. Because of their hypersaline nature and enrichment property, the selective enrichment broth media ensure 100% recovery of *C. auris* from specimens, especially body swabs [17].

The composition of various selective enrichment broth media used are as follows:

(a) *Salt Sabouraud Enrichment (Salt SAB) broth*:

5 g pancreatic digest of casein.

5 g peptic digest of animal tissue.

100 g sodium chloride (NaCl) dissolved in a liter of deionized (DI) water with 20 g of either dextrose, dulcitol/galactitol, or mannitol as the added carbon source. Mix the ingredients well and adjust the pH to 5.6.

Both gentamicin and chloramphenicol are added to a final concentration of 50 mg/L.

Cold sterilize by passing solution through a 0.2 micron filter.

(b) *Salt Yeast Nitrogen Base (Salt YNB) broth*:

6.8 g yeast nitrogen base.

100 g sodium chloride (NaCl) dissolved in a liter of deionized (DI) water with 20 g of either dextrose, dulcitol/galactitol, or mannitol as the added carbon source. Mix the ingredients well and adjust the pH to 5.6.

Both gentamicin and chloramphenicol are added to a final concentration of 50 mg/L.

Cold sterilize by passing solution through a 0.2 micron filter.

6 Identification of *Candida auris* by Commercial Biochemical-Based Systems

A vast range of biochemical-based commercial identification methods like VITEK 2 (bioMérieux), API 20C AUX (bioMérieux), BD Phoenix (Becton Dickinson), and MicroScan (Beckman Coulter) are used worldwide by clinical and public health laboratories to identify yeast isolates. These systems may incorrectly identify *C. auris* as other yeast species as their databases do not include *C. auris*. The most common misidentifications include *Candida haemulonii* and *Candida duobushaemulonii* by VITEK 2 YST, *Rhodotorula glutinis* (characteristic red color not present) by API 20C AUX, *Candida parapsilosis* by RapID Yeast Plus, and *Candida haemulonii* and *Candida catenulata* by BD Phoenix. MicroScan misidentifies *C. auris* as *Candida famata/Debaryomyces hansenii*, *Candida guilliermondii/Meyerozyma guilliermondii*, *Candida lusitaniae/Clavispora lusitaniae*, and *Candida parapsilosis* [23]. VITEK 2 automated identification system includes *C. auris* in its version 8.01 database. This updated version recognizes *C. auris* isolates belonging to Clade I and III isolates but misidentifies the *C. auris* Clade II isolates as *C. duobushaemulonii*. Even further updated versions of VITEK 2 show an accuracy of only 52% among tested *C. auris* strains [5], warranting also a cautious approach while identifying *C. auris* using the updated system.

7 MALDI–TOF Mass Spectrometry-Based Identification of *Candida auris*

The matrix-assisted laser desorption ionization time-of-flight mass spectrometry (MALDI–TOF MS) is a rapid and accurate identification system that depends on the analysis of an organic biomolecules spectrum. MALDI–TOF MS generates characteristic mass spectra, unique for each microorganism that can be used as their fingerprints. Lack of appropriate spectra in the databases has resulted in the misidentification of *C. auris* as *C. haemulonii* and *C. albicans*, among others, by MALDI–TOF MS [24, 25]. However, once research use-only (RUO) or manufacturer's databases are updated with *C. auris* and related species spectra, MALDI–TOF MS identification of *C. auris* appears to be accurate [26–28]. There are presently two FDA-approved, commercially available MALDI–TOF mass spectrometry systems available: the VITEK MS from bioMérieux and the MALDI Biotyper CA from Bruker Daltonics. Accurate identification of *C. auris* can be achieved by using the freely available MicrobeNet database validated by CDC experts or the following MALDI–TOF MS systems: Bruker Biotyper with FDA-approved MALDI Biotyper CA System library (Version Claim 4) or their "research use-only" libraries (Versions 2014 [5627] and more recent) and VITEK (MALDI–TOF) MS RUO (with Saramis Version 4.14 database and Saccharomycetaceae update). It is crucial that laboratories review the presence of aforementioned spectra in the database and confirm the laboratory detection capacity by testing a panel of reference strains, e.g., "*Candida auris* Panel" offered by CDC and FDA Antibiotic Resistance (AR) Isolate Bank, which contain *C. auris* isolates from all clades and other yeast species that are related to *C. auris* or are commonly misidentified as *C. auris*. MALDI–TOF MS is suitable for high-throughput and rapid microbial identification at low costs and is an alternative for conventional laboratory biochemical and molecular identification systems. Even though the running costs are significantly cheaper than those of conventional identification methods, the amount of money required to purchase a MALDI–TOF MS instrument itself is often an obstacle for laboratories, ultimately preventing implementation of the technique. Although MALDI–TOF MS improves timeliness of identification, the isolate must still be cultured, which may require up to a week for isolating *C. auris* from specimens.

8 Molecular-Based Identification Methods for *Candida auris*

Molecular techniques have greatly enhanced the identification and diagnosis of fungal agents of infectious disease, particularly when the fungal agent is difficult to culture and identify. An overview of significant advances made in the area of molecular diagnostics for *C. auris* is given below.

8.1 Sequence-Based Identification of Candida auris Isolates

Sequencing of rDNA genetic loci, namely, internal transcribed spacer (ITS) and D1/D2 region of large subunit (LSU) that are amplified with panfungal primers, provides accurate identification of *C. auris* and is considered the gold standard [21, 29–32]. However, sequencing is not routinely used to determine the species of isolates, since the cost, technical demands, and lengthy turnaround times for those without in-house sequencing capacity can make this technique unsuitable for some laboratories. It is also worth mentioning that comparative analysis of rDNA sequences should not be used to define relations between strains due to its insufficient discriminatory power [33]. In short, sequencing requires the isolate to be cultured, is expensive, and requires expertise.

8.2 Direct Detection of Candida auris DNA in Clinical Specimens

To facilitate a rapid turnaround time, many commercial and academic groups have designed various nucleic acid-based identification strategies to identify *C. auris* directly from patient samples. These assays deliver results within hours since the DNA can be isolated directly from a patient specimen without the need of obtaining a colony. Further, nucleic acid-based assays are exquisitely sensitive (with limits of detection in the range of 1–10 *C. auris* colony forming units (CFU)/PCR reaction) and specific, demonstrated by testing large panels of pathogens including closely related yeasts and other fungi on both clinical and screening samples. Nucleic acid-based identification strategies, especially real-time PCR assays are highly efficient tools for high-throughput screening of surveillance samples. Several platforms recently developed for identification of *C. auris* in clinical and surveillance samples as well as from yeast colonies are listed below.

8.3 Nucleic Acid-Based Amplification Testing Candida auris from Yeast Colonies

8.3.1 Method: PCR and Real-Time qPCR (SYBR Green)

Sample/target: yeast colonies/5.8-ITS-28S of rDNA

Assay specificity:	(1) *C. auris* (PCR and qPCR) (2) *C. auris*, *C. duobushaemulonii*, *C. haemulonii*, and *C. lusitaniae* (qPCR)

Detection limit:	(1) 10 CFU/rxn (2) 1000 CFU/rxn

Reference: Kordalewska et al. [34]

Note: This assay consists of two rapid, accurate, easy-to-perform and easy-to-interpret molecular diagnostic assays to identify *C. auris* and related species (*C. duobushaemulonii*, *C. haemulonii*, and *C. lusitaniae*) that overcome the deficiencies of existing phenotypic assays. Both conventional and real-time PCR assays allow specific identification of *C. auris* and related species (*C. duobushaemulonii*, *C. haemulonii*, and *C. lusitaniae*) within 2 (real-time PCR and colony PCR) to 2.5 (conventional PCR) hours. The differential specificities of the assays are obtained by highly specific primers.

8.3.2 Method: Duplex PCR

Sample/target: yeast colonies/species-specific GPI protein-encoding

Assay specificity: *C. auris*

Detection limit: N/A

Reference: Ruiz-Gaitan et al. [35]

Note: PCR approach that employs the uniqueness of certain *C. auris* glycosylphosphatidylinositol (GPI) protein-encoding genes which are unique and have no homology with sequences in other species. These are truly specific for *C. auris*, thus providing more flexibility in PCR amplification conditions and improving the accuracy of the species identification. This approach is primarily aimed at rapidly and unequivocally identifying *C. auris* culture isolates.

8.3.3 Method: Tetraplex PCR

Sample/target: yeast colonies/26s rDNA

Assay specificity: *C. auris*, *C. haemulonii*, *C. duobushaemulonii*, and *C. pseudohaemulonii*

Detection limit: N/A

Reference: Arastehfar et al. [36]

Note: End-point tetraplex PCR was designed to identify and differentiate the most medically important and related species to *C. auris*. Compared with the real-time PCR method, this end-point PCR is nonautomated and labor-intensive. The electrophoresis step in this assay requires 45 min, which is more time-consuming than real-time PCRs. This multiplex PCR system is the first assay for *C. auris* that has been validated by animal model testing. The overall PCR positive rates of mice blood and tissue were 28.6 and 92.9%, respectively, which showed slightly higher performance than culture-based identification.

8.3.4 Method: Multiplex PCR

Sample/target: yeast colonies, spiked serum sample/28s rDNA

Assay specificity: *C. auris*, *C. haemulonii*, *C. duobushaemulonii*, and *C. pseudohaemulonii*

Detection limit: N/A

Reference: Arastehfar et al. [37]

Note: This is the first multiplex PCR assay that can detect *C. auris* and its relatives directly from serum samples in a culture-independent manner. Serum, due to its applicability for downstream applications such as serology tests and its ease of DNA extraction procedure, has earned a reputation for being a suitable blood fraction for identification of fungemia. It has been shown that for identification of invasive candidiasis, serum samples showed the same performance as whole-blood samples when subjected to real-time PCR. Additionally, serum has been regarded as a suitable matrix for validation of PCR-based assays meant to detect *Candida* species from infected patients. Moreover, this multiplex qPCR assay utilizes the most basic and inexpensive qPCR chemistry, that is, SYBR Green I and EVA Green, and offers the first qPCR assay with the potential clinical utility for identification of *C. auris* and its relatives.

Moreover, despite the fact that *C. albicans* accounts for the majority of mixed infections, which could affect the specificity and sensitivity of PCR assays [38], this assay regardless of the DNA concentration of *C. albicans* (ten times more, ten times less and equal to target species) readily identified *C. auris*, *C. haemulonii*, *C. pseudohaemulonii*, and *C. duobushaemulonii* in serum samples spiked with a DNA mixture of target species and *C. albicans*.

8.3.5 Method: YEAST PANEL Multiplex PCR

Sample/target yeast colonies/species-specific primers and probes.

Assay specificity: 21 clinically important yeast species including *C. auris*, *Trichosporon*, *Rhodotorula*, *Cryptococcus*, and *Geotrichum*.

Detection limit: N/A

Reference: Arastehfar et al. [39]

Note: The YEAST PANEL multiplex PCR assay identifies and distinguishes the most clinically important opportunistic yeast species. This assay presents a novel stepwise strategy for identification of the clinically most important yeasts belonging to *Ascomycota* and *Basidiomycota*. The YEAST PANEL multiplex PCR assay can identify 99.6% of the causative agents of candidiasis including *C. auris* and 95% of infections caused by all yeast pathogens [40]. The consistency between YEAST PANEL multiplex PCR assay and MALDI–TOF MS/sequencing is 100%, and it is less expensive and requires less hands-on time when compared to sequencing. The turnaround time for identification, another inherent obstacle of conventional biochemical assays (up to 48 h), was improved with the YEAST PANEL multiplex PCR assay since identification was achieved in 3–6 h.

8.3.6 Method: GPS MONODOSE dtec-qPCR kit

Sample/target: yeast colonies/species-specific primers and probes

Assay specificity: *C. auris*

Detection limit: 5 CFU/rxn

Reference: Martinez-Murcia et al. [41]

Note: The MONODOSE GPS™ dtecq-PCR has paved the way to pathogen detection in a reduced amount of time with new technologies compared to the conventional methods. The MONODOSE GPS™ dtecq-PCR is a single ready-to-use qPCR test for specific pathogen detection that showed considerable advantages with respect to previously reported PCR methods. GPS™ MONODOSE dtec-qPCR is already dehydrated, making it easier for the technicians to just add their samples and run the PCR. This kit format can be transported at room temperature without the need of dry ice, a major breakthrough in the speed of transport and cost. Additionally, the enzyme is not frozen-thawed and, therefore, eliminates the risk of the enzyme degrading. Cross-contamination and fluorophore deterioration by UV light are minimized. In all, this is a time-saving process, which is key for a rapid response to nosocomial pathogens such as the emerging yeast *C. auris*. The method has been validated according to international standards and is available to track *C. auris* outbreaks.

8.4 Nucleic Acid Amplification Testing of *Candida auris* for Surveillance Studies

8.4.1 Method: Real-Time PCR (TaqMan)

Sample/target: swabs, environmental samples/ITS2-rDNA

Assay specificity: *C. auris*

Detection limit: 1 CFU/rxn

Reference: Leach et al. [42]

Note: A real-time PCR assay for rapid detection of *C. auris* from surveillance samples which provides rapid results within 4 h of sample processing. This assay is highly sensitive, with a detection limit of 1 *C. auris* CFU/PCR. The high sensitivity of this real-time PCR assay is not surprising since the multicopy ITS2 is being used as a gene target. More importantly, the assay is highly specific, as no cross-reactivity is observed with all known closely and distantly related yeasts and other pathogens. For culture-positive surveillance samples, a high sensitivity of 89% and 100% were recorded for swabs and sponges, respectively. Among the culture-negative surveillance samples, clinical specificities of 99% and 89% were recorded for swabs and sponges, respectively. This assay provides a new and superior method to increase accurate and rapid screening of *C. auris* in healthcare facilities.

8.4.2 Method: Real-Time PCR (SYBR Green)

Sample/target: swabs, environmental samples/ITS2-rDNA

Assay specificity: *C. auris*

Detection limit: 4 CFU/rxn

Reference: Sexton et al. [43]

Note: The *C. auris* SYBR green qPCR assay is an effective rapid alternative to culture for the identification of *C. auris* in clinical axilla/groin skin swabs. *C. auris* SYBR green qPCR shares many similarities with TaqMan qPCR assay in terms of workload and time needed to obtain results. Both assays can produce results in less than 5 h, delivering a significant improvement compared to the 2 weeks required for culture. Moreover, the SYBR green assay is more cost-effective, since it does not require TaqMan probes. The assay may also be more accessible because the emission wavelength (520 nm) of SYBR green dyes is universally detected across real-time PCR instruments. A technical shortcoming of the SYBR green method is that the nature of the inhibition control prevents quantitative data from being interpreted from the threshold cycle, although it is possible to remove the inhibition control for experiments seeking quantitative data. Compared to the *C. auris* T2MR assay (see Subheading 8.4.3), both the SYBR green and TaqMan (see Subheading 8.4.1) assays require more labor but are more affordable and offer higher throughput.

8.4.3 Method: T2 Magnetic Resonance (T2MR) System

Sample/target: swabs/species-specific primers and probes

Assay specificity: *C. auris*

Detection limit: 5 CFU/ml

Reference: Sexton et al. [44]

Note: The T2MR *C. auris* assay is an effective rapid alternative to culturing for the identification of *C. auris* in clinical axilla/groin skin swabs. T2 Biosystems (Lexington, MA) developed a new diagnostic for *C. auris* built into the established T2 Magnetic Resonance (T2MR) platform. T2MR technology uses target-specific probes linked to superparamagnetic particles that detect and subsequently amplify a pathogen-associated signal. Signal amplification occurs through clustering of the superparamagnetic particles in the presence of the molecular target, which can be detected by magnetic resonance technology [45]. T2 Biosystems has already used this technology to develop diagnostic panels for other pathogens including an FDA-approved T2 *Candida* panel that detects five clinically relevant *Candida* species including *C. albicans*, *C. tropicalis*, *C. parapsilosis*, *C. krusei*, and *C. glabrata* [46, 47]. Independent of performance, the T2MR *C. auris* assay is simple to operate and does not require labor-intensive steps like the DNA extraction required for qPCR. The T2MR *C. auris* assay also provides results quickly taking only 4–8 h depending on how many samples are being run simultaneously. The T2MR assay is performed on the T2Dx (a fully automated instrument); the potential for operator variability is also minimized. However, a considerable limitation is that the T2MR *C. auris* assay can only process up to seven samples simultaneously, whereas qPCR can be scaled up for high-throughput formats. In summary, the T2MR *C. auris* assay performed well in comparison with established culturing methods, indicating that it could be an effective rapid alternative to culturing for the identification of *C. auris* in clinical axilla/groin skin swabs.

8.4.4 Method: Loop-Mediated Isothermal Amplification (LAMP)

Sample/target: yeast colonies, environmental samples, swabs/ferredoxin oxidoreductase-encoding gene.

Assay specificity: *C. auris*

Detection limit: 20 CFU/rxn

Reference: Yamamoto et al. [48]

Note: The LAMP approach is reliably identifying *C. auris* strains and able to distinguish *C. auris* from other *Candida* species (even very closely related species) with a specificity of 100%. The LAMP assay is able to detect a template provided at concentrations as low as 2×10^1 copies of target DNA per reaction. Moreover, the results were obtained within a short time, without any technical complications regarding the use of the amplification instrument. Direct LAMP from a clinical specimen was demonstrated; thus, this technique is expected to save clinicians the time required for cultivation and DNA extraction, allowing an early diagnosis. Recently, portable LAMP amplification equipment has been made commercially available. This availability is expected to facilitate the use of the LAMP assay, enabling large-scale and field surveillance detection. However, care should be taken when handling the LAMPAuris product as opening the action tube could result in considerable contamination. Overall, this assay should be particularly valuable for *C. auris*, a pathogen that is an important target of environmental control in healthcare facilities.

9 Conclusion

Candida auris remains an unnoticed pathogen in routine microbiology laboratories due to misidentification as other phylogenetically related pathogens by commercial identification systems. As *C. auris* can colonize hospitalized patients and is a rapidly transmissible nosocomial yeast, public health laboratories providing diagnostic services to healthcare institutions should be capable of rapidly and accurately detecting *C. auris* in clinical samples. Osmotolerance (10% NaCl) and thermotolerance (42 °C) properties of *C. auris* help to differentiate it from other species of *C. haemulonii* complex, especially in resource-limited laboratories. The use of selective enrichment broth media and differential CHROMagar™ Candida Plus medium, optimized to *C. auris* growth characteristics achieves a faster recovery time in clinical specimens, along with improved sensitivity and specificity. Advances in spectral databases of MALDI–TOF MS Bruker

Biotyper systems have reduced misidentification. Nowadays nucleic acid-based identification strategies, especially real-time PCR assays, are highly efficient tools for high-throughput screening of clinical samples.

References

1. Du H, Bing J, Hu T, Ennis CL, Nobile CJ, Huang G (2020) *Candida auris*: epidemiology, biology, antifungal resistance, and virulence. PLoS Pathog 16:e1008921. https://doi.org/10.1371/journal.ppat.1008921
2. Adams E, Quinn M, Tsay S et al (2018) *Candida auris* in healthcare facilities, New York, USA, 2013–2017. Emerg Infect Dis 24:1816–1824. https://doi.org/10.3201/eid2410.180649
3. Yadav A, Singh A, Wang Y et al (2021) Colonisation and transmission dynamics of *Candida auris* among chronic respiratory diseases patients hospitalised in a chest hospital, Delhi, India: a comparative analysis of whole genome sequencing and microsatellite typing. J Fungi (Basel) 7:81. https://doi.org/10.3390/jof7020081
4. Pacilli M, Kerins JL, Clegg WJ et al (2020) Regional emergence of *Candida auris* in Chicago and lessons learned from intensive follow-up at 1 ventilator-capable skilled nursing facility. Clin Infect Dis 71:e718–e725. https://doi.org/10.1093/cid/ciaa435
5. Ambaraghassi G, Dufresne PJ, Dufresne SF et al (2019) Identification of *Candida auris* by use of the updated Vitek 2 yeast identification system, Version 8.01: a multilaboratory evaluation study. J Clin Microbiol 57:e00884–e00819. https://doi.org/10.1128/JCM.00884-19
6. Centre for Disease Prevention and Control Identification of *Candida auris*. Available online: https://www.cdc.gov/fungal/candida-auris/c-auris-patient-swab.html. Accessed on Sep 2017
7. Centre for Disease Prevention and Control Identification of *Candida auris*. Available online: https://www.cdc.gov/antibiotic-use/core-elements/collecting-cultures.html. Accessed on Aug 2019
8. Versalovic J, Carroll C, Funke G, Jorgensen JH, Landry L, Warnock W (2011) Manual of clinical microbiology, 10th edn. ASM Press American Society for Microbiology, Washington, DC
9. Vergidis P, Moore CB, Novak-Frazer L et al (2019) High-volume culture and quantitative real-time PCR for the detection of *Aspergillus* in sputum. Clin Microbiol Infect 26:935–940. https://doi.org/10.1016/j.cmi.2019.11.019
10. Eyre DW, Sheppard AE, Madder H et al (2018) A *Candida auris* outbreak and its control in an intensive care setting. N Engl J Med 379:1322–1331. https://doi.org/10.1056/NEJMoa1714373
11. Proctor DM, Dangana T, Sexton DJ et al (2021) Integrated genomic, epidemiologic investigation of *Candida auris* skin colonization in a skilled nursing facility. Nat Med. https://doi.org/10.1038/s41591-021-01383-w
12. Larkin E, Hager C, Chandra J et al (2017) The emerging pathogen *Candida auris*: growth phenotype, virulence factors, activity of antifungals, and effect of SCY-078, a novel glucan synthesis inhibitor, on growth morphology and biofilm formation. Antimicrob Agents Chemother 61:e02396–e02316. https://doi.org/10.1128/AAC.02396-16
13. Walsh TJ, Hayden RT, Larone DH (2018) Larone's medically important fungi: a guide to identification. ASM, Washington DC, pp 45–47,51
14. Sikora A, Zahra F (2021) *Candida auris*. In: StatPearls. StatPearls Publishing, Treasure Island (FL)
15. Fasciana T, Cortegiani A, Ippolito M et al (2020) *Candida auris*: an overview of how to screen, detect, test and control this emerging pathogen. Antibiotics (Basel) 9:778. https://doi.org/10.3390/antibiotics9110778
16. Wang X, Bing J, Zheng Q et al (2018) The first isolate of *Candida auris* in China: clinical and biological aspects. Emerg Microbes Infect 7:93. https://doi.org/10.1038/s41426-018-0095-0
17. Welsh RM, Bentz ML, Shams A et al (2017) Survival, persistence, and isolation of the emerging multidrug-resistant pathogenic yeast *Candida auris* on a plastic health care surface. J Clin Microbiol 55:2996–3005. https://doi.org/10.1128/JCM.00921-17
18. Chow NA, Gade L, Tsay SV et al (2018) Multiple introductions and subsequent transmission of multidrug-resistant *Candida auris* in

the USA: a molecular epidemiological survey. Lancet Infect Dis 18:1377–1384. https://doi.org/10.1016/S1473-3099(18)30597-8

19. Chow NA, Muñoz JF, Gade L et al (2020) Tracing the evolutionary history and global expansion of *Candida auris* using population genomic analyses. mBio 11:e03364–e03319. https://doi.org/10.1128/mBio.03364-19
20. Szekely A, Borman AM, Johnson EM (2019) *Candida auris* isolates of the Southern Asian and South African lineages exhibit different phenotypic and antifungal susceptibility profiles *in vitro*. J Clin Microbiol 57: e02055–e02018. https://doi.org/10.1128/JCM.02055-18
21. Chowdhary A, Sharma C, Duggal S et al (2013) New clonal strain of *Candida auris*, Delhi, India. Emerg Infect Dis 19: 1670–1673. https://doi.org/10.3201/eid1910.130393
22. Borman AM, Fraser M, Johnson EM (2021) CHROMagarTM *Candida* plus: a novel chromogenic agar that permits the rapid identification of *Candida auris*. Med Mycol 59: 253–258. https://doi.org/10.1093/mmy/myaa049
23. Keighley C, Garnham K, Harch SAJ et al (2021) *Candida auris:* diagnostic challenges and emerging opportunities for the clinical microbiology laboratory. Curr Fungal Infect Rep. https://doi.org/10.1007/s12281-021-00420-y
24. Kim TH, Kweon OJ, Kim HR, Lee MK et al (2016) Identification of uncommon *Candida* species using commercial identification systems. J Microbiol Biotechnol 26:2206–2213. https://doi.org/10.4014/jmb.1609.09012
25. Wattal C, Oberoi JK, Goel N, Raveendran R, Khanna S (2017) Matrix-assisted laser desorption ionization time of flight mass spectrometry (MALDI–TOF MS) for rapid identification of micro-organisms in the routine clinical microbiology laboratory. Eur J Clin Microbiol Infect Dis 36:807–812. https://doi.org/10.1007/s10096-016-2864-9
26. Girard V, Mailler S, Chetry M et al (2016) Identification and typing of the emerging pathogen *Candida auris* by matrix-assisted laser desorption ionisation time of flight mass spectrometry. Mycoses 59:535–538. https://doi.org/10.1111/myc.12519
27. Vatanshenassan M, Boekhout T, Meis JF et al (2019) *Candida auris* identification and rapid antifungal susceptibility testing against echinocandins by MALDI-TOF MS. Front Cell Infect Microbiol 9:20. https://doi.org/10.3389/fcimb.2019.00020
28. Bao JR, Master RN, Azad KN et al (2018) Rapid, accurate identification of *Candida auris* by using a novel matrix-assisted laser desorption ionization-time of flight mass spectrometry (MALDI-TOF MS) database (Library). J Clin Microbiol 56: e01700–e01717. https://doi.org/10.1128/JCM.01700-17
29. Prakash A, Sharma C, Singh A et al (2016) Evidence of genotypic diversity among *Candida auris* isolates by multilocus sequence typing, matrix-assisted laser desorption ionization time-of-flight mass spectrometry and amplified fragment length polymorphism. Clin Microbiol Infect 22:277.e1–9. https://doi.org/10.1016/j.cmi.2015.10.022
30. Chowdhary A, Anil Kumar V, Sharma C et al (2013) Multidrug-resistant endemic clonal strain of *Candida auris* in India. Eur J Clin Microbiol Infect Dis 33:919–926. https://doi.org/10.1007/s10096-013-2027-1
31. Kathuria S, Singh PK, Sharma C et al (2015) Multidrug-resistant *Candida auris* misidentified as *Candida haemulonii*: characterization by matrix-assisted laser desorption ionization-time of flight mass spectrometry and DNA sequencing and its antifungal susceptibility profile variability by Vitek 2, CLSI broth microdilution, and Etest method. J Clin Microbiol 53:1823–1830. https://doi.org/10.1128/JCM.00367-15
32. Tsay S, Kallen A, Jackson BR, Chiller TM, Vallabhaneni S (2018) Approach to the investigation and management of patients with *Candida auris*, an emerging multidrug-resistant yeast. Clin Infect Dis 66:306–311. https://doi.org/10.1093/cid/cix744
33. Jeffery-Smith A, Taori SK, Schelenz S et al (2017) *Candida auris*: a review of the literature. Clin Microbiol Rev 31:e00029–e00017. https://doi.org/10.1128/CMR.00029-17
34. Kordalewska M, Zhao Y, Lockhart SR, Chowdhary A, Berrio I, Perlin DS (2017) Rapid and accurate molecular identification of the emerging multidrug-resistant pathogen *Candida auris*. J Clin Microbiol 55: 2445–2452. https://doi.org/10.1128/JCM.00630-17
35. Ruiz-Gaitán AC, Fernández-Pereira J, Valentin E et al (2018) Molecular identification of *Candida auris* by PCR amplification of species-specific GPI protein-encoding genes. Int J Med Microbiol 308:812–818. https://doi.org/10.1016/j.ijmm.2018.06.014
36. Arastehfar A, Fang W, Badali H et al (2018) Low-cost tetraplex PCR for the global spreading multi-drug resistant fungus, *Candida auris*

and its phylogenetic relatives. Front Microbiol 9:1119. https://doi.org/10.3389/fmicb.2018.01119

37. Arastehfar A, Fang W, Daneshnia F et al (2019) Novel multiplex real-time quantitative PCR detecting system approach for direct detection of *Candida auris* and its relatives in spiked serum samples. Future Microbiol 14:33–45. https://doi.org/10.2217/fmb-2018-0227
38. Jensen J, Muñoz P, Guinea J, Rodríguez-Créixems M, Peláez T, Bouza E (2007) Mixed fungemia: incidence, risk factors, and mortality in a general hospital. Clin Infect Dis 44:e109–e114. https://doi.org/10.1086/518175
39. Arastehfar A, Fang W, Pan W et al (2019) YEAST PANEL multiplex PCR for identification of clinically important yeast species: stepwise diagnostic strategy, useful for developing countries. Diagn Microbiol Infect Dis 93: 112–119. https://doi.org/10.1016/j.diagmicrobio.2018.09.007
40. Pfaller MA, Diekema DJ, Gibbs DL et al (2009) Results from the ARTEMIS DISK Global Antifungal Surveillance Study, 1997 to 2007: 10.5-year analysis of susceptibilities of noncandidal yeast species to fluconazole and voriconazole determined by CLSI standardized disk diffusion testing. J Clin Microbiol 47: 117–123. https://doi.org/10.1128/JCM.01747-08
41. Martínez-Murcia A, Navarro A, Bru G, Chowdhary A, Hagen F, Meis JF (2018) Internal validation of GPS™ MONODOSE CanAur dtec-qPCR kit following the UNE/EN ISO/IEC 17025:2005 for detection of the emerging yeast *Candida auris*. Mycoses 61:877–884. https://doi.org/10.1111/myc.12834
42. Leach L, Zhu Y, Chaturvedi S (2018) Development and validation of a real-time PCR assay for rapid detection of *Candida auris* from surveillance samples. J Clin Microbiol 56: e01223–e01217. https://doi.org/10.1128/JCM.01223-17
43. Sexton DJ, Kordalewska M, Bentz ML, Welsh RM, Perlin DS, Litvintseva AP (2018) Direct detection of emergent fungal pathogen *Candida auris* in clinical skin swabs by SYBR green-based quantitative PCR Assay. J Clin Microbiol 56:e01337–e01318. https://doi.org/10.1128/JCM.01337-18
44. Sexton DJ, Bentz ML, Welsh RM, Litvintseva AP (2018) Evaluation of a new T2 Magnetic Resonance assay for rapid detection of emergent fungal pathogen *Candida auris* on clinical skin swab samples. Mycoses 61:786–790. https://doi.org/10.1111/myc.12817
45. Pfaller MA, Wolk DM, Lowery TJ (2016) T2MR and T2*Candida*: novel technology for the rapid diagnosis of candidemia and invasive candidiasis. Future Microbiol 11:103–117. https://doi.org/10.2217/fmb.15.111
46. Mylonakis E, Clancy CJ, Ostrosky-Zeichner L et al (2015) T2 magnetic resonance assay for the rapid diagnosis of candidemia in whole blood: a clinical trial. Clin Infect Dis 60: 892–899. https://doi.org/10.1093/cid/ciu959
47. Zervou FN, Zacharioudakis IM, Kurpewski J, Mylonakis E (2017) T2 Magnetic Resonance for fungal diagnosis. Methods Mol Biol 1508: 305–319. https://doi.org/10.1007/978-1-4939-6515-1_18
48. Yamamoto M, Alshahni MM, Tamura T et al (2018) Rapid detection of *Candida auris* based on Loop-Mediated isothermal amplification (LAMP). J Clin Microbiol 56: e00591–e00518. https://doi.org/10.1128/JCM.00591-18

Chapter 2

Candida auris Identification and Profiling by MALDI–ToF Mass Spectrometry

Alireza Abdolrasouli and Mark A. Fraser

Abstract

MALDI–ToF MS has become the standard method for routine identification of most medically important yeasts in clinical and public health laboratories and has largely replaced phenotypic identification methods as a first-line identification tool. Fungal identification is based on extensive and well-curated mass spectra libraries usually provided by the manufacturer of the MALDI–ToF MS platform; however, many centers do create specialized or in-house database collections to aid analysis. Most MALDI–ToF MS systems offer simple and standardized workflows for the identification of clinically relevant yeasts to species level with a high throughput, high accuracy, and a low overall cost per test. This makes MALDI-ToF MS an ideal platform for use in routine clinical, diagnostic, and research microbiology laboratories which may lack experience or expertise in the identification of pathogenic fungi.

In this chapter we review three standard protocols for the proteomic-based identification of *Candida auris* isolated from cultures of clinical or environmental surveillance samples in diagnostic and research laboratories.

Key words Mass spectrometry, MALDI–ToF, MALDI Biotyper, *Candida auris*

1 Introduction

Over the past decade, matrix-assisted laser desorption ionization–time of flight mass spectrometry (MALDI–ToF MS) has supplanted most of the biochemical and phenotypic methods used in clinical diagnostic laboratories for the routine identification of medically important yeasts and molds [1, 2]. The introduction of proteomic identification using commercially available MALDI–ToF MS platforms into microbiology laboratories has allowed the accurate and reliable species-level identification of most yeast species frequently encountered in clinical specimens. Once acquired, MALDI–ToF MS systems can provide inexpensive, high-throughput capacities with fast turnaround times, which suits the high demand of modern and centralized microbiology laboratories [3]. Current MALDI–ToF MS methods are mainly based on the

Alexander Lorenz (ed.), *Candida auris: Methods and Protocols*, Methods in Molecular Biology, vol. 2517,
https://doi.org/10.1007/978-1-0716-2417-3_2,

identification of yeasts grown on solid agar media; however, the direct detection and identification of many *Candida* species from positive automated blood culture systems has been demonstrated and may become the standard procedure in near future [4]. In addition, there is growing evidence that MALDI–ToF MS may provide a rapid and reliable means of detecting emerging antifungal resistance in fungal pathogens [5, 6] although the utility of this in a routine diagnostic microbiology setting is as yet not understood.

Early reports concerning *Candida auris* highlighted the difficulties encountered in accurately identifying this emerging pathogen using conventional biochemical and phenotypic methods such as VITEK 2 YST, API 20C, BD Phoenix yeast identification system, RapID Yeast Plus, and MicroScan [7–9]. Isolates were frequently misidentified as a variety of unrelated, yet often commonly identified yeast species, such as *Clavispora* (*Candida*) *lusitaniae* and *Saccharomyces cerevisiae*. Currently, accurate identification of *C. auris* can be performed using the Bruker Biotyper MALDI–ToF MS using the updated Bruker FDA-approved MALDI Biotyper CA System library (Version Claim 4) or their "research use-only" libraries (Versions 2014 [5627] and more recent) and using the bioMérieux VITEK (MALDI–ToF) MS using the FDA-approved IVD v3.2 or their "research use-only" libraries (with Saramis Ver 4.14 database and *Saccharomycetaceae* update).

Whole genome sequencing (WGS) was widely used to describe the population structure and genetic relatedness of *C. auris* [10, 11]. WGS analyses have revealed the existence of at least four phylogenetically separate clonal lineages of *C. auris*, each with firm associations with distinct geographic regions: clade I (South Asia), clade II (East Asia), clade III (South Africa), and clade IV (South America) [12]. Within each clade, WGS data indicate that there are very few single-nucleotide polymorphisms (SNPs), typically less than 70 SNPs, which is in sharp contrast to vast sequence divergence between different *C. auris* clades, consistent with their almost simultaneous and very recent emergence in multiple geographic areas [13]. Recently, a potential new *C. auris* clade (clade V) has been reported in Iran; interestingly, this clade is separated from the other clades by greater than 200,000 SNPs [14].

MALDI–ToF MS has been previously used in a limited number of studies to investigate the genetic diversity of *C. auris*. However, these applications remain under active development and do not currently provide the discriminatory power needed for epidemiologically relevant high-resolution strain typing. In one study, Prakash et al. [15] studied the population structure and genetic relatedness among 104 global *C. auris* isolates from India (clade II), South Africa (clade III), and Brazil (clade IV) using multilocus sequence typing (MLST), amplified fragment length polymorphism (AFLP) fingerprinting, and MALDI–ToF MS cluster analysis. Although MALDI–ToF MS spectral profiling was able to demonstrate evidence of geographical clustering, it did not correlate fully with the genotyping methods. Clade typing in this manner using proteomics provided by MALDI–ToF MS analysis would be beyond the remit of most diagnostic laboratories and would therefore not be achievable in a real-time diagnostic setting. Different clonal populations appear to possess different phenotypic and virulence traits and antifungal susceptibility profiles [16–19].

Supplemental MALDI–ToF MS databases that include additional *C. auris* strains from all four of the phylogenetic clades may enable users to overcome identification challenges by providing consistently higher MALDI identification scores.

In this chapter, we focus on the description of protocols for the standard identification of *C. auris* isolates obtained on solid media from clinical or environmental surveillance samples. Three different protein extraction methods that can be used in routine laboratory identification will be described. As a standard typing method using MALDI–ToF MS is not yet readily available for *C. auris*, we further explain a strategy to develop such profiling approach.

2 Materials

2.1 Laboratory Equipment

1. Pipettes, pipette tips.
2. A bench top microcentrifuge capable of reaching ~15,000 × g.
3. Vortex mixer.
4. Standard microfuge tube racks.
5. MALDI–ToF MS mass spectrometer and software for microbial species identification.
6. Clean MALDI target plates (reusable or disposable).

2.2 Basic Consumables and Chemicals

1. Media for cultivation of yeasts from clinical samples or subcultures of isolated colonies. Most laboratories use standard Sabouraud's dextrose agar supplemented with 50-mg/L chloramphenicol (SABC) to inhibit bacterial contamination. This medium is commercially available in different forms (ready prepared plates or dehydrated powder) from several manufacturers. The basic formulation consists of:

Glucose	40 g
Mycological peptone	10 g
Agar	15 g
Distilled water	1 L

Heat to dissolve. Autoclave at 121 °C for 15 min.

2. Microbiology inoculation loops (1 μL or 10 μL).
3. Deionized water.
4. High-grade ethanol, without denaturant.
5. Formic acid (MALDI-grade): add 700 μL of formic acid to 300-μL deionized water.
6. Acetonitrile (MALDI-grade).
7. α-cyano-4-hydroxy cinnamic acid (HCCA).
8. Trifluoroacetic acid (TFA).
9. Organic solvent (50% acetonitrile, 2.5% trifluoroacetic acid, 47.5% sterile water). To prepare 1-mL organic solvent, add 475-μL sterile water, 500-μL acetonitrile and 25-μL TFA. Mix by pipetting up and down several times.
10. Clean wooden sticks.
11. Sterile 1.5 mL-microfuge tubes.

3 Methods

MALDI–ToF MS protocols are generally based on transfering fungal cells from an isolated colony grown on solid media to a spot on a MALDI–ToF MS target plate, usually following a protein extraction procedure. Spots containing fungal cells, or fungal cell extractions, are then overlaid with an appropriate matrix. Most MALDI–ToF MS methods require some element of protein extraction, either in situ on target plate or through a separate extraction method, using formic acid, ethanol, and acetonitrile. Many extraction protocols have been developed and extensively evaluated in clinical diagnostic settings to simplify protein collection and maximize efficiency in the laboratory [20, 21]. Matching scores are generated related to the proteomic profile obtained and the proteomic profiles present in the database attached to the MALDI–ToF MS. This score can be used for identification unknown isolates and provide a level of confidence in the accuracy of the identification obtained.

3.1 Preanalytical Procedures

MALDI ToF MS-based identification of *C. auris* from clinical or environmental samples requires the primary isolation of yeast colonies in culture using a solid medium. This can be achieved in two ways: (1) direct inoculation of clinical specimens or environmental surveillance samples onto standard mycology media such as SABC agar, or (2) application of a liquid enrichment/selective culture medium (*see* **Note 1**) for selective isolation and growth of *C. auris* from mixed populations (screening environmental samples) which requires a final subculture into solid media. It is also important to obtain a "pure" culture of yeast prior to MALDI–ToF MS identification. Of note, many body sites like throat, axilla, and groin that are routinely screened for the presence of *C. auris* harbor mixed populations of yeasts that colonize the same anatomical site. Hence, application of commercially available chromogen-based media for screening of these samples is recommended. Application of chromogenic media such as CHROMagar™ Candida Plus allows the specific detection of *C. auris* in mixed cultures when two or more different species are present in the same sample [22]. Moreover, thermophilic growth of *C. auris* at 42–45 °C can be used for selection of this organism from mixed populations; however, other yeasts including *Candida albicans* can also grow in higher temperatures.

Most MALDI–ToF MS identification protocols are established based on the growth of yeast cells on SABC medium in pure culture. In addition to standard SABC agar, other common microbiological media like Columbia blood agar or chocolate agar (optional alternatives to SABC) have been used with acceptable results (*see* **Note 2**). These cultures are usually incubated at 28 °C ± 2 °C or 35 °C ± 2 °C for 24–48 h prior to MALDI–ToF MS identification.

3.2 Sample Preparation Procedures

Several different matrix substances have been described for diagnostic MALDI–ToF MS; in both commercial Bruker and bioMérieux MS protocols, α-cyano-4-hydroxy cinnamic acid (HCCA) is widely used to generate proteomic reference spectra and for the analysis of unknown yeast isolates in routine diagnostic procedures.

3.2.1 Preparation and Handling of Organic Solvent

1. Label a 1.5 mL tube with "organic solvent" and date (*see* **Note 3**).
2. Prepare organic solvent as described in Subheading 2, and add to the pre-labelled tube (*see* **Note 4**).
3. Organic solvent is stable at ambient temperature for ~2 weeks.

3.2.2 Preparation and Handling of Matrix

1. Take a pre-prepared vial of 25-mg HCCA supplied by MALDI–ToF MS manufacturer.
2. Transfer 250 μL of organic solvent into the matrix tube and mix. This can be done using a vortex mixer, or a "scratch-mix" achieved by running the tube along a tube rack. The organic solvent may take some time to dissolve, so it can be left for 5 min at ambient temperature before a second round of mixing is attempted.
3. Matrix can be stored in the dark at ambient temperature for up to 2 weeks; however, if any visible crystallization appears during this period, it must be discarded and a new matrix prepared.

3.2.3 Preparation and Handling of 70% Formic Acid

1. Label a 1.5 mL tube with "70% Formic Acid" and date (*see* **Note 5**).
2. Prepare 70% formic acid as described in Subheading 2.
3. 70% formic acid can be stored at ambient temperature for 1 week.

3.2.4 Full Extraction Protocol (FX)

1. Label one 1.5-mL conical microcentrifuge tube for each isolate to be tested.
2. Transfer 300-μL deionized water to each tube.
3. Pick several colonies from 24 to 48 h culture (*see* **Note 6**) using a plastic disposable microbiological loop. Use a new loop for each isolate.
4. Resuspend yeast cells in 300-μL water to a make a turbid suspension (approximate cell density of 6×10^8 CFU/mL or absorbance of 0.45 at wavelength of 600 nm). In clinical microbiology laboratories, turbidity equivalent to the McFarland standard number 2 is often used. This suspension needs to be homogenous and free of aggregates. Gently vortex the suspension. Any clumps may lead to poor results (*see* **Note** 7).
5. Add 900-μL absolute ethanol to each tube (creating a final concentration ~70%), mix by vortexing.
6. Place tubes in a bench top microfuge and centrifuge at maximum speed (~15,000 × g) for 2 min.
7. Remove the supernatant (ethanol) completely. Recentrifuge the pellet once more for 2 min.
8. Remove any remaining ethanol and allow the pellet to dry at ambient temperature (*see* **Note 8**). This step may take up to 30 min. Alternatively, this drying step can be skipped, and the biomass pellet can be immediately resuspended in 70% formic acid following the removal of residual ethanol.
9. Thoroughly resuspend the pellet in 10–60 μL of 70% formic acid. The appropriate volume of formic acid depends on the size of pellet; however, ~40-μL formic acid is usually sufficient (*see* **Note 9** and **10**).
10. Incubate the suspension for 5 min at ambient temperature.
11. Add an equal volume (as **step 9**) of pure acetonitrile and mix by pipetting.
12. Incubate the suspension for 5 min at ambient temperature.
13. Centrifuge the tubes at maximum speed (~15,000 × g) for 2 min to remove the cellular debris.
14. Place 1 μL of supernatant on MALDI target plate at the predecided position (*see* **Note 11** and **12**).
15. Allow the target plate to dry at ambient temperature. Do not use incubators or other procedures to speed up this step.

16. Overlay each spot with 1-μL matrix solution and allow to air dry at ambient temperature.
17. Target plate is ready for analysis on MALDI–ToF MS.

3.2.5 Partial Extraction Protocol (PX)

A modified partial extraction method has been developed by Fraser et al. [21] to simplify and expedite the identification of yeast isolates. The method was applied to 6343 isolates encompassing 71 yeast species and allowed the correct identification of 94% of isolates.

1. Prepare a labelled 1.5-mL sterile tube with 50-μL deionized water.
2. Emulsify the yeast cells in the water to a density roughly equivalent to 2 McFarland (OD of 0.45 at 600 nm).
3. Add 50-μL absolute ethanol and vortex for 30 s. If it is likely there is insufficient biomass, a centrifugation step can be inserted here to concentrate the cellular matter.
4. Take 1 μL of this suspension, and place directly onto the assigned spot on the target plate. Allow to air dry at ambient temperature.
5. Once dry, overlay the spot with 1 μL of 70% formic acid. Allow to air dry at ambient temperature.
6. Once dry, overlay the spot with 1-μL prepared HCCA matrix, and allow to air dry once more. The sample is now ready for analysis.

3.2.6 On-Target Plate Extraction Protocol (OTL)

This protocol (also known as short extraction or extended direct transfer) is widely used in many microbiology laboratories for routine identification of bacteria and yeasts isolated from clinical specimens. The on-target extraction method is easy and fast and follows the same steps as the direct deposit procedure used for identification of bacteria. However, an extra "lysis step" is added to the process (directly carried out on the cellular mass on the target plate) by adding 70% formic acid. In the bioMérieux Vitek MS system, 25% formic acid is used instead. This helps to increase the likelihood of an acceptable identification score.

1. Pick a single colony with a wooden toothpick (applicator) or a pipette tip.
2. Carefully spread the cellular mass onto the selected spot on the target plate.
3. Overlay the cells on the target plate with 1 μL of formic acid (70% v/v when using the Bruker system or 25% when using bioMérieux Vitek MS system), and let dry at ambient temperature.

4. Overlay the spot with 1-μL matrix solution and let air dry.
5. The plate is now ready to run on the MALDI–ToF MS.

3.2.7 Running the Target Plate on the MALDI–ToF MS Platform

In Bruker MALDI Biotyper spectra are acquired in a linear positive ion mode at a laser frequency of 60 Hz across a mass/charge ratio (*m/z*) of 2000–20,000 Da using the MicroFlex LT mass spectrometer (Bruker Daltonics). For each spectrum, 240 laser shots in 40-shot steps from different areas of the sample spot are accumulated and analyzed in automatic mode using default settings. The spectra are analyzed using the MALDI Biotyper software package with the reference database.

3.2.8 Result Interpretation

The MALDI Biotyper CA System Software electronically transforms the acquired spectrum of an unknown yeast into a peak list. Using a biostatistical algorithm, this peak list is compared to reference peak lists of all yeasts in the reference library, and a log (score) value between 0.00 and 3.00 is calculated. Interpretation is based on a "traffic light" color system (*see* Table 1). In the Bruker system, log scores above the threshold 2.00 indicate acceptable identification to species level.

3.2.9 *Candida auris* Mass Spectra Profiling

It is the experience of these authors and others [23] that it is not currently possible to strain type or clade cluster *C. auris* using the MALDI–ToF MS platform and methodologies currently on offer to the clinical diagnostic laboratory. While the addition of specific profiles representative of each clade group is possible, and may aid the accurate identification as *C. auris* overall, there is insufficient variation in proteomic profiles demonstrated between clades for them to be confidently distinguished using MALDI–ToF MS. Figure 1 shows an example of the spectra generated by examples of all four clade types of *C. auris*. Figure 2 demonstrates examples of four clades in a single image for comparison.

While strain typing methodologies for MALDI–ToF MS have been described, they have generally been performed using specifi-

Table 1
MALDI–ToF MS identification scores and their interpretations

Range	Interpretation	Colour
2.00–3.00	High confidence identification	Green
1.70–1.99	Low confidence identification	Yellow
<1.70	No organism identification possible	Red

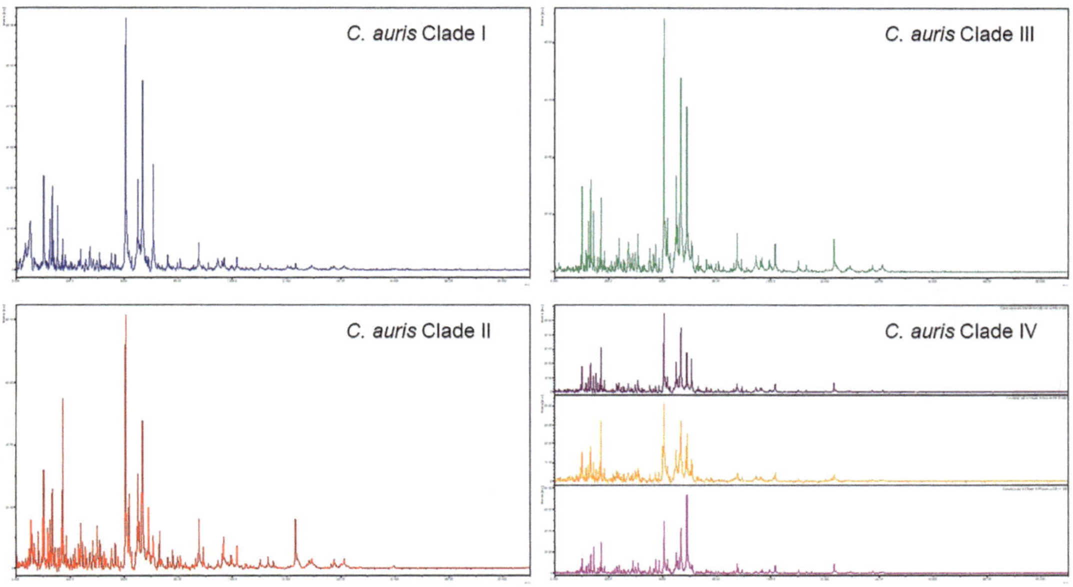

Fig. 1 Examples of proteomic spectra generated using a full extraction method with the Bruker Biotyper Sirius MALDI–ToF MS from each currently accepted clade of *C. auris*. Each isolate was grown on SABC agar at 30 °C ± 2 °C for 48 h. Isolates were provided by the National Collection of Pathogenic Fungi (NCPF), UK Health Security Agency (UKHSA), Bristol, UK. Clade IV represents examples from several geographic regions; purple is from Colombia, orange is from Israel, and pink is from Panama. The spectra are displayed using FlexAnalysis software from Bruker Daltonics GmBH, Bremen, Germany

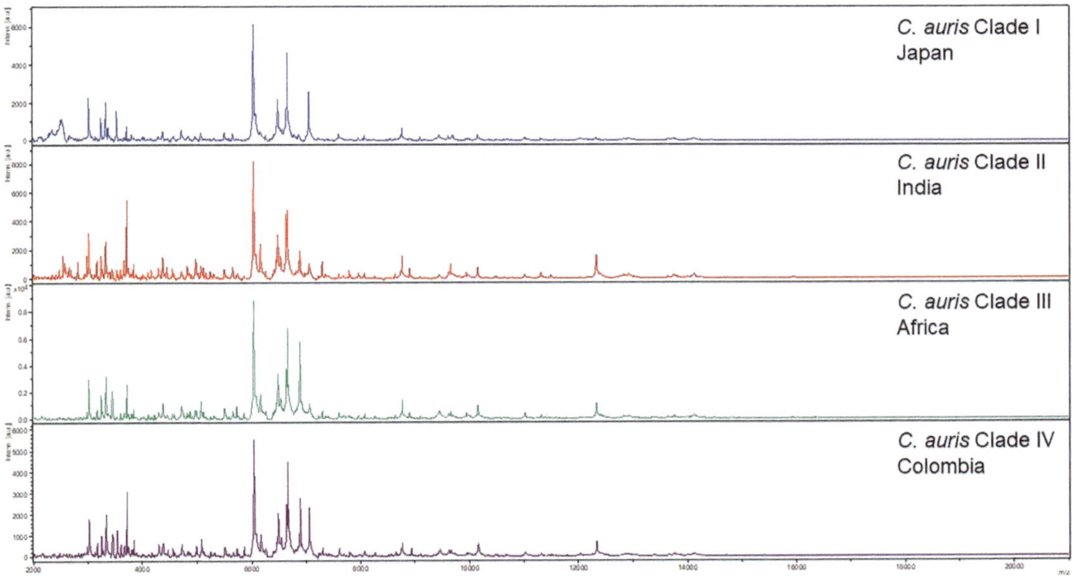

Fig. 2 All four currently accepted clades of *C. auris* in alignment. Proteomic spectra created from isolates provided by the NCPF grown at 30 °C ± 2 °C on SABC agar for 48 h. The similarities in all four spectral traces help to further demonstrate how it is not possible to separate clade grouping using MALDI–ToF MS and currently used diagnostic laboratory extraction/analysis techniques

cally designed, adapted methodologies and software algorithms which would not be conducive with the rapid identification required in clinical diagnostic laboratories. The improvement in identification that can be achieved by the addition of more *C. auris* isolates into commercially available and internally curated databases remains a critical component to the ongoing success of MALDI–ToF MS identification of *C. auris*.

4 Notes

1. Salt/dulcitol enrichment broth is commercially available from S2 media (Spokane Valley, WA) and Thermo Oxoid (Ontario, Canada). This broth contains 10% NaCl, which inhibits the growth of most, but not all, yeast species. *Candida glabrata* is one of the species that can sometimes grow at high salt, and therefore dulcitol is used as a carbon source rather than glucose. *C. glabrata* cannot assimilate or ferment dulcitol. Although CDC has primarily used this medium in broth form, it could also be used as a salt/dulcitol agar plate. Salt/dulcitol enrichment broth consists of 5-g pancreatic digest of casein, 5-g peptic digest of animal tissue, and 100-g sodium chloride (NaCl) dissolved in a liter of deionized (DI) water with 20 g of dulcitol as the added carbon source. The medium was mixed well, the pH was adjusted to 5.6 ± 2, and the filter sterilized through a 0.2-μm filter.
2. Columbia blood agar (containing horse or sheep blood) and chocolate agar are commercially available in pre-plated, ready-to-use form and can be purchased from various manufacturers in different countries.
3. Steps involved in the preparation of organic solvent must be performed inside a fume cabinet wearing chemical grade gloves. Chemicals used in organic solvent are harmful, and care must be taken in handling them.
4. Use only high-quality (MALDI-grade) chemicals.
5. Formic acid is caustic and irritant. Wear personal protective equipment (gloves and eye protection), and handle carefully.
6. MALDI–ToF MS should only be performed on pure cultures. Some yeasts may exhibit colonial variation in one culture; it is recommended to test each colonial variant separately.
7. If resuspension of yeast colonies with microbiological loop is insufficient, pipetting the solution up and down until a homogenous and aggregate-free suspension is achieved can be used.
8. Do not use incubators, hairdryers, hot blocks, or any direct heat source to speed up the drying process.

9. Ideally, formic acid needs to cover the pellet. Adding too much will dilute the sample, whereas adding too little will not allow the lysis of cell walls and release of intracellular proteins.
10. Usually pipetting up and down several times is needed to fully resuspend the pellet in formic acid.
11. A template can be used for each run to record the position of each sample on the target plate. This is particularly helpful when large numbers of isolates are tested on each run.
12. In some laboratories, it is preferred to test samples in duplicate (i.e., each sample is placed on two consecutive positions on the MALDI target plate).

References

1. Marklein G, Josten M, Klanke U et al (2009) Matrix-assisted laser desorption ionization-time of flight mass spectrometry for fast and reliable identification of clinical yeast isolates. J Clin Microbiol 47(9):2912–2917
2. Lau AF, Drake SK, Calhoun LB et al (2013) Development of a clinically comprehensive database and a simple procedure for identification of molds from solid media by matrix-assisted laser desorption ionization-time of flight mass spectrometry. J Clin Microbiol 51(3):828–834
3. Walsh TJ, McCarthy MW (2019) The expanding use of matrix-assisted laser desorption/ionization-time of flight mass spectroscopy in the diagnosis of patients with mycotic diseases. Expert Rev Mol Diagn 19(3):241–248
4. Spanu T, Posteraro B, Fiori B et al (2012) Direct MALDI-TOF mass spectrometry assay of blood culture broths for rapid identification of *Candida* species causing bloodstream infections: an observational study in two large microbiology laboratories. J Clin Microbiol 50(1):176–179
5. Vella A, De Carolis E, Vaccaro L et al (2013) Rapid antifungal susceptibility testing by matrix-assisted laser desorption ionization-time of flight mass spectrometry analysis. J Clin Microbiol 51(9):2964–2969
6. Paul S, Singh P, Shamanth A et al (2018) Rapid detection of fluconazole resistance in *Candida tropicalis* by MALDI-TOF MS. Med Mycol 56(2):234–241
7. Chowdhary A, Anil Kumar V, Sharma C et al (2014) Multidrug-resistant endemic clonal strain of *Candida auris* in India. Eur J Clin Microbiol Infect Dis 33(6):919–926
8. Govender NP, Magobo RE, Mpembe R et al (2018) *Candida auris* in South Africa, 2012–2016. Emerg Infect Dis 24(11): 2036–2040
9. Jeffery-Smith A, Taori SK, Schelenz S et al (2018) *Candida auris*: a review of the literature. Clin Microbiol Rev 31(1):1–18
10. Eyre DW, Sheppard AE, Madder H et al (2018) A *Candida auris* outbreak and its control in an intensive care setting. N Engl J Med 379(14):1322–1331
11. Rhodes J, Abdolrasouli A, Farrer RA et al (2018) Genomic epidemiology of the UK outbreak of the emerging human fungal pathogen *Candida auris*. Emerg Microbes Infect 7:43. https://doi.org/10.1038/s41426-018-0045-x
12. Lockhart SR, Etienne KA, Vallabhaneni S et al (2017) Simultaneous emergence of multidrug-resistant *Candida auris* on 3 continents confirmed by whole-genome sequencing and epidemiological analyses. Clin Infect Dis 64(2): 134–140
13. Borman AM, Johnson EM (2020) *Candida auris* in the UK: introduction, dissemination, and control. PLoS Pathog 16(7):e1008563
14. Chow NA, De Groot T, Badali H et al (2019) Potential fifth clade of *Candida auris*, Iran, 2018. Emerg Infect Dis 25(9):1780–1781
15. Prakash A, Sharma C, Singh A et al (2016) Evidence of genotypic diversity among *Candida auris* isolates by multilocus sequence typing, matrix-assisted laser desorption ionization time-of-flight mass spectrometry and amplified fragment length polymorphism. Clin Microbiol Infect 22(3):277.e1–277.e9
16. Borman AM, Szekely A, Johnson EM (2016) Comparative pathogenicity of United Kingdom isolates of the emerging pathogen *Candida auris* and other key pathogenic *Candida*

species. mSphere 1(4):e00189–16. https://doi.org/10.1128/mSphere.00189-16

17. Szekely A, Borman AM, Johnsona EM (2019) *Candida auris* isolates of the Southern Asian and South African lineages exhibit different phenotypic and antifungal susceptibility profiles in vitro. J Clin Microbiol 57(5): e02055–e02018. https://doi.org/10.1128/JCM.02055-18
18. Welsh RM, Sexton DJ, Forsberg K et al (2018) Insights into the unique nature of the East Asian clade of the emerging pathogenic yeast *Candida auris*. J Clin Microbiol 57(4): e00007–e00019. https://doi.org/10.1128/JCM.00007-19
19. Muñoz JF, Gade L, Chow NA et al (2018) Genomic insights into multidrug-resistance, mating and virulence in *Candida auris* and related emerging species. Nat Commun 9(1): 5346. https://doi.org/10.1038/s41467-018-07779-6
20. Fatania N, Fraser M, Savage M et al (2015) Comparative evaluation of matrix-assisted laser desorption ionisation-time of flight mass spectrometry and conventional phenotypic-based methods for identification of clinically important yeasts in a UK-based medical microbiology laboratory. J Clin Pathol 68(12): 1040–1042
21. Fraser M, Brown Z, Houldsworth M et al (2016) Rapid identification of 6328 isolates of pathogenic yeasts using MALDI-ToF MS and a simplified, rapid extraction procedure that is compatible with the Bruker Biotyper platform and database. Med Mycol 54(1):80–88
22. Borman AM, Fraser M, Johnson EM (2020) CHROMagar™ *Candida* Plus: a novel chromogenic agar that permits the rapid identification of *Candida auris*. Med Mycol. https://doi.org/10.1093/mmy/myaa049
23. Vatanshenassan M, Boekhout T, Mauder N et al (2020) Evaluation of microsatellite typing, ITS sequencing, AFLP fingerprinting, MALDI-TOF MS, and fourier-transform infrared spectroscopy analysis of *Candida auris*. J Fungi 6(3):146. https://doi.org/10.3390/jof6030146

Chapter 3

Molecular Tools for *Candida auris* Identification and Typing

Auke W. de Jong, Bert Gerrits van den Ende, and Ferry Hagen

Abstract

Identification of *Candida auris* by conventional identification methodologies can be challenging. While whole genome sequencing is seen as the golden standard to genotype *C. auris* at an inter- and intraspecies level, it is costly and time-consuming. Sequencing the transcribed spacer (ITS) region and microsatellite typing provide simple, fast, and inexpensive alternatives for identification and genotyping of *C. auris*. Here we will describe both molecular approaches.

Key words *Candida auris*, Molecular epidemiology, Genotyping, ITS sequencing, Microsatellite typing

1 Introduction

Candida auris was first described in 2009 as a novel species within the *Candida haemulonii* species complex [1]. Despite this description, identification of *Candida auris* remained challenging. Most conventional identification methodologies frequently misidentify *C. auris* as other related species [2, 3]. Moreover, *C. auris* consists of closely related but distinct clonal lineages divided into four, and potentially five, separate geographic-linked clades [4, 5]. The unique ability of *C. auris* to persistently colonize the skin makes it easily transmissible between patients [6]. Therefore, it is crucial to timely detect and identify *C. auris* to trace the source and prevent further spread. Various molecular typing methods can be applied to genotype *C. auris* at an inter- and intraspecies level. These methods include (a) whole genome sequencing (WGS), (b) sequencing of the internal transcribed spacer (ITS) region, and (c) microsatellite typing. While whole genome sequencing can be seen as a golden standard for identifying and genotyping *C. auris*, it is still laborious and costly. ITS sequencing and microsatellite typing provide simple, fast, and cheap alternatives for identification and genotyping of *C. auris*. Here, we will describe both methods. The evaluation of

Alexander Lorenz (ed.), *Candida auris: Methods and Protocols*, Methods in Molecular Biology, vol. 2517,
https://doi.org/10.1007/978-1-0716-2417-3_3,

six DNA regions as potential DNA barcodes for fungi showed that the nuclear ribosomal internal transcribed spacer (ITS) region has the highest probability of successful identification for the broadest range of fungi [7]. To this end the ITS region was proposed as the primary fungal barcode marker to the Consortium for the Barcode of Life. A quality-controlled ITS database was established containing 2800 ITS sequences representing 421 species, including *C. auris* [8]. Amplification of the ITS region in *C. auris* with primers ITS1 and ITS4 allows for identification but cannot be reliably used for genotyping all clades due to the fact that Clades I and III share an identical ITS sequence [9]. Microsatellite typing on the other hand is able to group *C. auris* isolates into clusters that correspond to the five recognized clades and identify distinct subgenotypes [9, 10]. While ITS identification is based on sequencing, microsatellite typing relies on fragment analysis of amplified microsatellite loci. Microsatellites are DNA repeat units of 1–6 nucleotides in length, while the larger repeat units have been coined as minisatellites. The panel developed for *C. auris* microsatellite typing consists of three dinucleotide repeat loci, six trinucleotide repeat loci, and three enea-nucleotide repeat loci [10]. ITS sequencing is a highly reliable method for the identification of *C. auris*, but when it comes to genotyping, it has poor performance compared to microsatellite typing. Therefore, microsatellite typing should be the preferred method to perform within an outbreak setting.

2 Materials

Prepare all solutions using ultrapure water with a sensitivity of 18MΩ cm at 25 °C and molecular biology grade reagents. Store the prepared reagents at room temperature unless stated otherwise.

2.1 Polymerase Chain Reaction

1. 1 M Tris–HCl, pH 8.0: Dissolve 121.1 g Tris (hydroxymethyl)-aminomethane in 800 mL ultrapure water. Adjust pH to 8.0 by adding 1 N HCl. Add ultrapure water to a final volume of 1000 mL.
2. 0.5 M Na–EDTA, pH 8.0: Add 186.1 g Na–EDTA to 800-mL ultrapure water, and stir on a magnetic stirrer. Add, while stirring, ± 20 g sodium hydroxide. The pH will be approximately 8.0. Adjust total volume to 1000 mL with ultrapure water.
3. 1 M DTT: Dissolve 1.54 g DL–dithiothreitol in 10 mL ultrapure water. Aliquot in portions of 1 mL and store at −20 °C.
4. 5 mM dNTP mix: Pipet 50 μL 25 mM dATP, 50 μL 25 mM dTTP, 50 μL 25 mM dCTP, 50 μL 25 mM dGTP, and 800 μL ultrapure sterile water into a sterile vial. Store aliquots of 50 μL

at −20 °C. Dilute this solution to 1 mM before use by adding 200-μL ultrapure sterile water to the aliquot.

5. DNA polymerase dilution buffer: Mix 200 μL 1 M Tris–HCl, pH 8.0, 200 μL 0.5-M EDTA pH 8.0, 100 μL 1 M DTT, 24.5 mL ultrapure water, and 25 mL 87% glycerol, and mix thoroughly. Store solution at +4 °C.
6. DNA polymerase solution: Dilute Taq polymerase to 0.5 U/μ L with DNA polymerase dilution buffer. Store aliquots at −20 °C.
7. PCR ammonium buffer (10×): 160 mM $(NH_4)_2SO_4$, 670 mM Tris–HCl, pH 8.8, and 0.1% (v/v) Tween 20. Store aliquots at −20 °C.
8. 50 mM $MgCl_2$ solution: 50 mM $MgCl_2$ in ultrapure water. Store aliquots at −20 °C.
9. TE buffer: 10 mM Tris–HCl, 1 mM EDTA, pH 8.0.
10. Primers: All primers should be diluted to 10 pmol/μL with TE buffer (*see* **Notes 1** and **2**).
11. Dimethyl sulfoxide solution (DMSO; *see* **Note 3**).
12. Genomic DNA (*see* **Note 4**).
13. Thermocycler.

2.2 Gel Electrophoresis

1. Ultrapure agarose.
2. TAE electrophoresis buffer (50×): 50 mM EDTA, pH 8.0, 2-M Tris base, 1 M glacial acid.
3. DNA stain: 10,000× GelRed in DMSO.
4. Sample loading buffer: Dissolve 50 mg bromophenol blue in 500 μL of 96% ethanol. Mix 5 mL 87% glycerol with 45 mL ultrapure water. Add the dissolved bromophenol blue. Mix both solutions and store at +4 °C. Before use prepare vials with 1 mL loading buffer and add 3 μL DNA stain. The mixture has a yellow brown color (*see* **Note 5**).
5. DNA size and concentration marker (*see* **Note 6**).

2.3 Sequencing (Sanger Method)

1. BigDye Terminator kit: BigDye™ Terminator v3.1 Cycle Sequencing Kit (Thermo Fisher Scientific, Bleiswijk, the Netherlands).
2. Primers: The same as PCR primers (see **step 10** in Subheading 2.1) but diluted to 4 pmol/μL with sterile ultrapure water.
3. 96-well skirted PCR plate.
4. Thermocycler.

2.4 Fragment Analysis

1. Orange 600 DNA Size Standard (600 bp).
2. A vortex mixer.
3. 96-well semi-skirted PCR plate ABI® style.

3 Methods

3.1 Internal Transcribed Spacer (ITS) Sequence-Based Identification

The multi-copy ribosomal DNA (rDNA) locus is used for identification and phylogenetic analysis of Fungi [7, 8]. The Internal Transcribed Spacer (ITS) 1 and 2, and the 5.8S gene that is located between these spacers are regarded as the gold standard in fungal identification. After a polymerase chain reaction, the obtained product needs to be Sanger-sequenced to identify the fungal isolate.

3.1.1 Polymerase Chain Reaction

The PCR is performed in a 25 μL reaction volume. It is preferred to prepare the master mix on ice.

1. Prepare a PCR master mix for N samples by mixing $(N+1) \times 14$ μL ultrapure water, $(N+1) \times 2.5$ μL PCR buffer, $(N+1) \times 2.5$ μL 1 mM dNTP mix, $(N+1) \times 1$ μL $MgCl_2$, $(N+1) \times 1$ μL of forward primer (ITS1 or ITS5), $(N+1) \times 1$ μL reverse primer (ITS4), $(N+1) \times 1$ μL DMSO, and $(N+1) \times 1$ μL DNA polymerase solution (*see* **Note 1**).
2. Make aliquots of 24 μL in 200 μL thin-walled PCR tubes. Last one is used as blank (negative template control, NTC).
3. Add 1 μL of genomic DNA sample to the aliquoted PCR mix.
4. Start the PCR procedure: initial denaturation at 95 °C for 5 min, 35 cycles of 95 °C for 45 s, 52 °C for 30 s, and 72 °C for 1 min, followed by a final elongation at 72 °C for 6 min. Cool the reaction to 15 °C and store at 4 °C for further processing.
5. Check PCR result by electrophoresis on a 1% agarose gel.

3.1.2 Gel Electrophoresis

1. Mix 200 mL 50× TAE with 9800 mL ultrapure water to make 1× TAE buffer.
2. Prepare a 1% agarose gel for electrophoresis with 1× TAE buffer (*see* **Note 7**).
3. Mix 2.5 μL DNA size marker with 3 μL sample loading buffer.
4. Mix 2.5 μL of each PCR amplicon with 3 μL sample loading buffer.
5. Transfer all samples to one well each.
6. Run electrophoresis in 1× TAE buffer at 100 V (6–10 V/cm) for 20 min.
7. View the agarose gel under UV light (*see* **Note 8**).

3.1.3 Sequencing

1. Prepare master mixes for N samples for each primer, if sense and antisense strand has to be analyzed: $(N+1) \times 4.3$ μL molecular grade water, $(N+1) \times 3$ μL BigDye buffer,

$(N + 1) \times 1$ μL primer (4 pmol/μL), $(N + 1) \times 0.7$ μL BigDye Terminator.

2. Add 1 μL (20–50× diluted) PCR amplicon to 9 μL master mix.
3. Run sequencing program in a thermal cycler using the following program: 95 °C 1 min, 30 cycles that consist of 95 °C for 10 s, 50 °C for 5 s, and 60 °C for 4 min, followed cooling at 15 °C. Sequencing PCR reaction can be stored at −20 °C until further use.
4. Prior to analysis on a capillary sequencer, the sequencing PCR products need to be purified. This approach is largely depending on the sequencing facility used. Hence, we do not describe this step nor the analysis on the sequencing platform.
5. Raw sequencing files (.ab1-extension) can be checked by commercial software packages or by freely available ones, like MEGA version 7 or newer versions (https://www.megasoftware.net/) [11].
6. The obtained plain sequence file (FASTA format) can then be uploaded to either the NCBI Blast website (https://blast.ncbi.nlm.nih.gov/) [12] or, the most reliable option, the curated ISHAM–ITS database (https://its.mycologylab.org/) [8].

3.2 Microsatellite Typing

Microsatellites are DNA repeat units of 1–6 nucleotides in length, while the larger repeat units have been coined as minisatellites. The panel developed for *C. auris* microsatellite typing consists of three dinucleotide repeat loci, six trinucleotide repeat loci, and three enneanucleotide repeat loci [10]. These 12 loci result after fragment analysis in a microsatellite profile that can be interpret in repeat units per locus (total fragment length minus the repeat unit flanking sites divided by the repeat unit size) [10] or as the total fragment length in base pairs [9]. The obtained profiles can be compared and visualized in commercially available software packages, like Biolomics (BioAware, Hannut, Belgium) and BioNumerics (Applied Maths, Sint-Martens-Latem, Belgium).

3.2.1 Polymerase Chain Reaction

The PCR is performed in a 25 μL reaction volume. It is preferred to prepare the master mix on ice.

1. Prepare a PCR master mix for N samples by mixing $(N + 1) \times 15$ μL ultrapure water, $(N + 1) \times 2.5$ μL ammonium buffer, $(N + 1) \times 2.5$ μL 1 mM dNTP mix, $(N + 1) \times 1$ μL $MgCl_2$, $(N + 1) \times 1$ μL of forward primer, $(N + 1) \times 1$ μL reverse primer, and $(N + 1) \times 1$ μL DNA polymerase solution.

   Make aliquots of 24 μL in 200 μL thin-walled PCR tubes (*see* Table 1 for the primer pairs).
2. Add 1 μL of genomic DNA sample to the aliquoted PCR mix.

Table 1
Primers used for microsatellite typing of *C. auris* [10]

Target	Forward primer sequence (5′–3′) *Labelled 5′ with 6-FAM*	Reversed primer sequence (5′–3′)	Repeat unit
M2a	GCAACATCCTGAGCAGTATCAC	GGTGTTGACGTGCCCAAATA TGC	AG
M2b	CCACTCCGTTTTGGGTCTG	AGAGAATCTACAAATGTGTCGC	AG
M2c	CTGTTTCTGTGGCAGGCTTCC	GCCACGTTTCACYGCYACCAT	AG
M3Ia	GCATGGATCAACAGCTAACAG	AGTGCCAGGCTGTGTACTTTTG	CAA
M3Ib	CATCCTAACGCTGGCTCTTC	GGYTTTGAGGYTGCCCTAGC	CAA
M3Ic	GCAACTACGCATTGTGTATTC	CTAACAGAGGATTTCAATTGCC	TTA
M3IIa	GTTCAAAATCGCTGACGGTC	GAGATGATGATGGCACTTGC	CTA
M3IIb	GTGAATGGAGCACCACAACCAG	GCGCAAATGACTGGCCCATG	GTA
M3IIc	GTGATGAGCGCACTACACAGG	GGCGAAGAAACGGTGAGTAC	CAA
M9a	CTTGTCTAGTTTGCGATC TACGC	GAGACTGCCAAGCCAAGC	GATGA TGAA
M9b	CTGCTTACTGGAGACTCTTCC	GATGAGGAGGACGAGGACG	TCATCG TCA
M9c	GTACGAAATGGGGATAATTGGG	ACCAACCGTGCTATTCTC	TCCTTC TTC

3. Transfer the thin-walled PCR tubes containing the PCR mix into a thermocycler, and start the PCR procedure. Initial denaturation at 95 °C for 5 min, 35 cycles of 95 °C for 30 s, 60 °C for 30 s, and 72 °C for 1 min, followed by a final elongation at 72 °C for 5 min. Cool the reaction to 15 °C and store at 4 °C for further processing.
4. Check PCR result by electrophoresis on 1% agarose gel.

3.2.2 Gel Electrophoresis

See Subheading 3.1.2 to follow the practical steps of gel electrophoresis.

3.2.3 Fragment Analysis

Prior the actual fragment analysis on a capillary sequencer, the PCR products need to be diluted 100–300× as otherwise the detector returns a saturated signal that cannot be analyzed.

1. To dilute the PCR products 100×, aliquot 198 μL ultrapure water in a 96-well semi-skirted PCR plate.
2. Add 2 μL of PCR product to each well.
3. Next, prepare a dilution of the size standard. Calculate 8.9 μL ultrapure water and 0.1 μL Orange 600 DNA Size Standard per sample (for a full 96-well plate, calculate for 100 samples). Mix using a vortex mixer.

4. Aliquot 9 μL of the diluted size standard into each well of a new 96-well semi-skirted PCR plate.
5. Transfer 1 μL diluted PCR product (from **step 2** above) to the corresponding well of the plate filled with diluted Orange 600 DNA Size Standard.
6. Seal the plate and centrifuge to get rid of droplets and bubbles.
7. Place the 96-well plate in the PCR machine, and run the following denaturation program: 96 °C for 1 min followed by 4 °C for 1 min.
8. Centrifuge the plate to get rid of droplets and bubbles.
9. Transfer the plate to the sequence facility to perform a fragment analysis.
10. Raw fsa-format files can be checked using a commercial software package, like BioNumerics (Applied Maths, Sint-Martens-Latem, Belgium).

4 Notes

1. Primers used for sequencing the internal transcribed spacer region are ITS1 (5′-TCC GTA GGT GAA CCT GCG G-3′), ITS4 (5′-TCC TCC GCT TAT TGA TAT GC-3′), and ITS5 (5′-GGA AGT AAA AGT CGT AAC AAG G-3′). Primers ITS1 and ITS5 are situated at the end of the 18S of the ribosomal DNA locus. Some fungal genera contain an intron directly after the ITS5 site but before the ITS1 site. The choice of one of the primers depends on whether the intron, if present, should be amplified or not. Primer ITS4 is situated at the beginning of the 28S of the ribosomal DNA locus.
2. The primers needed for microsatellite loci are listed in Table 1. All forward primers have a fluorescent 6-FAM label attached at the 5′-side [9]. This differs from the original publication of De Groot et al., which used multiple fluorescent labels to enable multiplex microsatellite typing [10].
3. DMSO acts as a PCR enhancer, which increases yield and specificity.
4. Genomic DNA samples are checked for quality using Nano-Drop 1000 and/or running on a 0.8% agarose gel. Although many DNA extraction kits are available on the market, they do not always have the same effectivity in extracting DNA from a wide range of fungal genera.
5. The sample loading buffer contains bromophenol blue that shifts color at pH 6.2. The buffer itself has a pH lower than 6.2 and changes into a dark blue color after adding samples, which have a pH higher than 6.2.

6. The DNA size and concentration marker should cover a fragment range of 200–10,000 nucleotides and DNA concentrations between 10 and 100 ng.
7. How much agarose gel has to be prepared depends on the system that is used. The agarose gel does not contain a DNA stain. Samples are mixed with sample loading buffer containing GelRed DNA stain.
8. The length of the amplicon varies and depends on the genus/species. Normally the length varies between 400 and 600 bp except when an intron is present; then the amplicon could be 1000–1100 bp long. No intron has yet been detected in this region in *Candida auris*.

References

1. Satoh K, Makimura K, Hasumi Y, Nishiyama Y, Uchida K, Yamaguchi H (2009) *Candida auris* sp. nov., a novel ascomycetous yeast isolated from the external ear canal of an inpatient in a Japanese hospital. Microbiol Immunol 53(1): 41–44. https://doi.org/10.1111/j.1348-0421.2008.00083.x
2. Kathuria S, Singh PK, Sharma C, Prakash A, Masih A, Kumar A, Meis JF, Chowdhary A (2015) A. Multidrug-resistant *Candida auris* misidentified as *Candida haemulonii*: characterization by Matrix-Assisted Laser Desorption Ionization-Time of Flight Mass Spectrometry and DNA sequencing and its antifungal susceptibility profile variability by Vitek 2, CLSI broth microdilution, and Etest method. J Clin Microbiol 53(6):1823–1830. https://doi.org/10.1128/JCM.00367-15
3. Mizusawa M, Miller H, Green R, Lee R, Durante M, Perkins R, Hewitt C, Simner PJ, Carroll KC, Hayden RT, Zhang SX (2017) Can multidrug-resistant *Candida auris* be reliably identified in clinical microbiology laboratories? J Clin Microbiol 55(2):638–640. https://doi.org/10.1128/JCM.02202-16
4. Chow NA, de Groot T, Badali H, Abastabar M, Chiller TM, Meis JF (2019) Potential fifth clade of *Candida auris*, Iran, 2018. Emerg Infect Dis 25(9):1780–1781. https://doi.org/10.3201/eid2509.190686
5. Lockhart SR, Etienne KA, Vallabhaneni S, Farooqi J, Chowdhary A, Govender NP, Colombo AL, Calvo B, Cuomo CA, Desjardins CA, Berkow EL, Castanheira M, Magobo RE, Jabeen K, Asghar RJ, Meis JF, Jackson B, Chiller T, Litvintseva AP (2017) Simultaneous emergence of multidrug-resistant *Candida auris* on 3 continents confirmed by whole-genome sequencing and epidemiological analyses. Clin Infect Dis 64(2):134–140. https://doi.org/10.1093/cid/ciw691
6. de Jong AW, Hagen F (2019) Attack, defend and persist: how the fungal pathogen *Candida auris* was able to emerge globally in healthcare environments. Mycopathologia 184(3): 353–365. https://doi.org/10.1007/s11046-019-00351-w
7. Schoch CL, Seifert KA, Huhndorf S, Robert V, Spouge JL, Levesque CA, Chen W, Fungal Barcoding Consortium, Fungal Barcoding Consortium Author List (2012) Nuclear ribosomal internal transcribed spacer (ITS) region as a universal DNA barcode marker for Fungi. Proc Natl Acad Sci U S A 109(16):6241–6246. https://doi.org/10.1073/pnas.1117018109
8. Irinyi L, Serena C, Garcia-Hermoso D, Arabatzis M, Desnos-Ollivier M, Vu D et al (2015) International Society of Human and Animal Mycology (ISHAM)-ITS reference DNA barcoding database--the quality controlled standard tool for routine identification of human and animal pathogenic fungi. Med Mycol 53(4):313–337. https://doi.org/10.1093/mmy/myv008
9. Vatanshenassan M, Boekhout T, Mauder N, Robert V, Maier T, Meis JF, Hagen F (2020) Evaluation of microsatellite typing, ITS sequencing, AFLP fingerprinting, MALDI-TOF MS, and Fourier-transform infrared spectroscopy analysis of *Candida auris*. J Fungi 6(3):146
10. de Groot T, Puts Y, Berrio I, Chowdhary A, Meis JF (2020) Development of *Candida auris* short tandem repeat typing and its application

to a global collection of isolates. mBio 11(1): e02971–19. https://doi.org/10.1128/mBio.02971-19

11. Kumar S, Stecher G, Tamura K (2016) MEGA7: molecular evolutionary genetics analysis version 7.0 for bigger datasets. Mol Biol Evol 33(7):1870–1874. https://doi.org/10.1093/molbev/msw054

12. Altschul SF, Gish W, Miller W, Myers EW, Lipman DJ (1990) Basic local alignment search tool. J Mol Biol 215(3):403–410. https://doi.org/10.1016/S0022-2836(05)80360-2

Chapter 4

qPCR Detection of *Candida auris* Using the GPS™ CanAur MONODOSE dtec–qPCR Test

Antonio Martínez-Murcia, Gema Bru, and Aaron Navarro

Abstract

Candida auris is a multidrug-resistant pathogenic ascomycete yeast of increasing health concern. *C. auris* colonizes patient's skin and can persist for weeks on surfaces, so it can be transmitted within and between hospitals. The most common diagnostic platforms in microbiology use reference databases that have not yet incorporated *C. auris*, misidentifying it. This chapter describes how to detect *C. auris* by qPCR with the GPS™ *CanAur MONODOSE dtec–qPCR Test* (Alicante, Spain) in less than 45 min, using ready-to-use tubes with all the components dehydrated. This commercial kit was subjected to validation following the guidelines of the UNE–EN ISO/IEC 17025:2005 and French Standard NF T90-471:2010.

Key words DNA extraction, qPCR, *Candida auris* diagnosis, UNE–EN ISO/IEC 17025:2005, Validation

1 Introduction

Candida auris is a multidrug-resistant pathogenic ascomycete yeast of increasing health concern, which was first described by Japan in 2009 [1–4]. *C. auris* colonizes patient's skin and can persist for weeks on surfaces, so it has a high potential of environmental contamination and can be transmitted within and between hospitals [5–7]. For this reason, an early detection allows to reduce transmission, initiate the appropriate antifungal therapy, and implement infection control measures [8–10]. The most common diagnostic platforms prevalent in microbiology laboratories, such as the MicroScan (Beckman Coulter), RapID Yeast Plus (Thermo Scientific), and BD Phoenix (Becton Dickinson Diagnostics), use reference databases that have not yet incorporated *C. auris*, misidentifying it with *C. haemulonii*, *C. duobushaemulonii*, *C. lusitaniae*, *C. parapsilosis*, and *Rhodotorula glutinis* [11–16]. Kordalewska et al. [17] developed in 2017 a conventional end-point PCR and melting temperature PCR analysis to identify

Alexander Lorenz (ed.), *Candida auris: Methods and Protocols*, Methods in Molecular Biology, vol. 2517,
https://doi.org/10.1007/978-1-0716-2417-3_4,

C. auris, *C. haemulonii*, *C. duobushaemulonii*, and *C. lusitaniae*. Moreover, end-point PCR and real-time PCR using hydrolysis probes have been reported for detection of *C. auris* from surveillance samples [18–20]. However, a simple, fast, and low-cost quantitative method to detect this nosocomial pathogen is still not commercially available or has not been validated according to international standards [21]. The GPS™ *CanAur MONODOSE dtec–qPCR Test* (Alicante, Spain) was subjected to validation following the guidelines of the UNE–EN ISO/IEC 17025:2005 and French Standard NF T90-471:2010 [22, 23]. The diagnostic validation was performed using extracted DNA from JCM15448 strain and 117 clinical isolates of *Candida auris* from India, Oman, South Africa, Spain, the United Kingdom, and Venezuela. Samples were subjected to qPCR by using this commercial kit, and the results showed Ct values within the interval of 8–17 (very strong positive signal). No signal was detected (negative result) for all strains of the other species tested, *C. duobushaemulonii* (9 strains), *C. haemulonii* (8 strains), *C. lusitaniae* (2 strains), and *C. pseudohaemulonii* (6 strains) [21].

This chapter describes how to detect *C. auris* by qPCR using the GPS™ *CanAur MONODOSE dtec–qPCR Test* (Genetic Analysis Strategies SL, Alicante, Spain) in less than 45 min. It uses ready-to-use tubes with all the components dehydrated, where it is only needed to add the extracted DNA before thermo-cycling.

2 Materials

All components from *GPStraction DNA colonies* kit and GPS™ *CanAur MONODOSE dtec–qPCR Test* are stable at room temperature for transport but should be stored at −20 °C if not immediately used. Keep a linear workflow and components refrigerated in a cooling block.

2.1 DNA Extraction Using the GPStraction DNA Colonies Kit

1. Colonies Extraction Buffer.
2. 1.5-ml Eppendorf tubes.
3. Vortex.
4. Freezer.

2.2 GPS™ CanAur MONODOSE dtec–qPCR Test for C. auris Detection

1. TargetSpecies MONODOSE dtec–qPCR (24 or 96 individual tubes). They contain a dehydrated mixture of specific primers and labelled probe, deoxynucleotides (dNTPs), bovine serum albumin (BSA), polymerase, and buffer at optimal concentrations.
2. Internal Control. It is in MONODOSE tubes and includes primers, probe, and DNA template for a PCR internal control.

3. Standard Template. It contains dehydrated target copies for positive control.
4. Standard Buffer. This is exclusively for the standard template resuspension (500 μl).
5. DNA isolation kit, as *GPStraction DNA colonies.*
6. DNase-/RNase-free water, to prepare standard curve dilution and qPCR.
7. Sterile pipette tips with filters, to prevent contamination of micropipettes.
8. qPCR tubes, strips, or plates.
9. Micropipettes.
10. Vortex mixer.
11. Spinner centrifuge.
12. Cooling block.
13. Real-time PCR device.

3 Methods

3.1 GPStraction DNA Colonies Kit

1. Pipette 200 μl of Colony Extraction Buffer in a 1.5-ml tube.
2. Resuspend some biomass from a single colony in the buffer.
3. Vortex for 10 s.
4. Keep at −20 °C for at least 5 min (Fig. 1).
5. Use in PCR.

3.2 qPCR for *C. auris* Detection

3.2.1 Preparation of Standard Curve Dilution Series

1. Pipette 900 μl of DNAse-/RNAse-free water into five tubes, and label them as 2–6 (*see* **Note 1**).
2. Pulse-spin for 5 s the standard template, reconstitute with 120 μl of standard buffer, and vortex thoroughly, label as number 1 (*see* **Note 2**).
3. Pipette 100 μl of diluted standard template into tube 2.
4. Vortex thoroughly and pulse-spin for 5 s.
5. Change tip and pipette 100 μl from tube 2 into tube 3.
6. Vortex thoroughly and pulse-spin.
7. Repeat **steps 5** and **6** with the tubes 4–6 to complete serial dilution (Fig. 2 and Table 1).
8. Pipette 5 μl of template into each well for the standard curve according to your plate setup (*see* **Note 3**).

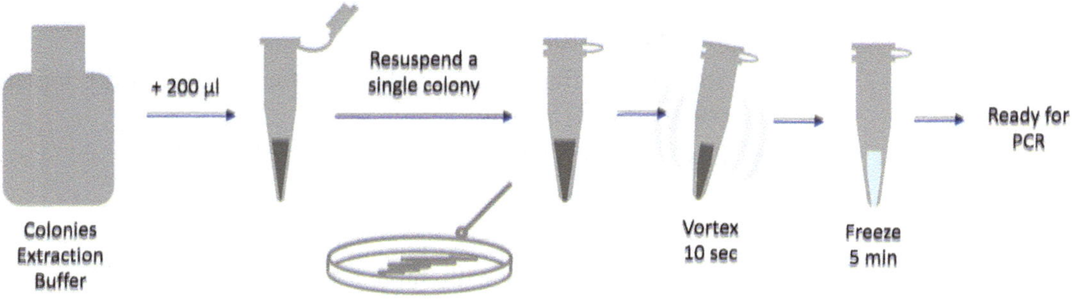

Fig. 1 Process of *C. auris* DNA extraction using the *GPStraction DNA colonies* kit

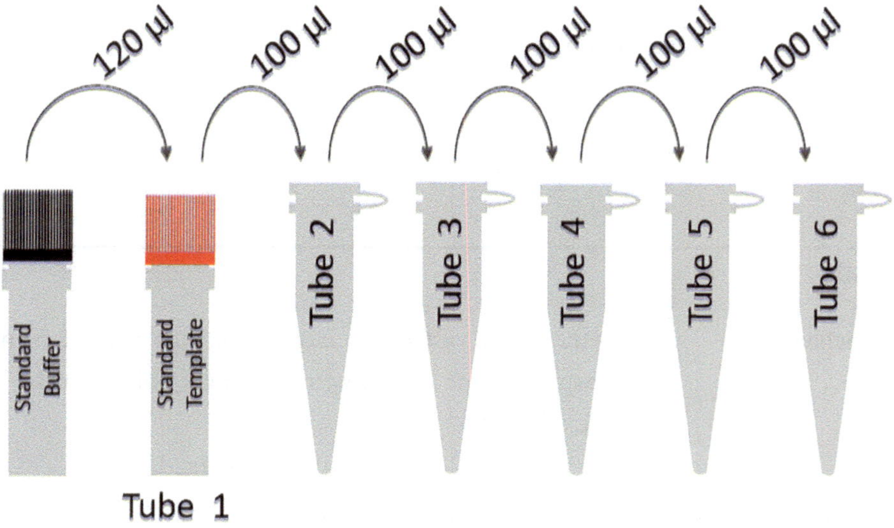

Fig. 2 Scheme of the standard curve dilution series preparation

Table 1
Concentration of each tube of the standard curve dilution series as copies/µl and the number of copies used in qPCR (5 µl)

Standard curve dilution series	Copies/µl	Copies in 5 µl
Standard template	$2{\cdot}10^5$	10^6
Tube 2	$2{\cdot}10^4$	10^5
Tube 3	$2{\cdot}10^3$	10^4
Tube 4	$2{\cdot}10^2$	10^3
Tube 5	$2{\cdot}10$	10^2
Tube 6	2	10

Table 2
Amplification regime for *C. auris* detection by qPCR using the GPS™ *CanAur MONODOSE dtec–qPCR Test*

Step		Time	Temperature
Activation		1 min	95 °C
40 cycles	Denaturation	10 s	95 °C
	Hybridization/extension and data collection	1 min	60 °C

3.2.2 qPCR

GPS™ reagents contain BSA and are compatible with all real-time PCR thermal cyclers, glass capillary or plate-based (*see* **Note 4**).

1. Add the desired volume of sample, and, when needed, complete this final volume (20 μl) by adding DNase-/RNase-free water (*see* **Note 5**).
2. Vortex thoroughly and pulse-spin for 5 s.
3. Put samples in the real-time PCR thermal cyclers following the Amplification Regime shown in Table 2 (*see* **Note 6**).
4. There are some recommended reaction controls to ensure the quality system of the procedure (*see* **Note 7**).

3.2.3 Recommended Reaction Controls

Negative Control (Ctrl−)

1. Add 20 μl of DNase-/RNase-free water to one tube (*see* **Note 8**).

Positive Control (Ctrl+)

1. Prepare a standard curve dilution series as described above from the standard template.
2. Add 5 μl of the standard template dilution (i.e., $2 \cdot 10^2$ copies/μl) to 15 μl of water (*see* **Note 9**).

Matrix Inhibition Control (M-Inh)

1. Add a known amount of standard template (i.e., $2 \cdot 10^2$ copies/μl) together with the sample (Table 3 and *see* **Note 10**).

Extraction Negative Control (ExtCtrl−)

1. Perform an extraction according to *GPStraction DNA colonies* kit, or similar, without adding the sample.
2. Add 5 μl of extraction negative sample to 15 μl of water (*see* **Note 11**).

Table 3
Example of a matrix inhibition control (IC) reaction

Reagent	Volume
Standard template dilution (i.e., $2{\cdot}10^2$ copies/μl)	5 μl
DNase-/RNase-free water	10 μl
Sample	5 μl
Final reaction volume	**20 μl**

Table 4
Interpretation of the possible results obtained in qPCR

Ctrl+	Sample	Ctrl−	ExtCtrl−	M-Inh	Interpretation
−	+/−	+/−	+/−	+/−	Experiment failed
+	+	+	+/−	+/−	PCR reagents are contaminated
		−	+	+/−	Possible contamination at extraction step
			−	+	Positive sample
+	−	−	−	+	Negative sample
				−	PCR inhibition

Key symbols + and −: amplification does or not occur, respectively

Extraction Positive Control (ExtCtrl+)

1. Perform an extraction according to *GPStraction DNA colonies* kit, or similar, adding the standard template, or DNA extracted from pure cultures, into the first extraction buffer (*see* **Note 12**).

*3.2.4 Interpretation of Results (see **Note 13**)*

1. The linear regression of copy number logarithm versus Ct gives the constants Y-intercept and slope of the standard curve (Eq. 1).
2. The number of copies in the sample can be calculated based on the regression (Eq. 2).

$$\mathrm{Ct} = \mathrm{Yinter} + \mathrm{Slope}\tilde{n}\log(\text{copy number}) \quad (1)$$

$$\text{Copy number} = 10^{\frac{(\mathrm{Ct}-\mathrm{Yinter})}{\mathrm{Slope}}} \quad (2)$$

3. The possible interpretations of results could be experiment fails, PCR reagents are contaminated, possible contamination at extraction step, positive sample, negative sample, and PCR inhibition (Table 4).

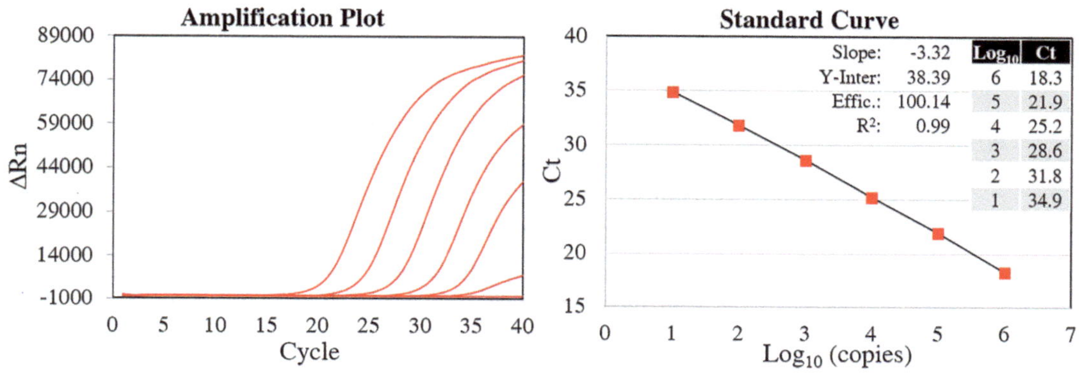

Fig. 3 Quality control of the GPS™ *CanAur MONODOSE dtec–qPCR Test* using tenfold serial dilutions of standard template from 10 to 10^6 DNA copies and negative control. Amplification plot (left) and calibration curve with stats (right)

3.2.5 Validation Methods

1. All batches are calibrated with a standard curve from 10^6 to 10 copies with our standard template (Fig. 3). Diverse parameters are evaluated: Ct, slope, R^2, and efficiency.

4 Notes

1. Micropipettes used to dilute the standard template should not be used for other PCR reagents. Also, extract, store, and prepare positive materials (samples, positive controls, and PCR products) in a separate laboratory environment.
2. The concentration of the standard template will be $2{\cdot}10^5$ copies/µl after adding 120 µl of standard buffer. Be sure to resuspend the dehydrated standard template completely.
3. The final volume in each qPCR reaction well is 20 µl.
4. Plastic of the *Generic Tube* (provided in the GPS™ *CanAur MONODOSE dtec–qPCR Test*) is compatible with: StepOne™, StepOnePlus™, ABI 7500 Fast, LightCycler® 96, LightCycler® Nano, CFX96™, PikoReal™ 24 well, DNA Engine® systems, MiniOpticon™ 48-12, and Opticon® 2. For other devices, please transfer the content of the MONODOSE (20 µl) to appropriate tubes.
5. The volume of sample should range from 5 µl to a maximum qPCR volume of 20 µl. To determine the sample volume, please take into account the possible presence of inhibitors. Protect the primer/probe from prolonged exposure to light.
6. After preparing MONODOSE tubes, run reactions immediately. Fluorogenic signal should be collected during extension and data collection step by using the FAM channel for the target and the HEX channel for the internal control.

7. Some reaction controls are recommended when considering the guidelines of ISO-IEC 17025:2005 Standard. When setting up your qPCR protocol, select the controls considered better suits your quality system.
8. This reaction should be negative. A positive result may be considered as a symptom of DNA contamination in the water, making the test inconclusive. Water must be replaced.
9. To avoid cross-contamination with the positive control, pipette it after closing reaction tubes with negative control and samples. A positive result indicates that qPCR setup is correct and works. If negative, the test should be carefully repeated after checking the thermal protocol.
10. An optimal result should show a positive signal, equal, or higher (same or lower Ct) than these found for the positive control alone (in this example, tube 4, with 10^3 copies in the qPCR reaction). Inhibition may be total (negative result) or partial when a considerable increase in the Ct compared to the standard template dilution is observed. If there is inhibition, a sample dilution to 1/10 is recommended (as long as concentration is not close to detection limit). The matrix inhibition control is external, allowing to check the inhibition on the main target of interest.
11. In this case, the test for DNA includes the reagents used in the extraction steps. If positive, when the negative control is negative, a contamination occurred during the extraction process. Extraction reagents must be discarded.
12. The positive extraction control indicates the effectiveness of the extraction method used. A positive result should be expected. If negative, extraction must be carefully repeated or the extraction method replaced.
13. To obtain the sample quantification directly from the device, the standard dilution series must be defined in the software of your qPCR device as standard with the specified copies for each dilution (see Subheading 3.2.1). To refer the values obtained with qPCR to the sample material, please take into account the elution volume after extraction, the sample volume processed and any dilution performed.

Acknowledgments

We are grateful to Genetic Analysis Strategies SL laboratories for providing this validated method to detect *Candida auris*.

References

1. Satoh K, Makimura K, Hasumi Y et al (2009) *Candida auris* sp. nov., a novel ascomycetous yeast isolated from the external ear canal of an inpatient in a Japanese hospital. Microbiol Immunol 53:41–44
2. Arikan-Akdagli S, Mahmoud G, Meis JF (2018) Antifungal resistance: specific focus on multidrug resistance in *Candida auris* and secondary azole resistance in *Aspergillus fumigatus*. J Fungi 4:129
3. De Jong A, Hagen F (2019) Attack, defend and persist: how the fungal pathogen *Candida auris* was able to emerge globally in healthcare environments. Mycopathologia 184:353–365
4. Kordalewska M, Perlin DS (2019) Identification of drug resistant *Candida auris*. Front Microbiol 10:1918
5. Welsh RM, Bentz ML, Shams A et al (2017) Survival, persistence, and isolation of the emerging multidrug-resistant pathogenic yeast *Candida auris* on a plastic health care surface. J Clin Microbiol 55:2996–3005
6. Chowdhary A, Voss A, Meis JF (2016) Multidrug-resistant *Candida auris*: "new kid on the block" in hospital-associated infections? J Hosp Infect 94:209–212
7. Schelenz S, Hagen F, Rhodes JL et al (2016) First hospital outbreak of the globally emerging *Candida auris* in a European hospital. Antimicrob Resist Infect Control 5:35
8. Kordalewska M, Perlin DS (2019) Molecular diagnostics in the times of surveillance for *Candida auris*. J Fungi 5:77
9. Lockhart SR (2019) *Candida auris* and multidrug resistance: defining the new normal. Fungal Genet Biol 131:103243
10. Ong CW, Chen SC, Clark JE et al (2019) Diagnosis, management and prevention of *Candida auris* in hospitals: position statement of the Australasian Society for Infectious Diseases. Intern Med J 49:1229–1243
11. Kathuria S, Singh PK, Sharma C et al (2015) Multidrug-resistant *Candida auris* misidentified as *Candida haemulonii*: characterization by matrix-assisted laser desorption ionization-time of flight mass spectrometry and DNA sequencing and its antifungal susceptibility profile variability by Vitek 2, CLSI broth microdilution, and Etest method. J Clin Microbiol 53:1823–1830
12. Mizusawa M, Miller H, Green R et al (2016) Can multidrug-resistant *Candida auris* be reliably identified in clinical microbiology laboratories? J Clin Microbiol 55:638–640
13. Sexton DJ, Kordalewska M, Bentz ML et al (2018) Direct detection of emergent fungal pathogen *Candida auris* in clinical skin swabs by SYBR green-based quantitative PCR assay. J Clin Microbiol 56:e01337–e01318
14. Lone SA, Ahmad A (2019) *Candida auris* – the growing menace to global health. Mycoses 62:620–637
15. Caceres DH, Forsberg K, Welsh RM et al (2019) *Candida auris*: a review of recommendations for detection and control in healthcare settings. J Fungi 5:111
16. Sabino R, Verissimo C, Pereira AA et al (2020) *Candida auris*, an agent of hospital-associated outbreaks: which challenging issues do we need to have in mind? Microorganisms 8:181
17. Kordalewska M, Zhao Y, Lockhart SR et al (2017) Rapid and accurate molecular identification of the emerging multidrug resistant pathogen *Candida auris*. J Clin Microbiol 55: 2445–2452
18. Theill L, Dudiuk C, Morales-Lopez S et al (2018) Single-tube classical PCR for *Candida auris* and *Candida haemulonii* identification. Rev Iberoam Micol 32:110–112
19. Leach L, Zhu Y, Chaturvedi S (2018) Development and validation of a real-time PCR assay for rapid detection of *Candida auris* from surveillance samples. J Clin Microbiol 56: e01223–17
20. Mahmoudi S, Afshari S, Kuchak A et al (2019) Methods for identification of *Candida auris*, the yeast of global public health concern: a review. J Mycol Med 29:174–179
21. Martínez-Murcia A, Navarro A, Bru G et al (2018) Internal validation of GPS™ MONODOSE CanAur dtec-qPCR kit following the UNE-EN ISO/IEC 17025:2005 for detection of the emerging yeast *Candida auris*. Mycoses 61:877–884
22. UNE-EN ISO/IEC 17025:2005. Conformity assessment. General requirements for the competence of testing and calibration laboratories. https://www.iso.org/standard/39883.html. Accessed 08 Feb 2021
23. NF T90-471 April 2010. Water quality – detection and quantification of *Legionella* and/or *Legionella pneumophila* by concentration and genic amplification by real time polymerase chain reaction (qPCR). https://nf-validation.afnor.org/en/water-analysis/legionella-spp-legionella-pneumophila-pcr-all-water/. Accessed 08 Feb 2021

Part II

Testing Susceptibility of *Candida auris* to Antifungal Drugs

Chapter 5

Susceptibility Testing of *Candida auris* Isolates

Elizabeth M. Johnson and Andrew M. Borman

Abstract

Susceptibility testing of isolates of *Candida auris* is helpful as a guide to the selection of the most appropriate antifungal agent for treatment as different clades and strains within clades often demonstrate markedly different susceptibility profiles. Some strains are relatively susceptible to all antifungal drugs, but most demonstrate innate resistance to fluconazole, many are cross-resistant to other azoles and others demonstrate resistance to other classes of antifungal. The finding of multi-drug resistance, where an isolate is resistant to two or more classes of antifungal agent, is not uncommon, and development of resistance during a course of treatment has also been documented. This chapter describes a reference broth microdilution method for susceptibility testing and a commercially available gradient strip method.

Key words Susceptibility testing, Resistance, MIC, *Candida auris*

1 Introduction

Susceptibility testing methods and the minimum inhibitory concentration (MIC) results generated by their application play an increasingly pivotal role in the selection of the most appropriate antifungal agents for treatment especially with yeast infections. As methods have evolved and become more sophisticated, there has been the demonstration of species-specific differences in antifungal susceptibility profiles for key species. The correlation of MIC of a given antifungal agent with outcome of treatment with that agent has enabled the development and introduction of clinical breakpoints for many yeast–drug interactions [1–7].

There are two main reference methods for the antifungal susceptibility testing of yeast, those produced by the Clinical and Laboratory Standards Institute (CLSI) and the European Committee on Antimicrobial Susceptibility Testing (EUCAST) [8, 9]. Both are based on broth microdilution testing in which suspensions of the test organism are inoculated into doubling dilutions of the antifungal agents in microtiter plates to enable high throughput of testing. Differences in the inoculum concentration employed,

Alexander Lorenz (ed.), *Candida auris: Methods and Protocols*, Methods in Molecular Biology, vol. 2517,
https://doi.org/10.1007/978-1-0716-2417-3_5,

medium composition, and endpoint determination mean that different absolute MIC values are obtained; thus the two organizations have also developed antifungal susceptibility testing breakpoints that can be applied to results obtained by their test methods with given drug–organism combinations [10, 11]. Although the absolute MICs and thus the associated breakpoints differ, the overall categorization of organisms should be uniform irrespective of the method employed. Thus, isolates can be classified as susceptible or likely to respond to standard dosing, intermediate which incorporates those organisms which are susceptible–dose-dependent and are likely to respond with increased exposure or resistant which describes those organisms that are unlikely to respond to treatment with that agent.

There is also a plethora of commercial methods that have been developed including colorimetric variations of the microtiter plate methodologies incorporating chromogenic substrates for ease of reading, disc tests, gradient strip tests, breakpoint testing, and automated systems such as VITEK. Each has been developed to produce results consistent with one or other of the reference methods, and thus it is imperative to apply the appropriate breakpoints as specified by the manufacturers of these methods [6].

Published clinical breakpoints for the CLSI and EUCAST reference methods have been validated in the clinical setting and are available for some drug–organism combinations [10, 11]. However, in the absence of interpretable clinical outcome data to establish clinical breakpoints, wild-type distributions for a given species can be used to establish epidemiological cutoff values (ECOFF/ECV) [12–16]. This provides an indication of the usual susceptibility patterns encountered with a given drug–organism combination and thus suggests whether the test isolate conforms to the predicted wild-type susceptibility profile or is displaying a less susceptible or non-wild-type phenotype. This is the current situation with *C. auris* for which there are no established clinical breakpoints. However, the wild-type profile for *C. auris* with fluconazole suggests that the majority of isolates have very high fluconazole MICs which exceed predicted blood concentrations of the drug on standard dosing and for most *Candida* species would be considered resistant. With fluconazole EUCAST has suggested a non-species-related breakpoint for *Candida* of 2.0 mg/L [11], which is the same as the EUCAST and CLSI breakpoint for *Candida albicans*, and when assessed by this criterion, the vast majority of *C. auris* isolates would be considered fluconazole-resistant. Moreover, genetic pointers to azole resistance have been found in *C. auris* strains; two-point mutations in the *ERG11* gene leading to fluconazole resistance have been documented and cells have high ABC–efflux pump activity and the presence of *CDR1* and *MPS* pump genes all of which can contribute to azole drug resistance (for a review *see* [17]). Echinocandin resistance, described in as many as 30% of *C. auris* strains, has also been correlated with the finding of mutations in the *FKS1* gene which encodes the target enzyme [18, 19]. A proportion of strains (10–17%) also demonstrate resistance to amphotericin B, and 5 SNPs in different genomic loci leading to membrane changes or increased catalase activity have been implicated in reduced susceptibility to this agent [20]. Finally, some strains have demonstrated high 5-fluorocytosine (flucytosine) MICs or developed resistance during treatment and an amino acid substitution in *FUR1*, one of the genes involved in metabolism of flucytosine to its toxic form 5-flurouracil, has been documented [18].

This chapter describes application of the CLSI reference method for the susceptibility testing of *C. auris* isolates as well as a commercial gradient strip method which is easier to perform. We recommend the gradient strip method for caspofungin which does not perform uniformly well in broth microdilution testing methods, but this method is it also suitable for testing all the systemic antifungal agents.

2 Materials

2.1 Culture Media

1. Sabouraud (SAC) medium (glucose–peptone agar) containing 0.05-mg/mL chloramphenicol, prepared both as slopes and plates (*see* Subheading 3.1).
2. Chloramphenicol solution: 0.5-g chloramphenicol in 100-mL 96% ethanol.
3. RPMI 1640 medium (with glutamine, without bicarbonate, and with a pH indicator) with 2% glucose, buffered to pH 7.0 at 25 °C with MOPS (final concentration 0.165-mol/L) (*see* Subheading 3.2 for details).

2.2 Buffers

1. 0.165 M MOPS (3-[N-morpholino] propanesulfonic acid) buffer, pH 7.0.
2. 1-mol/L sodium hydroxide: 20 g NaOH in 500 mL of sterile distilled water.

2.3 Antifungal Agents

The antifungal agents required to be tested should be obtained from pharmaceutical companies or other commercial sources and stored in powder form in a desiccator at 4 °C. Caspofungin should be stored at −80 °C.

Appropriate antifungal ranges to be tested:

Antifungal agent	Concentration range (mg/L)	Solvent
Amphotericin B	0.03–16.0	DMSO
Echinocandin	0.016–8.0	DMSO

(continued)

Antifungal agent	Concentration range (mg/L)	Solvent
Fluconazole	0.12–64.0	Water
Itraconazole	0.03–16.0	DMSO (*see* **Note 1**)
Isavuconazole	0.03–16.0	DMSO
Posaconazole	0.03–16.0	DMSO (*see* **Note 1**)
Voriconazole	0.03–16.0	DMSO
Flucytosine	0.12–64.0	Water

2.4 Solvents

1. Dimethyl sulfoxide (DMSO) (*see* **Note 2**).
2. Polyethylene glycol 400 (PEG 400) (*see* **Note 2**).

2.5 Suspension Fluid

1. Sterile 0.85% saline: 8.5 g NaCl in 1 L of sterile distilled water.

2.6 Equipment

1. PH meter.
2. Vacuum pump.
3. Screw-topped 500-mL flasks.
4. Vortex mixer.
5. Spectrophotometer with a wavelength set to 530 nm.
6. Reading mirror.
7. Pipettes and tips, including multichannel pipette.
8. Sterile loops.
9. Sterile saline (4 mL) in capped glass tubes.
10. Sterile distilled water.
11. Universal tubes.
12. 0.22 micron filters for sterilization of medium.
13. 96-well, round-bottomed microtiter plates with lids (*see* **Note 3**).
14. Conical flasks (4 L).
15. Volumetric flasks (2 L).
16. Desiccator for storage of antifungal agents.
17. Delicate balance for weighing powders (10 mg).
18. Heat block to heat to 56 °C.
19. Petri dishes.
20. 35 °C incubator.
21. Plate shaker.
22. Gradient strips for systemic antifungal agents.
23. Ready poured RPMI agar plates.
24. Glass bijou containers.

25. 0.5 McFarland standard.
26. Sterile swabs.
27. Sterile forceps.
28. Incubator at 35 °C.
29. Freezers at −20 °C and − 80 °C.

2.7 Fungal Isolates

1. Pure cultures of all yeast isolates to be tested (*see* Subheading 3.1).
2. Control (QC) strains to be included on each run:

   Candida parapsilosis NCPF 8334/ATCC 22019.

   Candida krusei (*Pichia kudriavzevii*) NCPF 3953/ ATCC 6258.

These strains should be maintained on SAC agar slopes at 4 °C and subcultured every 2 months (*see* **Note 4**).

The day before a test run, prepare a SAC culture plate for single colonies of each control organism, and incubate at 35 °C for 18–24 h.

3 Methods

3.1 Preparation of SAC Media and Culturing of Yeast Isolates

1. Dissolve 65-g Sabouraud agar in 990 mL of sterile water.
2. Autoclave at 115 °C (10 lbs) for 20 min, and then add 10 mL/ L of chloramphenicol solution.
3. Pour into sterile petri dishes or sterile 20-mL tubes (for slopes). Allow to solidify and store at 4 °C.
4. Prepare pure cultures of each yeast isolate by inoculating onto SAC medium to obtain single colonies, and incubate plates inverted at 35 °C for 18–24 h (*see* **Note 5**).

3.2 MIC Testing

3.2.1 Preparation of RPMI Medium

1. Dissolve two vials of pre-weighed RPMI medium (10.4 g each) in 1600-mL distilled water in a 4-L conical flask.
2. Add 69.09-g MOPS buffer and stir until dissolved (*see* **Note 6**).
3. Transfer to a 2-L volumetric flask.
4. Adjust the PH to 7.0 while stirring at 25 °C by adding 1 M NaOH solution.
5. Add additional sterile distilled water to bring the volume to 2.0 L.
6. Filter sterilize and dispense in 200-mL aliquots in sterile screw-topped flasks.
7. Label with date of preparation, batch number, and expiry date. Each batch can be stored for up to 1 month at 4 °C.

8. To perform sterility testing, remove 10-mL aliquots of RPMI from each sterile container into each of two separate sterile universal containers. Incubate one aliquot at 35 °C for 5 days and the other at room temperature for 2 days, and record the results. If there is turbidity in the sterility tests after the incubation period, the batch is contaminated and should be discarded.

3.2.2 Preparation of Drug Dilutions (see ***Note 7****)*

Fluconazole and Flucytosine

1. Weigh out 10 mg of powder into a glass vial, and add 1 mL of sterile distilled water to give a 10,000-mg/L stock solution. Heat to 56 °C to dissolve.
2. Add 6.8 mL of RPMI to give a 1280-mg/L solution.
3. Label the tops of ten universal containers with numbers from 1 to 10 and the sides with the final concentration outlined in Table 1.
4. Dispense the 1280-mg/L stock concentration into the first three labelled tubes (2 mL in tube 1 and 1 mL each into tubes 2 and 3), and then follow the rest of procedure outlined in Table 1.

Table 1
Preparation of drug concentrations for fluconazole and flucytosine

	Amount to be added to container (mL):				
Top label	**Stock drug conc. (mg/L)**	**Stock drug solution**	**RPMI medium**	**Intermediate conc. (mg/L)**	**Side label**
1	1280	2	2	640	128
2	1280	1	3	320	64
3	1280	1	7	160	32
Transfer 2 mL from container 3 to container 4 and 1 mL each to containers 5 and 6					
4	160	2	2	80	16
5	160	1	3	40	8
6	160	1	7	20	4
Transfer 2 mL from container 6 to container 7 and 1 mL each to containers 8 and 9					
7	20	2	2	10	2
8	20	1	3	5	1
9	20	1	7	2.5	0.5
Transfer 2 mL from container 9 to 10. Discard 2 mL from container 9					
10	2.5	2	2	1.25	0.25

Table 2
Preparation of drug concentrations for amphotericin B, isavuconazole, itraconazole, posaconazole, and voriconazole

	Amount to be added to container (mL):				
Top label	Stock drug conc. (mg/L)	Stock drug solution	RPMI medium	Intermediate conc. (mg/L)	Side label
1	320	2	2	160	32
2	320	1	3	80	16
3	320	1	7	40	8
Transfer 2 mL from container 3 to container 4 and 1 mL each to containers 5 and 6					
4	40	2	2	20	4
5	40	1	3	10	2
6	40	1	7	5	1
Transfer 2 mL from container 6 to container 7 and 1 mL each to containers 8 and 9					
7	5	2	2	2.5	0.5
8	5	1	3	1.25	0.25
9	5	1	7	0.62	0.12
Transfer 2 mL from container 9 to 10. Discard 2 mL from container 9					
10	0.62	2	2	0.31	0.06

5. Add 16 mL of RPMI medium to each universal to give final concentrations ranging from 128 to 0.25 mg/L; when diluted with the inoculum during testing this will produce the final desired concentration range of 64–0.125 mg/L.

Amphotericin B, Isavuconazole, Itraconazole, Posaconazole, and Voriconazole

1. Look at the data sheet provided by the manufacturer regarding % activity of the drug, and calculate the weight to be added to 1 mL of DMSO to give a 10,000-mg/L solution equivalent to 100% activity; this will vary from batch to batch of each drug.
2. Add 500 μL of this stock solution to 15.125 mL of RPMI medium to give a 320-mg/L stock solution.
3. Label the tops of ten universal containers with numbers from 1 to 10 and the sides with the final concentration shown in Table 2.
4. Dispense the 320-mg/L stock concentration into the first three labelled tubes (2 mL in tube 1 and 1 mL each into tubes 2 and 3), and then follow the rest of procedure outlined in Table 2.

Table 3
Preparation of drug concentrations for anidulafungin and micafungin

	Amount to be added to container (mL):				
Top label	Stock drug conc. (mg/L)	Stock drug solution	RPMI medium	Intermediate conc. (mg/L)	Side label
1	320	1	3	80	16
2	320	1	7	40	8
Transfer 2 mL from container 2 to container 3 and 1 mL each to containers 4 and 5					
3	40	2	2	20	4
4	40	1	3	10	2
5	40	1	7	5	1
Transfer 2 mL from container 5 to container 6 and 1 mL each to containers 7 and 8					
6	5	2	2	2.5	0.5
7	5	1	3	1.25	0.25
8	5	1	7	0.62	0.12
Transfer 2 mL from container 8 to 9 and 1 mL to container 10. Discard 1 mL from container 8					
9	0.62	2	2	0.31	0.06
10	0.62	1	3	0.16	0.03

5. Add 16 mL of RPMI medium to each universal to give final concentrations ranging from 32 to 0.06 mg/L; when diluted with the inoculum during testing, this will produce the final desired concentration range of 16–0.03 mg/L.

Anidulafungin and Micafungin (Echinocandins) (*see* Note 8)

1. Look at the data sheet provided by the manufacturer regarding % activity of the drug and calculate the weight to be added to 1 mL of DMSO to give a 10,000-mg/L solution equivalent to 100% activity; this will vary from batch to batch of each drug.
2. Add 500 μL of this stock solution to 15.125 mL of RPMI medium to give a 320-mg/L stock solution.
3. Label the tops of ten universal containers with numbers from 1 to 10 and the sides with the final concentration shown in Table 3.
4. Dispense the 320-mg/L stock concentration into the first two labelled tubes (1 mL each into tubes 1 and 2), and then follow the rest of procedure outlined in Table 3.
5. Add 16 mL of RPMI medium to each universal to give final concentrations ranging from 16 to 0.03 mg/L; when diluted with the inoculum during testing, this will produce the final desired concentration range of 8–0.015 mg/L.

3.2.3 Preparation of Plates for the Microdilution Test

Plates should be prepared in batches of 21 using the pre-prepared drug solutions in the ten universal containers in each dilution series described in Subheading 3.2.2 and Tables 1, 2, and 3.

1. Label 21 round-bottom microtiter plates with the drug name, dilution series, batch number, date of preparation, and expiry (6 months from the date of preparation).
2. Place 100 μL of drug-free RPMI solution into column 11 and 200 μL into column 12 (*see* **Note 9**).
3. Pour the 20-mL contents of the universal containing the lowest drug concentration (labelled 10) into a sterile petri dish to facilitate use of a multichannel pipette.
4. By means of the multichannel pipette dispense 100 μL into column 10 of each of the microtiter plates.
5. Repeat with the remaining drug dilutions in ascending order, placing 100 μL of each of the appropriately labelled universal contents into each of the wells of the corresponding labelled column of the plate, such that column ten contains the lowest concentration and column 1 the highest.

3.2.4 Preparation of Inoculum

1. The spectrophotometer, set to wavelength 530 nm and in % transmission mode (T), should be allowed to warm up for 15 min before use.
2. Label one sterile capped tube of 4-mL 0.85% saline, suitable for fitting into the cell holder of the spectrophotometer, and two plastic universals for each strain to be tested and the two control strains.
3. Aliquot 4.9-mL and 9.5-mL RPMI medium into the two universals labelled 1 and 2.
4. Place an uninoculated tube of sterile saline into the cell holder of the spectrophotometer and use to calibrate to 100% transmission.
5. With a sterile loop touch five colonies on a SAC agar plate, and mix the colonies on an uninoculated area of the plate surface. With the same loop, take a light inoculum from this area, and suspend in the sterile saline in the capped tube, and mix thoroughly on a vortex mixer (*see* **Note 10**).
6. Place the tube in the light path of the spectrophotometer and read the % transmission. To achieve the desired cell concentration of 1–5 × 10^6 CFU/mL, the transmission should be between 85–90%. To increase the transmission, add saline; to decrease it add more cell suspension.
7. A 1000-fold dilution of this is required as the working suspension. To achieve this, add 0.1 mL of the stock solution prepared in saline to the universal containing 4.9 mL of RPMI (1:50

dilution), vortex mix, and then aliquot 0.5 mL of this diluted stock to the universal containing 9.5 mL of RPMI (1:20 dilution). This is now the inoculum suspension at twice the final inoculum concentration of 2×10^3 CFU/mL, and the 10 mL suspension should be sufficient for inoculating nine rows on microtiter plates, i.e., sufficient for tests with nine different drugs.

8. Repeat **steps 4–7** above for each strain to be tested and the two control yeast strains.

3.2.5 Inoculation and Reading of Microtiter Plates and Interpreting the Results

1. Add 100 μL of the cell suspension in RPMI medium to the 100 μL of drug-containing RPMI medium in the wells in a row (columns 1–10). Also add 100 μL of cell suspension to the drug-free control well (column 11), but add nothing to column 12 which is the inoculum-free sterility control (*see* **Note 11**).
2. Incubate lidded plates at 35 °C for 24 h.
3. Shake the plates on a plate shaker before reading, and compare the growth in each well with that in the control well (column 11) with the aid of a reading mirror (Fig. 1). There should be no growth in the inoculum-free control (column 12) (*see* **Note 12**).

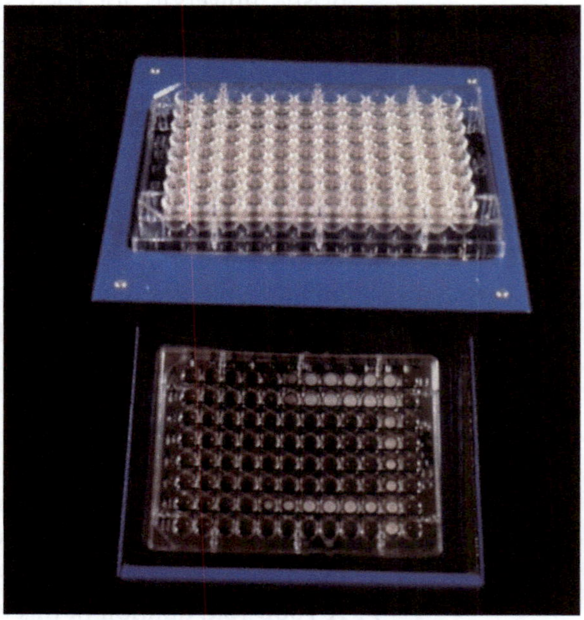

Fig. 1 Photograph of MIC reading using a 96-well round-bottomed plate and plate-reading mirror

Table 4
Acceptable MIC value ranges for the QC organisms and the 10 major systemic antifungal drugs

Drug	MIC range (mg/L)	
	NCPF 8334 *Candida parapsilosis*	NCPF 3953 *Candida krusei* (*Pichia kudriavzevii*)
Amphotericin B	0.25–2.0	0.5–2.0
Anidulafungin	0.25–2.0	0.03–0.12
Caspofungin	0.25–1.0	0.12–1.0
Fluconazole	0.5–4.0	8.0–64
Flucytosine	0.06–0.25	4.0–16
Isavuconazole	0.015–0.06	0.06–0.5
Itraconazole	0.06–0.5	0.12–1.0
Micafungin	0.5–2.0	0.12–0.5
Posaconazole	0.03–0.25	0.06–0.5
Voriconazole	0.03–0.12	0.06–0.5

4. As amphotericin is cidal for yeast, there should be a clear endpoint with 100% inhibition, the lowest concentration at which there is complete inhibition should be recorded as the MIC.
5. For all other drugs, the MIC is recorded as the minimum concentration at which there is a prominent reduction in growth (>50% reduction) as assessed by eye.
6. The control (QC) strains must produce results within the ranges shown in Table 4 or the test run is invalid.
7. There are currently no established clinical breakpoints for *Candida auris*, and it is hard to establish epidemiological cutoff values when quite a large proportion of the population can demonstrate high MICs with several of the classes of drugs. Instead we would suggest applying the CLSI breakpoints for *Candida albicans* as suggested in several publications [21, 22].

3.3 Gradient Strip Methodology for Assessing Antifungal Susceptibility of Candida auris Isolates (see Note 13)

3.3.1 Inoculum Preparation

1. With a sterile loop touch five colonies on a SAC agar plate, and mix the colonies on an uninoculated area of the plate surface. With the same loop, take a light inoculum from this area, and emulsify in a bijou containing 4 mL of 0.85% sterile saline, and mix thoroughly on a vortex mixer (*see* **Note 10**).
2. Adjust the turbidity to correspond to a 0.5 McFarland standard by adding more cells or more saline depending on whether the suspension is too light or too heavy.

3.3.2 Setting Up the Gradient Strip Test

1. Dry the surface of the RPMI agar plates by placing in the 35 °C incubator inverted and resting slightly open on the lids for about 10 min or longer, as required.
2. Dip a sterile swab in the prepared cell suspension in the bijou container and squeeze out excess fluid on the inside wall of the bijou container as the tip of the swab is withdrawn.
3. Use the swab to inoculate the RPMI agar plate passing over the surface lightly in three directions to obtain even growth on the entire surface of the agar.
4. Allow the moisture to absorb for 30 min before applying the test strip centrally on the surface of the plate with sterile forceps ensuring there are no air bubbles (*see* **Note 14**).
5. Incubate the plates at 35 °C in a moist incubator for 24 or 48 h depending on when the inhibition zone ellipse is clearly visible, usually 24 h for *C. auris* (*see* **Note 15**).
6. One or both of the QC strains documented in Subheading. 2.6, **item 2** should be tested initially on first set up and then periodically depending on the volume of the testing.

3.3.3 Interpretation of Gradient Strip Test Results

1. The MIC is read where the inhibition zone ellipse intersects with the scale on the strip (Fig. 2).
2. *Candida auris* may produce diffuse endpoints with the azole drugs in which a zone of very small colonies is seen in the background. In this case the MIC should be read at about 80% inhibition.

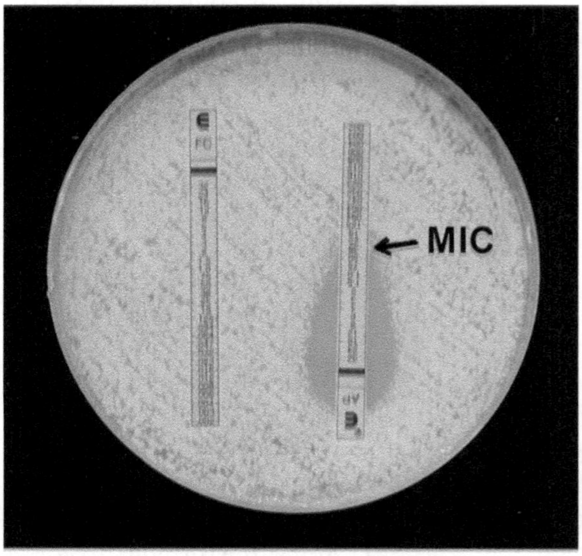

Fig. 2 Gradient strip test. The MIC is read at the point at which the inhibition ellipse intersects the scale on the strip (*see* arrow). This photograph depicts an amphotericin B (AP) susceptible, flucytosine (FC) resistant isolate

3. If with any drug there are large colonies formed within the zone, especially if these are adjacent to the strip, they should be taken with a loop and used as the inoculum for a further test with the same drug to establish if there is a resistant subpopulation, often described as hetero-resistance.
4. The MIC should be read against the strip and then allocated to the next doubling dilution higher in the CLSI dilution series and interpreted against the CLSI breakpoints for *C. albicans* as discussed above.

4 Notes

1. Posaconazole and itraconazole have poor solubility in DMSO, and you may need to use PEG for these drugs. Heat 10-mL PEG 400 to 70 °C in a beaker, and then add 16 mg of the drug powder to give a stock solution of 1600 mg/L. The temperature will rise, but do not let it go above 80 °C; remove from heat source if necessary. Once dissolved, add 1 mL of the 1600 mg/L to 4 mL of 50% PEG (2 mL of PEG and 2-mL sterile distilled water) to give a concentration of 320 mg/L, and then follow the dilution series as specified for DMSO with these drugs. However, when preparing the plates, you will need to have a column for a PEG control as some strains are inhibited by PEG (*see* **Notes 9**, **11**, and **12**).
2. This compound is potentially toxic; consult the material safety data sheets provided by the manufacturer.
3. Some plastics will adsorb the more lipophilic drugs thus decreasing the accuracy of the tests; we prefer Corning Costar® round-bottomed tissue culture-treated non-pyrogenic polystyrene plates.
4. No more than four serial passages should be made for the control strains. On receipt, multiple aliquots of control strains should be frozen at −80 °C or lower. After 8 months and four passages, fresh cultures should be prepared from new aliquots.
5. If you are working with primary cultures, then purity of the culture can be confirmed by plating onto CHROMAgar™ Candida Plus, a novel chromogenic agar that permits the rapid identification of *C. auris* [23].
6. Addition of 40 g of glucose at this point so that the final medium contains 20 g/L (2%) glucose is an accepted, and in our hands, highly recommended modification of the CLSI methodology as it allows better growth of the organisms leading to easier endpoint interpretation.
7. Preparation of microtiter plates by simple doubling dilution from one column of a microtiter plate to the next has been shown to produce highly inaccurate concentrations especially

at the lower concentrations. Instead, a series of concentrations should be prepared in universal containers which can then be aliquoted into the appropriate columns of microtiter plates to produce batches of plates which can then be frozen at −20 °C for up to 6 months.

8. Caspofungin does not perform uniformly well in microtiter plate testing for reasons that have yet to be fully established. Anidulafungin can be used as a sentinel echinocandin, but for specific caspofungin testing, we would recommend gradient strip testing as described in Subheading 3.3.

9. In the microtiter plates, column 11 should contain 100 μL of 5% PEG in RPMI medium (2 mL of 50% PEG added to 18-mL RPMI medium). Column 12 should contain 100 μL of drug-free RPMI medium.

10. To prepare an inoculum for heavily aggregating strains, such as those in the South African clade [22], it is necessary to initially attempt to disrupt the aggregates on a vortex mixer. Prepare a heavy suspension in 5 mL of sterile saline, and homogenize vigorously with a vortex mixer. This will help to break up some of the larger aggregates, but many will remain intact. Allow the suspension to settle for 10 min to allow any aggregates to sink to the bottom, and remove the top 3 mL of single-celled suspension into a fresh tube to prepare the inoculum. If aggregated cells are used as an initial inoculum, it will mean that antifungal drugs are unable to penetrate as effectively and may lead to erroneously elevated MICs (Fig. 3).

11. If you have used PEG to dissolve itraconazole and posaconazole, all wells in each row of these plates should receive 100 μL of cell suspension as there is no sterility control; instead there is a PEG control to ensure isolates grow well in the presence of 5% PEG.

12. If you have used PEG as the solvent for itraconazole and posaconazole instead, ensure there is good growth in the PEG control in column 11 comparable to the drug-free control in column 12.

13. Once opened the packs of gradient strips should be stored at −20 °C in a sealed container with a desiccant.

14. Once placed gradient strips must not be moved as the drug will start to diffuse into the agar immediately. Make sure the side with the drug concentrations easily visible is uppermost not placed face down; otherwise the drug will not diffuse and the MIC cannot be read. Two gradient strips can be placed per plate provided the concentration gradients are facing in opposite directions (Fig. 2).

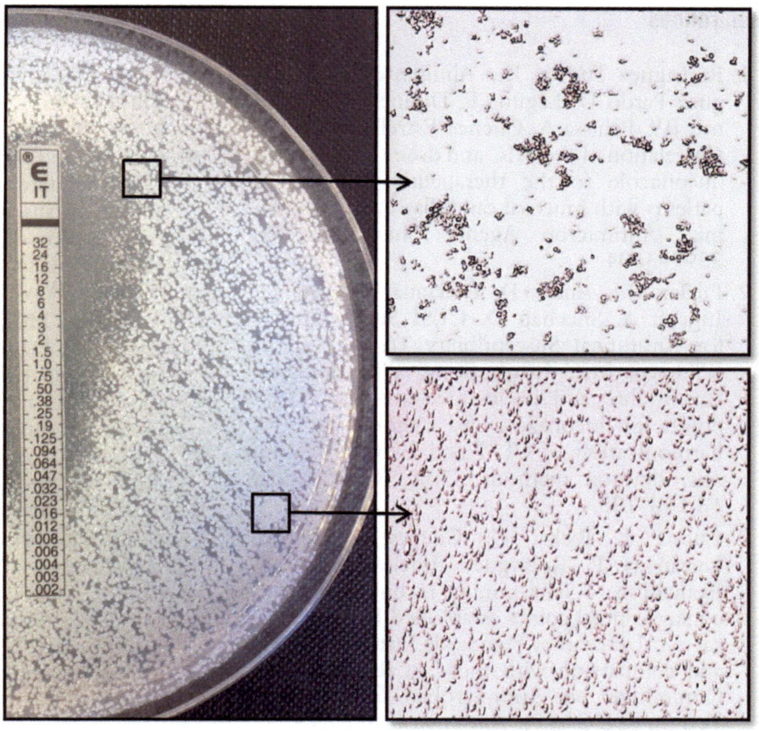

Fig. 3 Exposure to azole or echinocandin antifungal drugs induces a reversible aggregative phenotype in *C. auris* isolates of certain clades [22]. Gradient strip test performed with an itraconazole (IT) strip (left) and microscopic appearance of suspensions prepared from yeast cells adjacent to the zone of inhibition (right, top) or distant from the strip (right, bottom)

15. If there is no growth visible after 48 h, it may indicate a really susceptible strain or that the strain will not grow on RPMI medium, test this by inoculating an RPMI agar plate as above, but do not add a gradient strip.

Acknowledgments

We thank the staff of the UK National Mycology Reference Laboratory for their expertise in conducting these methods, for providing much of the data helping to validate the methods for this species and for preparing detailed SOPs. Thank you also to Associate Professor Annette Fothergill from The Fungus Testing laboratory, San Antonio, Texas, USA, who participated in the introduction of the CLSI plate dilution preparation method into our laboratory.

References

1. Rodríguez-Tudela JL, Almirante B, Rodríguez-Pardo D, Laguna F, Donnelly JP, Mouton JW, Pahissa A, Cuenca-Estrella M (2007) Correlation of the MIC and dose/MIC ratio of fluconazole to the therapeutic response of patients with mucosal candidiasis and candidemia. Antimicrob Agents Chemother 51: 3599–3604
2. Pfaller MA, Andes D, Diekema DJ, Espinel-Ingroff A, Sheehan D, CLSI Subcommittee for Antifungal Susceptibility Testing (2010) Wild-type MIC distributions, epidemiological cutoff values and species-specific clinical breakpoints for fluconazole and Candida: time for harmonization of CLSI and EUCAST broth microdilution methods. Drug Resist Updat 13:180–195
3. Cuesta I, Bielza C, Cuenca-Estrella M, Larrañaga P, Rodríguez-Tudela JL (2010) Evaluation by data mining techniques of fluconazole breakpoints established by the Clinical and Laboratory Standards Institute (CLSI) and comparison with those of the European Committee on Antimicrobial Susceptibility Testing (EUCAST). Antimicrob Agents Chemother 54:1541–1546
4. Pfaller MA, Diekema DJ, Andes D, Arendrup MC, Brown SD, Lockhart SR, Motyl M, Perlin DS, CLSI Subcommittee for Antifungal Testing (2011) Clinical breakpoints for the echinocandins and Candida revisited: integration of molecular, clinical, and microbiological data to arrive at species-specific interpretive criteria. Drug Resist Updat 14:164–176
5. Shields RK, Nguyen MH, Press EG, Updike CL, Clancy CJ (2013) Caspofungin MICs correlate with treatment outcomes among patients with Candida glabrata invasive candidiasis and prior echinocandin exposure. Antimicrob Agents Chemother 57:3528–3535
6. Johnson EM, Arendrup MC (2019) Susceptibility test methods: yeasts and filamentous fungi. In: Carroll KC, Pfaller MA, Landrt ML, McAdam AJ, Patel R, Richter SS, Warnock DW (eds) Manual of clinical microbiology, 12th edn. ASM Press, Washington, DC, pp 2351–2376
7. Borman AM, Muller J, Walsh-Quantick J, Szekely A, Patterson Z, Palmer MD, M. Fraser M, Johnson EM. (2020) MIC distributions for amphotericin B, fluconazole, itraconazole, voriconazole, flucytosine and anidulafungin for thirty-five uncommon pathogenic yeasts species from the UK determined using CLSI broth microdilution method. J Antimicrob Chemother 75:1194–1205. https://doi.org/10.1093/jac/dkz568
8. Clinical and Laboratory Standards Institute (2008) Reference method for broth dilution antifungal susceptibility testing of yeasts, 3rd ed. Approved standard. CLSI document M27-A3. Clinical and Laboratory Standards Institute, Wayne
9. Arendrup MC, Meletiadis J, Mouton JW, Lagrou K, Hamal P, Guinea J, Subcommittee on Antifungal Susceptibility Testing (AFST) of the ESCMID European Committee for Antimicrobial Susceptibility Testing (EUCAST) (2020) EUCAST definitive document E.DEF 7.3.2: method for the determination of broth dilution minimum inhibitory concentrations of antifungal agents for yeasts. http://www.eucast.org
10. Clinical and Laboratory Standards Institute (2017) Performance standards for antifungal susceptibility testing of yeast. CLSI Suppl. M60. Clinical and Laboratory Standards Institute, Wayne
11. https://www.eucast.org/fileadmin/src/media/PDFs/EUCAST_files/AFST/Clinical_breakpoints/AFST_BP_v10.0_200204_updatd_links_200924.pdf
12. Pfaller MA, Espinel-Ingroff A, Boyken L, Hollis RJ, Kroeger J, Messer SA, Tendolkar S, Diekema DJ (2011) Comparison of the broth microdilution (BMD) method of the European Committee on Antimicrobial Susceptibility Testing with the 24-hour CLSI BMD method for testing susceptibility of Candida species to fluconazole, posaconazole, and voriconazole by use of epidemiological cutoff values. J Clin Microbiol 49:845–850
13. Pfaller M, Boyken L, Hollis R, Kroeger J, Messer S, Tendolkar S, Diekema D (2011) Use of epidemiological cutoff values to examine 9-year trends in susceptibility of Candida species to anidulafungin, caspofungin, and micafungin. J Clin Microbiol 49:624–629
14. Espinel-Ingroff A, Pfaller MA, Bustamante B, Canton E, Fothergill A, Fuller J, Gonzalez GM, Lass-Flörl C, Lockhart SR, Martin-Mazuelos E, Meis JF, Melhem MS, Ostosky-Zeichner L, Pelaez T, Szeszs MW, St-Germain G, Bonfietti LX, Guarro J, Turnidge J (2014) A multilaboratory study of epidemiological cutoff values for detection of resistance in eight Candida spp. to fluconazole,

posaconazole, and voriconazole. Antimicrob Agents Chemother 58:2006–2012

15. Clinical and Laboratory Standards Institute (2018) Epidemiological cutoff values for antifungal susceptibility testing, 2nd ed. CLSI Suppl. M59. Clinical and Laboratory Standards Institute, Wayne
16. https://www.eucast.org/fileadmin/src/media/PDFs/EUCAST_files/AFST/Clinical_breakpoints/EUCAST_BP_ECOFF_v2.0_20-09-24.pdf
17. Chaabane F, Graf A, Jequier L, Coste AT (2019) Review on antifungal resistance mechanisms in the emerging pathogen Candida auris. Front Microbiol 10:2788. https://doi.org/10.3389/fmicb.2019.02788
18. Rhodes J, Abdolrasouli A, Farrer RA, Cuomo CA, Aanensen DM, Armstrong-James D, Fisher MC, Schelenz S (2018) Genomic epidemiology of the UK outbreak of the emerging human fungal pathogen Candida auris. Emerg Microbes Infect 7:43. https://doi.org/10.1038/s41426-018-0045-x
19. Kordalewska M, Lee A, Park S, Berrio I, Chowdhary A, Zhao Y, Perlin DS (2018) Understanding echinocandin resistance in the emerging pathogen Candida auris. Antimicrob Agents Chemother 62:e00238–e00218. https://doi.org/10.1128/AAC.00238-18
20. Escandón P, Chow NA, Caceres DH, Gade L, Berkow EL, Armstrong P, Rivera S, Misas E, Duarte C, Moulton-Meissner H, Welsh RM, Parra C, Pescador LA, Villalobos N, Salcedo S, Berrio I, Varón C, Espinosa-Bode A, Lockhart SR, Jackson BR, Litvintseva AP, Beltran M, Chiller TM (2019) Molecular epidemiology of Candida auris in Colombia reveals a highly related, countrywide colonization with regional patterns in amphotericin B resistance. Clin Infect Dis 68:15–21. https://doi.org/10.1093/cid/ciy411
21. Arendrup MC, Prakash A, Meletiadis J, Sharma C, Chowdhary A (2017) Comparison of EUCAST and CLSI reference microdilution MICs of eight antifungal compounds for Candida auris and associated tentative epidemiological cutoff values. Antimicrob Agents Chemother 61:e00485–e00417. https://doi.org/10.1128/AAC.00485-17
22. Szekely A, Borman AM, Johnson EM (2019) Candida auris isolates of the Southern Asian and South African lineages exhibit different phenotypic and antifungal susceptibility profiles in vitro. J Clin Microbiol 57:e02055–e02018. https://doi.org/10.1128/JCM.02055-18
23. Borman AM, Fraser M, Johnson EM (2020) CHROMagar™ Candida Plus: a novel chromogenic agar that permits the rapid identification of Candida auris. Med Mycol. https://doi.org/10.1093/mmy/myaa049

Chapter 6

Deciphering *Candida auris* Paradoxical Growth Effect (Eagle Effect) in Response to Echinocandins

Milena Kordalewska and David S. Perlin

Abstract

The paradoxical growth effect (PGE; also known as Eagle effect) is an in vitro phenomenon observed during antifungal susceptibility testing (AFST). In PGE, some fungal isolates grow in medium containing high concentrations of an echinocandin, above the minimal inhibitory concentration (MIC), despite being fully susceptible at lower concentrations. The presence of PGE complicates the assignment of isolates to susceptible or resistant category, especially in the case of newly emerged pathogens like *Candida auris*, for which susceptibility breakpoints are not established.

Here we describe a protocol aiding in the determination of whether a given *C. auris* isolate is echinocandin-resistant or echinocandin-susceptible but exhibiting paradoxical growth.

Key words *Candida*, *Candida auris*, Paradoxical growth effect, Eagle effect, Echinocandins, Caspofungin, Micafungin, Antifungal susceptibility testing, Antifungal drug resistance

1 Introduction

Echinocandins are lipopeptide antifungal drugs that act by noncompetitive inhibition of β-1,3-D-glucan synthase, which catalyzes the biosynthesis of β-1,3-D-glucan—one of the key carbohydrates in fungal cell wall [1]. Echinocandins exhibit broad fungicidal (killing) activity against most pathogenic *Candida* species [2, 3] and are recommended as a first-line empiric therapy of invasive candidiasis in non-neutropenic patients [4]. Pharmacodynamic resistance to echinocandins occurs only when a fungal strain develops mutations in the *FKS* genes encoding the catalytic subunits of β-1,3-D-glucan synthase. In vitro, presence of amino acid substitutions in Fks subunits manifests as elevated minimal inhibitory concentration (MIC) values [1]. In clinical isolates of *Candida auris*, the only amino acid changes associated with echinocandin resistance occur at position S639 of Fks1 [5–8].

Alexander Lorenz (ed.), *Candida auris: Methods and Protocols*, Methods in Molecular Biology, vol. 2517,
https://doi.org/10.1007/978-1-0716-2417-3_6,

Antifungal susceptibility testing (AFST) to determine whether a given fungal isolate is echinocandin-susceptible or echinocandin-resistant is most often performed using Clinical and Laboratory Standards Institute (CLSI) and European Committee on Antimicrobial Susceptibility Testing (EUCAST) broth microdilution methods [9, 10]. These phenotypic methods are standardized, but reading of their results, MIC values, can be complicated by the presence of paradoxical growth effect (PGE, also known as Eagle effect). This phenomenon presents as reduced fungicidal activity at high concentrations of antifungals, above the MIC, meaning that some fungal isolates grow in medium containing higher concentrations of echinocandin, while being fully susceptible (no growth) at lower concentrations [11]. PGE is not associated with a genetically stable resistant fungal subpopulation, where fungal growth would be expected at all concentrations above the MIC, even at the lower concentration range [12]. Moreover, the effect is not due to mutations in *FKS* and is not related to the upregulated activity of β-1,3-D-glucan synthase in the presence of the drug. PGE observed in vitro has no translation into in vivo response to the echinocandin treatment, as was shown in *C. auris* and *C. dubliniensis* invasive murine candidiasis models [6, 13].

Among currently approved echinocandins, caspofungin is most notably associated with PGE. In case of the newly emerged *C. auris*, we noticed [6] that previously reported 37% caspofungin resistance rate, without supporting molecular data [14], may possibly be due to the fact that isolates exhibiting high PGE were categorized as resistant, thereby overestimating echinocandin resistance in *C. auris*. Thus, as with many *Candida* species [15], using

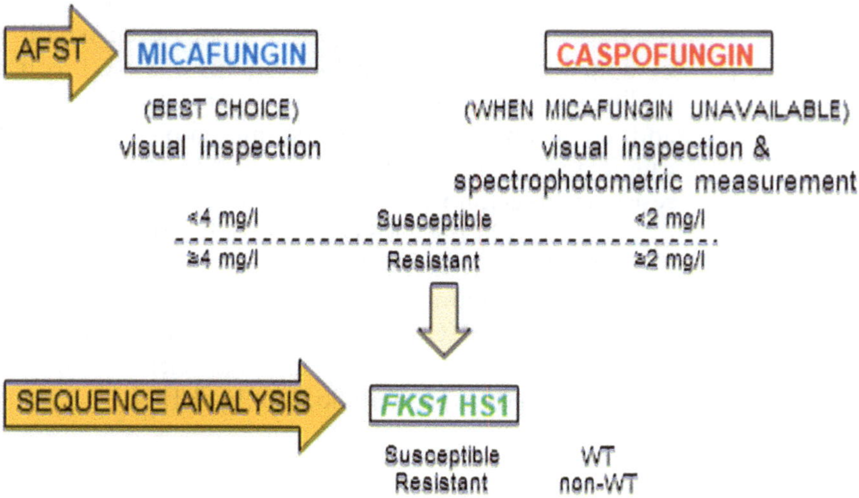

Fig. 1 Overview of methods enabling accurate echinocandin susceptibility determination of *Candida auris* isolates

solely caspofungin AFST to determine *C. auris* echinocandin resistance is unreliable. Accurate results are obtained with micafungin AFST and/or *FKS1* sequencing, where resistance is reflected by MIC ≥4 mg/L and presence of characteristic mutation, respectively (Fig. 1).

2 Materials

2.1 Fungal Culture

1. YPDA medium (1% yeast extract, 2% peptone, 2% glucose, 2% agar): place a 1000 mL autoclavable media bottle on a magnetic stirrer plate. Add 800 mL of double-distilled water (ddH_2O) and a magnetic stir bar. Then, add 8 g yeast extract, 16 g peptone, 16 g glucose, and 16 g Bacto agar. Dissolve all ingredients by stirring. Autoclave to sterilize. Place autoclaved medium on a magnetic stirrer plate, and stir for at least 30 min to cool it down. While still warm, pour 20 mL of medium into 100 × 15 mm Petri dishes by using a 25 mL serological pipette. Allow agar medium to solidify. Store at 4 °C.
2. Disposable inoculating loops of 1 μL volume.
3. Frozen glycerol stocks (−80 °C) of *C. auris* isolates.

2.2 Broth Microdilution

Broth microdilution according to CLSI standard M27 and supplement M60 [9, 16] (*see* **Note 1**)

1. 2× RPMI-1640: place a 1000 mL beaker on a magnetic stirrer plate. Add 900 mL of ddH_2O and a magnetic stir bar. Then, add 20.8 g RMPI-1640 (with glutamine, without sodium bicarbonate, and with phenol red as a pH indicator), and 69.06 g MOPS (3-(N-morpholino)propanesulphonic acid). Stir until components are completely dissolved. While stirring, adjust the pH to 7.0 at 25 °C with NaOH. Add ddH_2O to a final volume of 1 L. Filter sterilize using a 0.22 μm vacuum filter/storage bottle system. Store at 4 °C.
2. Stocks of echinocandin drugs in DMSO prepared as described in CLSI standard M27 and supplement M60 [9, 16]. Store at −80 °C.
3. *C. auris* isolates streaked on YPDA plates.
4. Autoclaved ddH_2O.
5. 0.5 McFarland standard.
6. 96 U-shaped well microdilution plates made with untreated polystyrene.
7. Cuvette spectrophotometer and appropriate cuvettes.
8. Microplate reader enabling the optical density measurement at 600 nm.

9. Vortex mixer.
10. Data visualization and analysis software, e.g., Microsoft Excel or GraphPad Prism.

2.3 DNA Extraction [17]

1. Buffer A: 60 mM sodium bicarbonate [$NaHCO_3$], 250 mM potassium chloride [KCl], and 50 mM Tris, pH 9.5. Autoclave to sterilize. Store at 4 °C.
2. Buffer B: 2% BSA in ddH_2O. Filter sterilize using a 0.22 μm filter. Store at 4 °C.

2.4 Amplification of Hotspot 1 (HS1) of FKS1 Gene of C. auris

1. PCR primers.
   - Cau_FKS1_HS1_F: 5′-CTA CTG GGC TGG TTT GTC TGC C-3′.
   - Cau_FKS1_HS1_R: 5′-GAC AAT GGC ATT CCA CAC CTG C-3′.
2. PCR Master Mix (*see* **Note 2**).
3. PCR-grade water.
4. Thermal cycler (*see* **Note 2**).

2.5 Gel Electrophoresis

1. Low electroendosmosis (LE) agarose.
2. 50× TAE buffer: 242 g of Tris base, 18.61 g disodium EDTA, 57.1 mL of glacial acetic acid, volume adjusted to 1 L with ddH_2O.
3. 1× TAE buffer prepared by mixing 49 parts of ddH_2O and 1 part of 50× TAE buffer.
4. Microwave (*see* **Note 3**).
5. Nucleic acid stain.
6. DNA size standard.
7. Gel electrophoresis system (buffer chamber, lid with attached power supply leads, gel tray, combs) and electrophoresis power supply.
8. Gel imaging and documentation system.

2.6 Sanger Sequencing

Prepare PCR products for Sanger sequencing following the guidelines of your service provider (in-house or outsourced). If PCR purification is needed, use a commercially available kit.

2.7 FKS1 HS1 Sequence Analysis

1. *C. auris* reference sequence of *FKS1* HS1 wild-type genotype (B9J08_000964): TTCTTGACTTTGTCCTTGAGAGATCCT (nucleotide position: 1903–1929) which translates into FLTLSLRDP (amino acid position: 635–643).
2. Multiple sequence alignment software, e.g., Clustal Omega available at https://www.ebi.ac.uk/Tools/msa/clustalo/.

3 Methods

3.1 Antifungal Susceptibility Testing

1. Streak out *C. auris* isolates from glycerol stock on YPDA plates by using a loop (*see* **Note 4**). Incubate the plates at 35 °C in ambient air for 24 h. Store at 4 °C.
2. Set up broth microdilution test following the guidelines provided in CLSI standard M27 and supplement M60 [9, 16]. Incubate the microdilution plates at 35 °C in ambient air for 24 h.
3. Read the AFST results. View microdilution plates under normal laboratory lightning and determine MIC (Figs. 2, 3, 4, and 5)—wells with prominent decrease (~50%) in visible growth (*see* **Note 5**).
4. Measure absorbance at 600 nm by using a microplate reader.

   Calculate growth in each well, relative to the growth control:

   - A_{neg}, absorbance of the negative control (medium); A_{gc}, absorbance of the growth control; A_x, absorbance in the well containing *x* mg/L of the drug
   - $\text{Growth}\% = \frac{A_x - A_{\text{neg}}}{A_{\text{gc}} - A_{\text{neg}}} \times 100\%$

   Plot the results (Figs. 2, 3, 4, and 5):

   - *x* axis—drug concentration on log2 scale
   - *y* axis—growth % on linear scale

   Determine, the spectrophotometric measurement-based MIC—a drug concentration closest to the point where the growth % is $\leq$50%.
5. Compare MIC results obtained through visual inspection and spectrophotometric measurement to determine susceptibility profile of a given *C. auris* isolate (*see* **Note 6**). To categorize isolates as susceptible or resistant, use tentative MIC breakpoints developed by the Centers for Disease Control and Prevention [18].

3.2 FKS1 *HS1* Sequence Analysis

1. Streak out *C. auris* isolates from glycerol stock on YPDA plates by using a loop (*see* **Note 4**). Incubate the plates at 35 °C in ambient air for 24 h. Store at 4 °C.
2. DNA extraction: aliquot 100 μL of Buffer A into a sterile 1.5 mL microcentrifuge tube. Use a pipette tip to pick one colony from YPDA plate. Mix the colony into Buffer A, and incubate for 10 min at 95 °C. Then, add 100 μL Buffer B, and mix intensively by using a vortex mixer. Store at 4 °C.
3. PCR: Prepare Master Mix as outlined in Table 1. Run PCR with time-temperature profile presented in Table 2.

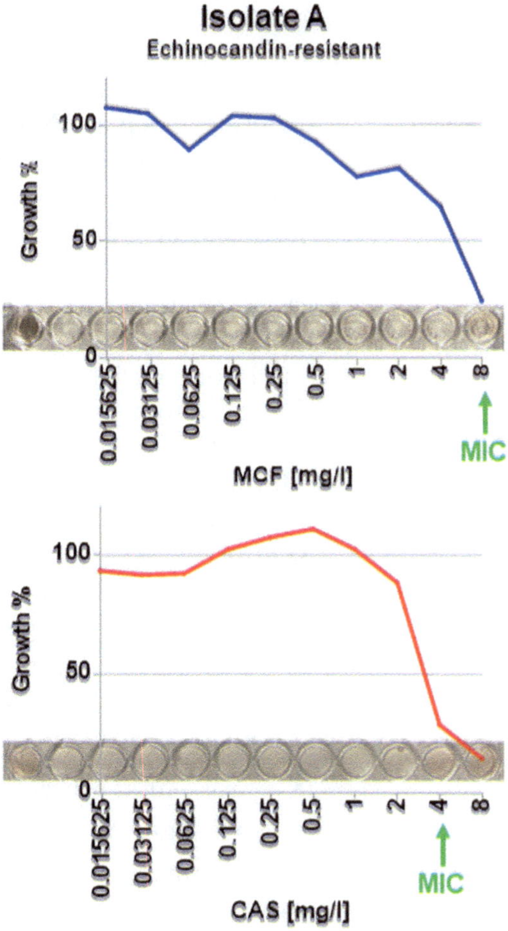

Fig. 2 Micafungin (top panel) and caspofungin (bottom panel) AFST results for Isolate A (echinocandin-resistant) presented as photograph of the row of a microdilution plate (visual inspection) and a graph presenting growth % as a function of drug concentration. MCF, micafungin; CAS, caspofungin; MIC, minimal inhibitory concentration

4. Gel electrophoresis: prepare a 2% agarose gel. Load 5 μL of the samples and DNA size marker. Run electrophoresis (10 V per 1 cm of a gel) until the front of the loading dye is 1 cm from the bottom of the gel. Inspect the gel for the presence of correct amplicons (625 bp) by using a gel imaging and documentation system (Fig. 6).
5. Sanger sequencing of PCR products: Perform Sanger sequencing by using an in-house or outsourced service. If needed, purify PCR products by using a commercially available kit.
6. *FKS1* HS1 sequence alignment: Paste reference sequence and sequences obtained for PCR products into a multiple sequence alignment software, and run alignment. Determine the

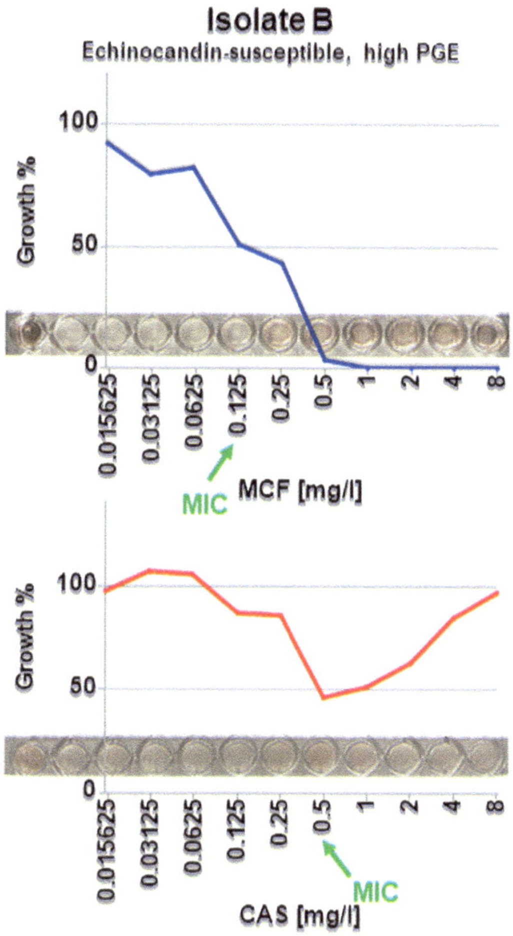

Fig. 3 Micafungin (top panel) and caspofungin (bottom panel) AFST results for Isolate B (echinocandin-susceptible, presenting high level of caspofungin PGE) presented as photograph of the row of a microdilution plate (visual inspection) and a graph presenting growth % as a function of drug concentration. MCF, micafungin; CAS, caspofungin; MIC, minimal inhibitory concentration

genotype of the *FKS1* HS1 fragment: wild-type (WT) or non-WT genotype with a single-nucleotide polymorphism (SNP) present (Fig. 7).

Determine the type of the SNP: synonymous or non-synonymous (*see* **Note 7**). Screen the literature for information on whether this *FKS1* HS1 alteration has been shown to confer resistance to echinocandins in *C. auris* or other *Candida* spp. (*see* **Note 7**).

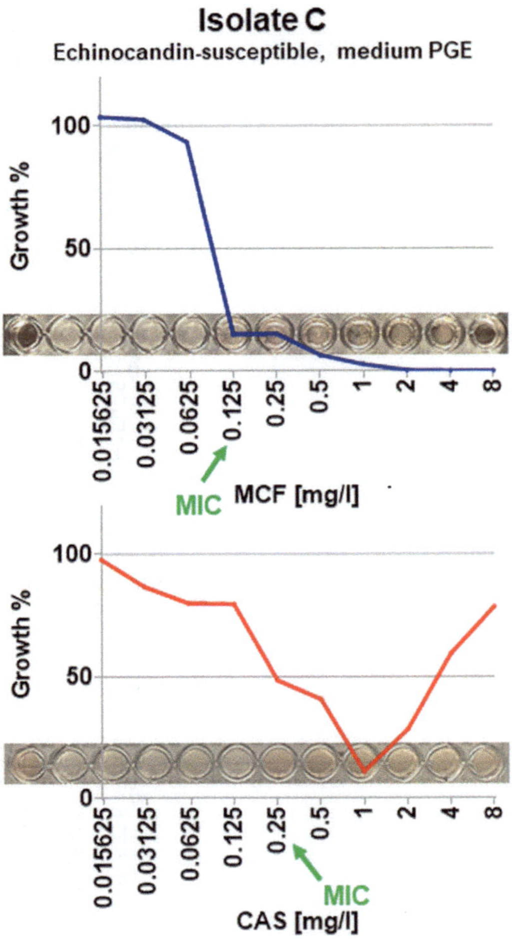

Fig. 4 Micafungin (top panel) and caspofungin (bottom panel) AFST results for Isolate C (echinocandin-susceptible, presenting medium level of caspofungin PGE) presented as photograph of the row of a microdilution plate (visual inspection) and a graph presenting growth % as a function of drug concentration. MCF, micafungin; CAS, caspofungin; MIC, minimal inhibitory concentration

4 Notes

1. Antifungal susceptibility testing should be performed exactly like described in the CLSI or EUCAST guidelines. In this chapter, we followed the CLSI methodology.
2. The *FKS1* HS1 PCR protocol was optimized for EmeraldAmp MAX PCR Master Mix (Takara Bio) and C1000 Touch Thermal Cycler (Bio-Rad) and would need to be adapted if other products and/or instrument are used.
3. In the absence of a microwave, a hot plate can be used for an agarose gel preparation.

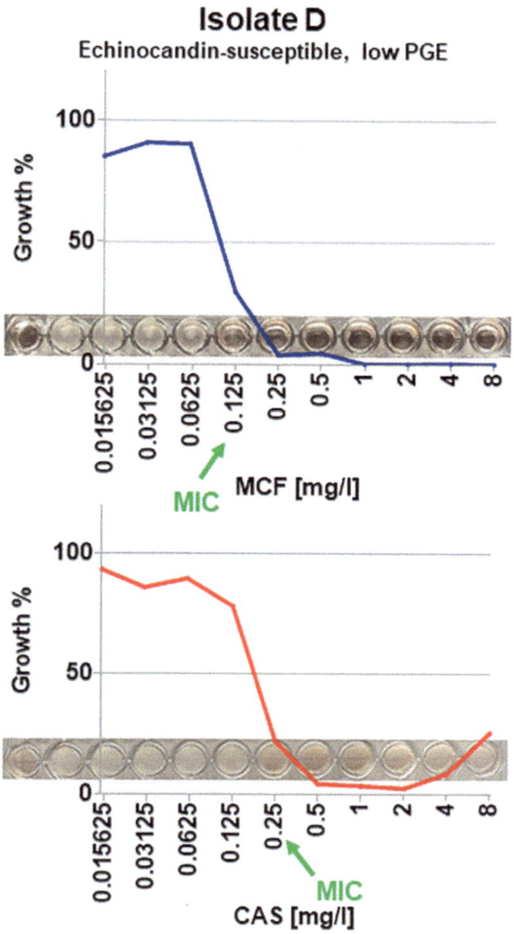

Fig. 5 Micafungin (top panel) and caspofungin (bottom panel) AFST results for Isolate D (echinocandin-susceptible, presenting low level of caspofungin PGE) presented as photograph of the row of a microdilution plate (visual inspection) and a graph presenting growth % as a function of drug concentration. MCF, micafungin; CAS, caspofungin; MIC, minimal inhibitory concentration

Table 1
***FKS1* HS1 PCR Master Mix composition**

Component	Volume per reaction (μL)
Cau_FKS1_HS1_F (10 μM)	1
Cau_FKS1_HS1_R (10 μM)	1
PCR Master Mix (2×)	15
Water	11
DNA template	2
Total volume	30

Table 2
Time–temperature profile of *FKS1* HS1 PCR

Temperature (°C)	Time	Repeats
95	3 min	
95	30 s	×35
60	30 s	
72	30 s	
72	5 min	
12	∞	

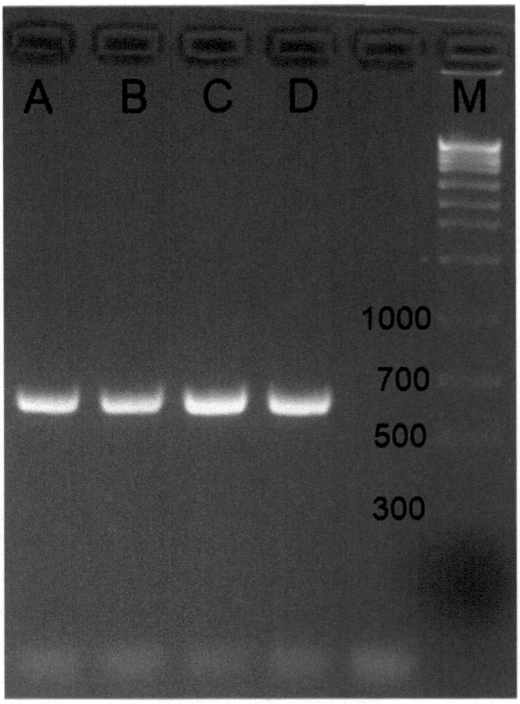

Fig. 6 PCR results of the *C. auris FKS1* HS1 amplification for *C. auris* isolates A, B, C, and D. M: DNA size marker

4. Apply streak plate technique. To obtain single colonies, especially from glycerol stocks of high cell concentration, use a separate YPDA plate for each isolate.

5. Isolates exhibiting high level of PGE, e.g., Isolate B, might not seem to have ~50% of growth inhibited by any concentration of caspofungin when inspected visually (Fig. 3). Caspofungin MIC might be read as higher than the maximum concentration tested, leading to an incorrect categorization of the isolate as an echinocandin-resistant one. However, when such isolates are

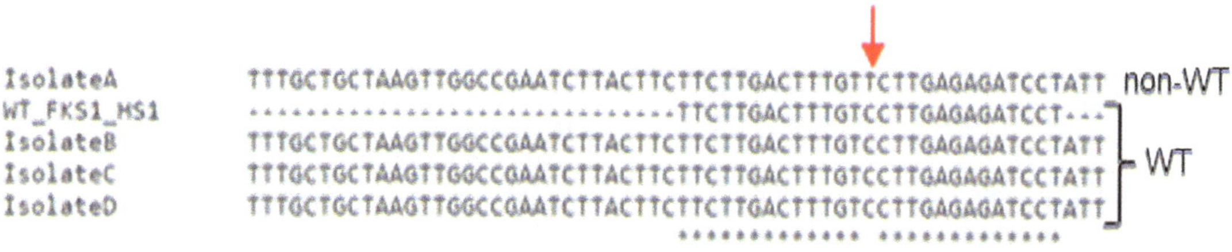

Fig. 7 Results of the alignment of the reference *FKS1* HS1 sequence (extracted from B9J08_000964) with sequences obtained by Sanger sequencing of PCR products

tested with micafungin, the assessment of growth inhibition is much easier and MIC read-out results in a value below the tentative MIC breakpoint (Fig. 3).

If micafungin is not available, caspofungin MIC value reading should be accompanied by spectrophotometric measurement and growth inhibition plot (*see* **Note 6**).

6. Isolate A (Fig. 2) is an example of an echinocandin-resistant *C. auris* isolate. The curves of Isolate A's growth inhibition for both micafungin and caspofungin have one sharp drop (and no rebound) resulting in MIC values higher than the tentative MIC breakpoints. Isolates B–D (Figs. 3, 4, and 5) exhibit different levels of PGE for caspofungin, complicating visual assessment of the growth inhibition. The curves of growth inhibition of such isolates have one sharp drop (MIC) followed by a rebound (growth % increases with an increasing drug concentration). If the caspofungin MIC value resulting from visual assessment is very high but the shape of the growth inhibition curve is as described above, the isolate should be categorized as echinocandin-susceptible but exhibiting high level of PGE.
7. Isolate A harbors the C1916T SNP leading to the amino acid change S639F, which is one of the *FKS1* HS1 alterations proven to confer echinocandin resistance in *C. auris* (Table 3) [5–8].

Table 3
***Candida auris* WT and resistance-conferring *FKS1* HS1 genotypes with corresponding Fks1 HS1 phenotypes**

Phenotype		Genotype
WT	F L T L S L R D P	TTCTTGACTTTGTCCTTGAGAGATCCT
S639F	F L T L **F** L R D P	TTCTTGACTTTG<u>TTC</u>TTGAGAGATCCT
S639P	F L T L **P** L R D P	TTCTTGACTTTG<u>CCC</u>TTGAGAGATCCT
S639Y	F L T L **Y** L R D P	TTCTTGACTTTG<u>TAC</u>TTGAGAGATCCT

References

1. Perlin DS (2015) Mechanisms of echinocandin antifungal drug resistance. Ann N Y Acad Sci 1354:1–11. https://doi.org/10.1111/nyas.12831
2. Barchiesi F, Spreghini E, Tomassetti S, Arzeni D, Giannini D, Scalise G (2005) Comparison of the fungicidal activities of caspofungin and amphotericin B against *Candida glabrata*. Antimicrob Agents Chemother 49(12):4989–4992. https://doi.org/10.1128/AAC.49.12.4989-4992.2005
3. Ernst EJ, Klepser ME, Ernst ME, Messer SA, Pfaller MA (1999) *In vitro* pharmacodynamic properties of MK-0991 determined by time-kill methods. Diagn Microbiol Infect Dis 33(2):75–80. https://doi.org/10.1016/s0732-8893(98)00130-8
4. Pappas PG, Kauffman CA, Andes DR, Clancy CJ, Marr KA, Ostrosky-Zeichner L, Reboli AC, Schuster MG, Vazquez JA, Walsh TJ, Zaoutis TE, Sobel JD (2016) Clinical practice guideline for the management of candidiasis: 2016 update by the Infectious Diseases Society of America. Clin Infect Dis 62(4):e1–e50. https://doi.org/10.1093/cid/civ933
5. Chowdhary A, Prakash A, Sharma C, Kordalewska M, Kumar A, Sarma S, Tarai B, Singh A, Upadhyaya G, Upadhyay S, Yadav P, Singh PK, Khillan V, Sachdeva N, Perlin DS, Meis JF (2018) A multicentre study of antifungal susceptibility patterns among 350 *Candida auris* isolates (2009-17) in India: role of the *ERG11* and *FKS1* genes in azole and echinocandin resistance. J Antimicrob Chemother 73(4):891–899. https://doi.org/10.1093/jac/dkx480
6. Kordalewska M, Lee A, Park S, Berrio I, Chowdhary A, Zhao Y, Perlin DS (2018) Understanding echinocandin resistance in the emerging pathogen *Candida auris*. Antimicrob Agents Chemother 62(6):e00238-00218. https://doi.org/10.1128/AAC.00238-18
7. Berkow EL, Lockhart SR (2018) Activity of CD101, a long-acting echinocandin, against clinical isolates of *Candida auris*. Diagn Microbiol Infect Dis 90(3):196–197. https://doi.org/10.1016/j.diagmicrobio.2017.10.021
8. Woodworth MH, Dynerman D, Crawford ED, Doernberg SB, Ramirez-Avila L, Serpa PH, Nichols A, Li LM, Lyden A, Tato CM, Miller S, Derisi JL, Langelier C (2019) Sentinel case of *Candida auris* in the Western United States following prolonged occult colonization in a returned traveler from India. Microb Drug Resist 25(5):677–680. https://doi.org/10.1089/mdr.2018.0408
9. CLSI (ed) (2017) Reference method for broth dilution antifungal susceptibility testing of yeast, CLSI Standard M27, 4th edn. Clinical and Laboratory Standards Institute, Wayne
10. Arendrup MC, Meletiadis J, Mouton JW, Lagrou K, Hamal P, Guinea J, the Subcommittee on Antifungal Susceptibility Testing (AFST) of the ESCMID European Committee for Antimicrobial Susceptibility Testing (EUCAST) (2020) EUCAST definitive document E.DEF 7.3.2 Method for the determination of broth dilution minimum inhibitory concentrations of antifungal agents for yeasts
11. Vanstraelen K, Lagrou K, Maertens J, Wauters J, Willems L, Spriet I (2013) The Eagle-like effect of echinocandins: what's in a name? Expert Rev Anti-Infect Ther 11(11):1179–1191. https://doi.org/10.1586/14787210.2013.841543
12. Stevens DA, Espiritu M, Parmar R (2004) Paradoxical effect of caspofungin: reduced activity against *Candida albicans* at high drug concentrations. Antimicrob Agents Chemother 48(9):3407–3411. https://doi.org/10.1128/AAC.48.9.3407-3411.2004
13. Marine M, Pastor FJ, Sahand IH, Ponton J, Quindos G, Guarro J (2009) Paradoxical growth of *Candida dubliniensis* does not preclude *in vivo* response to echinocandin therapy. Antimicrob Agents Chemother 53(12):5297–5299. https://doi.org/10.1128/AAC.00980-09
14. Kathuria S, Singh PK, Sharma C, Prakash A, Masih A, Kumar A, Meis JF, Chowdhary A (2015) Multidrug-resistant *Candida auris* misidentified as *Candida haemulonii:* characterization by matrix-assisted laser desorption ionization-time of flight mass spectrometry and DNA sequencing and its antifungal susceptibility profile variability by Vitek 2, CLSI broth microdilution, and Etest method. J Clin Microbiol 53(6):1823–1830. https://doi.org/10.1128/JCM.00367-15
15. Espinel-Ingroff A, Arendrup MC, Pfaller MA, Bonfietti LX, Bustamante B, Canton E, Chryssanthou E, Cuenca-Estrella M, Dannaoui E, Fothergill A, Fuller J, Gaustad P, Gonzalez GM, Guarro J, Lass-Florl C, Lockhart SR, Meis JF, Moore CB, Ostrosky-Zeichner L, Pelaez T, Pukinskas SR, St-Germain G, Szeszs MW, Turnidge J (2013) Interlaboratory variability of caspofungin MICs for *Candida* spp. using CLSI and EUCAST methods: should the clinical

laboratory be testing this agent? Antimicrob Agents Chemother 57(12):5836–5842. https://doi.org/10.1128/AAC.01519-13

16. CLSI (ed) (2020) Performance standards for antifungal susceptibility testing of yeasts, CLSI Supplement M60, 2nd edn. Clinical and Laboratory Standards Institute, Wayne
17. Brillowska-Dabrowska A, Nielsen SS, Nielsen HV, Arendrup MC (2010) Optimized 5-hour multiplex PCR test for the detection of tinea unguium: performance in a routine PCR laboratory. Med Mycol 48(6):828–831. https://doi.org/10.3109/13693780903531579
18. Centers for Disease Control and Prevention (2020) *Candida auris*. Information for laboratorians and health professionals. Antifungal susceptibility testing and interpretation. https://www.cdc.gov/fungal/candida-auris/c-auris-antifungal.html. Accessed 23 Apr 2021

Part III

Genetic Manipulation of *Candida auris*

Chapter 7

An Optimized Electroporation Procedure for Genetic Transformation of *Candida auris*

Dalal Kasir, Marwan Osman, Monzer Hamze, Jean-Philippe Bouchara, Vincent Courdavault, and Nicolas Papon

Abstract

Candida auris is responsible for recent outbreaks with significant mortality in hospitalized and long-term care patients. As a highly transmissible and multidrug-resistant fungal pathogen, genetic tools are urgently needed for deciphering mechanisms involved in the host–pathogen interactions and potentially identifying new fungal targets for therapeutic development. Here, we provide a reliable transformation protocol based on an efficient electroporation procedure and the use of a mycophenolic acid resistance marker.

Key words Transgenesis, Drug resistance marker, Fluorescent protein, Mycophenolic acid, *Candida auris*

1 Introduction

Candida auris has emerged globally as a multidrug-resistant healthcare-associated fungal pathogen. This ascomycetous yeast belongs to the *Candida haemulonii* complex of the *Metschnikowiaceae* clade [1]. Despite the implementation of proactive infection prevention and control measures, this species frequently leads to nosocomial outbreaks in intensive care settings with serious healthcare-related invasive infections [2]. After its first description in the ear canal of a Japanese patient in 2009 [3], *C. auris* now represents a public health concern since its overall mortality rate reaches up to 39% [4]. Indeed, the high rates of resistance to current antifungal drugs such as fluconazole, amphotericin B, and caspofungin associated with the rapid clonal transmission within and across healthcare facilities; ongoing challenges due to organism misidentification in the clinical laboratory and significant patient mortality have placed this yeast species under the research spotlight [5–7].

Alexander Lorenz (ed.), *Candida auris: Methods and Protocols*, Methods in Molecular Biology, vol. 2517,
https://doi.org/10.1007/978-1-0716-2417-3_7,

To decipher the host–pathogen interactions [8, 9], functional genomic analysis is considered for a long time a relevant tool since it provides a clear view on the virulence and drug-resistance genes [10]. Importantly, as *Candida albicans*, *C. auris* belongs to the *Candida* CTG clade that decodes CTG codons as serine rather than leucine [11]. Therefore, as observed for a broad series of non-*albicans Candida* species in the past two decades, appropriate molecular tools had to be developed for genetic manipulation [12–14]. Here, we present an optimized transformation protocol based on an electroporation procedure and mycophenolic acid resistance conferred by the recently developed resistance gene *IMH3.2* from *Meyerozyma guilliermondii* (Fig. 1) [15].

2 Material

1. Micropipette set.
2. Sterile pipette tips.
3. Eppendorf (1.5 mL) and falcon (50 mL) tubes.
4. 250-mL flask (Erlenmeyer).
5. DNA electrophoresis apparatus.
6. Spectrophotometer.
7. Electroporation system.
8. 0.2-cm gap width electroporation cuvette.
9. Incubators: 32 °C with rotary shaker (150 × *g*), 37 °C.
10. Centrifuge (1000–13,000 × *g*).
11. Epifluorescence microscope equipped with a digital camera, imaging software, and an appropriate filter (excitation filter 545–580 nm and emission filter 610-nm Long Pass).
12. Sterilized deionized water.
13. *C. auris* genome-sequenced strain VPCI479/P/13 [16] (*see* **Note 1**).
14. YPD liquid medium: 0.5% yeast extract, 1% peptone, 2% dextrose.
15. Selective medium (YNBD solid medium containing mycophenolic acid): 0.67% Yeast Nitrogen Base with ammonium sulfate and without amino acids, 2% dextrose, 2% agar, supplemented with 50-μg/mL mycophenolic acid before plating.
16. pAYCU257 purified plasmid (1 μg/μL) (Fig. 1a) [15].
17. *Apa*I endonuclease and adequate digestion buffer.
18. DNA purification kit.

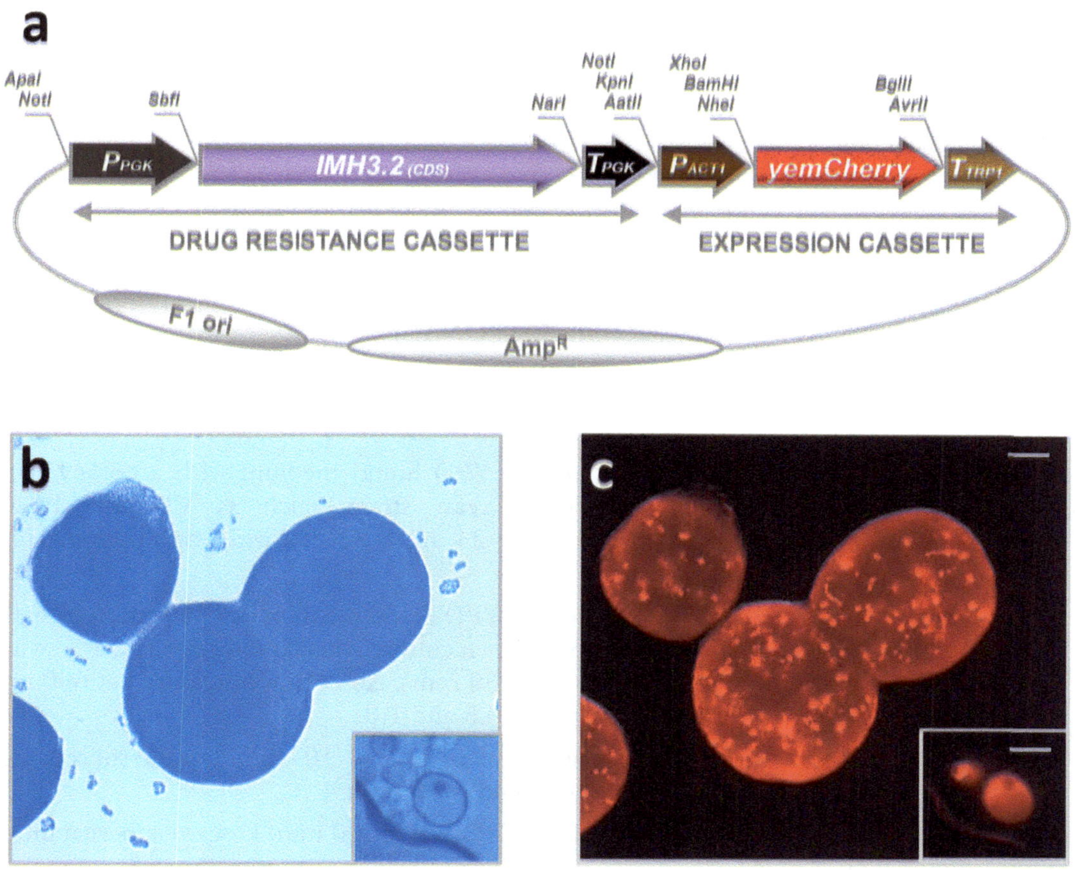

Fig. 1 pAYCU257 synthetic construct and efficient transformation and expression in *C. auris*. (**a**) pAYCU257 vector construct with P_{PGK}-*IMH3.2*-T_{PGK} drug resistance and P_{ACT1}-*yemCherry*-T_{TRP1} expression cassettes. (**b**, **c**) Colonies and individual cells (small inserts) of *C. auris* strain VPCI479/P/13 transformed with the linearized pAYCU257 plasmid showing (**b**) no signal under bright field microscopy and (**c**) mCherry fluorescence under epifluorescence microscopy. Scale bars in (**c**) large picture, 300 μm; insert, 3 μm apply to (**b**) as well

19. 1% agarose gel in 0.5X TAE buffer (20-mM Tris–HCl, 0.6-g/L acetic acid, 0.5-mM EDTA), 2-μg/mL ethidium bromide.
20. 10-mg/mL mycophenolic acid stock solution in sterilized deionized water.
21. 0.1-M lithium acetate, 10-mM Tris–HCl, 1-mM EDTA pH 7.5.
22. 1-M DTT stock solution.
23. 1-M sorbitol in sterilized deionized water.

3 Methods

3.1 Preparation of the Transforming DNA

1. Digest 10 μg of freshly purified pAYCU257 (Fig. 1a) [15] with endonuclease *Apa*I in a final volume of 200 μL (2 h, 37 °C) (*see* **Note 2**).
2. Electrophorese the linearized plasmid on a 1% agarose gel.
3. Purify DNA with a purification kit.
4. Electrophorese an aliquot (2 μL) of the purified product on 1% agarose gel to check DNA integrity.
5. Determine DNA concentration with a spectrophotometer and set final concentration at 500 ng/mL in sterile deionized water.

3.2 Preparation of C. auris Electro-Competent Cells

1. Inoculate 2 mL of YPD liquid medium with a colony of a freshly subcloned yeast strain, such as VPCI479/P/13 [16]. Incubate for 24 h at 32 °C with vigorous shaking (150 × *g*).
2. Inoculate 50 mL of YPD liquid medium in a 250-mL flask with 2 mL of the pre-grown culture. Incubate for 5 h at 30 °C with vigorous shaking (150 rpm). Absorbance measured at 600 nm should be between 0.5 and 1.0.
3. Chill cell suspension in ice for 10–15 min. All subsequent manipulations should be done at 4 °C.
4. Pellet the cells at 3000 × *g* for 10 min. Discard supernatant.
5. Resuspend the cells in 20 mL of 0.1-M lithium acetate, 10-mM Tris–HCl, 1-mM EDTA pH 7.5, 10-mM DTT (*see* **Note 3**). Incubate at 25 °C for 60 min.
6. Pellet the cells at 3000 × *g* for 10 min. Discard supernatant.
7. Resuspend the cells in 50 mL of ice-cold sterile deionized water. Pellet the cells at 3000 × *g* for 10 min. Discard supernatant.
8. Gently resuspend the cells by pipetting up and down in 20-mL ice-cold 1-M sorbitol. Pellet the cells at 3000 × *g* for 10 min. Discard supernatant.
9. Gently resuspend the cells 20 mL in ice-cold 1-M sorbitol (20 mL). Pellet the cells at 3000 × *g* for 10 min. Discard supernatant.
10. Gently resuspend the cells in 2-mL ice-cold 1-M sorbitol (*see* **Note 4**).

3.3 Electroporation of C. auris Cells

1. Place 190 μL of *C. auris* electro-competent cells in a cold 1.5-mL Eppendorf tube.
2. Add 5 μg (10 μL) of purified transforming DNA (here, *Apa*I--digested pAYCU257) (Fig. 1a) to the *C. auris* cells.

3. Mix gently.
4. Transfer the *C. auris*/transforming DNA mix (200 μL) in a 0.2-cm gap width electroporation cuvette.
5. Apply an electric pulse (1.8 kV, 25 μF, 200 ohms) on the cuvette by using an electroporation machine (*see* **Note 5**).
6. Immediately after electroporation, washed out the cells from the cuvettes with 800 μL YPD liquid medium and place them in a 50-mL Falcon tube containing 4-mL YPD liquid medium. Incubate for 6 h at 32 °C with vigorous shaking (150 rpm) (*see* **Note 6**).
7. Centrifuge (3800 × *g*, 2 min) and resuspend the cells in 200-μL YPD liquid medium.
8. Plate on selective medium.
9. Incubate the agar plates at 32 °C for 3–5 days.
10. Fluorescence signal can be directly observed on an epifluorescence microscope (Fig. 1b, c).

4 Notes

1. In the present protocol, the *C. auris* genome-sequenced strain VPCI479/P/13 isolated from a candidemia case in 2013 [16] is used. This strain is known to be susceptible to mycophenolic acid (minimum inhibitory concentration (MIC) < 10 μg/mL). Other *C. auris* strains may be used after checking their susceptibility to mycophenolic acid (MIC <10 μg/mL).
2. pAYCU257 does not contain any *C. auris* homologous sequence. The plasmid is routinely linearized with the single site *Apa*I for promoting random integration into the *C. auris* genome.
3. Importantly, 10 mM DTT (from a freshly prepared stock solution of 1 M DTT) must be added just before use to prepare electro-competent cells to the buffer containing 0.1-M lithium acetate, 10-mM Tris–HCl, 1-mM EDTA pH 7.5.
4. This suspension of *C. auris* electro-competent cells can be either stored at −80 °C (for several months) or directly used for the electroporation procedure.
5. The pulse given is for a Bio-Rad gene pulser and that conditions for other systems might be different.
6. This recovery step is of high importance because it allows the expression of the *IMH3.2* resistance gene and subsequently confers mycophenolic acid resistance.

References

1. Ninan MM, Sahni RD, Chacko B, Balaji V, Michael JS (2020) *Candida auris*: clinical profile, diagnostic challenge and susceptibility pattern: experience from a tertiary-care Centre in South India. J Glob Antimicrob Resist 21: 181–185
2. Du H, Bing J, Hu T, Ennis CL, Nobile CJ, Huang G (2020) *Candida auris*: epidemiology, biology, antifungal resistance, and virulence. PLoS Pathog 16(10):e1008921
3. Satoh K, Makimura K, Hasumi Y, Nishiyama Y, Uchida K, Yamaguchi H (2009) *Candida auris* sp. nov., a novel ascomycetous yeast isolated from the external ear canal of an inpatient in a Japanese hospital. Microbiol Immunol 53(1): 41–44
4. Chen J, Tian S, Han X, Chu Y, Wang Q, Zhou B et al (2020) Is the superbug fungus really so scary? A systematic review and meta-analysis of global epidemiology and mortality of *Candida auris*. BMC Infect Dis 20(1):827
5. O'Brien B, Liang J, Chaturvedi S, Jacobs JL, Chaturvedi V (2020) Pan-resistant *Candida auris*: New York subcluster susceptible to antifungal combinations. Lancet Microbe 1(5): e193–e1e4
6. Papon N, Sanglard D (2020) Tracking the origin and evolution of multidrug resistance in *Candida auris*. Lancet Microbe. 1(6):e237
7. Chow NA, Muñoz JF, Gade L, Berkow EL, Li X, Welsh RM, Forsberg K, Lockhart SR, Adam R, Alanio A, Alastruey-Izquierdo A, Althawadi S, Araúz AB, Ben-Ami R, Bharat A, Calvo B, Desnos-Ollivier M, Escandón P, Gardam D, Gunturu R, Heath CH, Kurzai O, Martin R, Litvintseva AP, Cuomo CA (2020) Tracing the evolutionary history and global expansion of *Candida auris* using population genomic analyses. MBio 11:e03364–e03319
8. Bruno M, Kersten S, Bain JM, Jaeger M, Rosati D, Kruppa MD, Lowman DW, Rice PJ, Graves B, Ma Z, Jiao YN, Chowdhary A, Renieris G, van de Veerdonk FL, Kullberg BJ, Giamarellos-Bourboulis EJ, Hoischen A, Gow NAR, Brown AJP, Meis JF, Williams DL, Netea MG (2020) Transcriptional and functional insights into the host immune response against the emerging fungal pathogen *Candida auris*. Nat Microbiol 5:1516–1531
9. Chaturvedi V, Linhardt RJ, Papon N (2020) *Candida auris* mannans and pathogen-host interplay. Trends Microbiol 28:954–956
10. Muñoz J, Gade L, Chow N, Loparev V, Juieng P, Berkow E et al (2018) Genomic insights into multidrug-resistance, mating and virulence in *Candida auris* and related emerging species. Nat Commun 9(1):5346
11. Defosse TA, Le Govic Y, Courdavault V, Clastre M, Vandeputte P, Chabasse D, Bouchara JP, Giglioli-Guivarc'h N, Papon N (2018) Les levures du clade CTG (clade Candida): biologie, incidence en santé humaine et applications en biotechnologie [yeasts from the CTG clade (*Candida* clade): biology, impact in human health, and biotechnological applications]. J Mycol Med 8(2):257–268
12. Papon N, Courdavault V, Clastre M, Simkin AJ, Crèche J, Giglioli-Guivarc'h N (2012) *Deus* ex *Candida* genetics: overcoming the hurdles for the development of a molecular toolbox in the CTG clade. Microbiology 158: 585–600
13. Papon N, Courdavault V, Clastre M (2014) Biotechnological potential of the fungal CTG clade species in the synthetic biology era. Trends Biotechnol 32(4):167–168
14. Defosse TA, Courdavault V, Coste AT, Clastre M, de Bernonville TD, Godon C, Vandeputte P, Lanoue A, Touzé A, Linder T, Droby S, Rosa CA, Sanglard D, d'Enfert C, Bouchara JP, Giglioli-Guivarc'h N, Papon N (2018) A standardized toolkit for genetic engineering of CTG clade yeasts. J Microbiol Methods 144:152–156
15. Defosse TA, Le Govic Y, Vandeputte P, Courdavault V, Clastre M, Bouchara JP et al (2018) A synthetic construct for genetic engineering of the emerging pathogenic yeast *Candida auris*. Plasmid 95:7–10
16. Sharma C, Kumar N, Meis JF, Pandey R, Chowdhary A (2015) Draft genome sequence of a fluconazole-resistant *Candida auris* strain from a candidemia patient in India. Genome Announc 3(4):00722–00715

Chapter 8

Genetic Transformation of *Candida auris* via Homology-Directed Repair Using a Standard Lithium Acetate Protocol

Gustavo Bravo Ruiz and Alexander Lorenz

Abstract

Reverse genetics is a particularly powerful tool in non-model organisms with known whole-genome sequences enabling the characterization of gene and, thus, protein function via a mutant phenotype. Reverse genetic approaches require genetic manipulation techniques which often need to be specifically developed for non-model organisms; this can be fraught with difficulties. Here, we describe a genetic transformation protocol for the recently emerged human pathogen *Candida auris* to target the integration of DNA constructs into genomic locations via homology-directed repair using long flanking homologous sequences (>1 kb). We detail the generation of DNA constructs for gene deletion with dominant drug resistance markers via fusion PCR, the transformation of these constructs into chemically competent *C. auris* cells, and the confirmation of correct integration by PCR. This strategy can be adapted to deliver DNA constructs other than deletion cassettes, including promoter exchanges and protein tags.

Key words *Candida auris*, Gene deletion, Gene targeting, Genetic manipulation, Genetic transformation, Dominant drug resistance markers

1 Introduction

Candida auris emerged about a decade ago as a multidrug-resistant fungal pathogen of humans. Since its first identification, this yeast has been found across all continents, causing clonal outbreaks in intensive care settings [1, 2]. Our ability to explain the life cycle and unique phenotypic characteristics of *C. auris* is currently impeded by a lack of information about its basic biology [3]. Development of molecular tools for *C. auris* to make it genetically tractable and traceable will be key for elucidating which mechanisms underpin its (multi)drug resistance, virulence, pathogenicity, and other relevant traits.

To date, genetic manipulation of *C. auris* has been successfully performed by several laboratories delivering genome alteration

Alexander Lorenz (ed.), *Candida auris: Methods and Protocols*, Methods in Molecular Biology, vol. 2517, https://doi.org/10.1007/978-1-0716-2417-3_8,

cassettes by various methods [4–13]. Also, a CRISPR (clustered regularly interspaced short palindromic repeats)–Cas9 system was developed for non-*albicans Candida* species [4] and proved successful in genetically manipulating *C. auris* [8–10]. However, this system relies on the delivery of the genome-editing tool as a ribonucleotide protein. Additionally, we and others have started to generate deletion mutants by introducing dominant drug resistance markers flanked by sequence homologies to the target locus [5, 11–13]. The targeting method via homology-directed repair can be used more widely for applications, such as gene deletion, placing genes under the control of heterologous promoters, protein engineering, or introducing point mutations. Currently, there are three dominant drug resistance marker (DDRMs) cassettes available for *Candida auris*; these cassettes confer resistance to the antibiotics nourseothricin (NAT or clonNAT), hygromycin B, and mycophenolic acid [6]. Other (model) yeasts such as *Saccharomyces cerevisiae* and *Candida albicans* only need short flanking regions (30–40 bp) to carry out homologous recombination [14, 15]. Although short flanking regions have been used successfully to delete *HOG1* in *C. auris* [5], our experience indicates that *C. auris* is less efficient at homology-directed targeting and requires homologous flanking regions in excess of 1 kb for correct integration of a DNA construct at a reasonable frequency [11].

Here, we detail all the steps necessary for homology-directed targeting of a DDRM cassette to a desired genomic region of *C. auris*. The protocol described here uses chemically competent cells subjected to heat-shock to deliver a DDRM cassette generated by a fusion PCR method [16] to delete the target region. However, this protocol can easily be adapted to deliver other genetic manipulation constructs (promoter exchange, protein tags, etc.).

2 Materials

Prepare all solutions using ultrapure H_2O (prepared by purifying deionized H_2O, to attain a sensitivity of 18 MΩ cm at 25 °C) and analytical grade reagents. Autoclaving is performed at 121 °C for 15 min. Prepare and store all reagents at room temperature (unless indicated otherwise). Diligently follow local regulations for disposal of waste materials.

2.1 General Materials

1. Thermocycler.
2. Heat block (or water bath with temperature regulation).
3. Static and shaking incubators.
4. Standard and NanoDrop spectrophotometers.

5. Agarose gel electrophoresis system: gel tray, electrophoresis chamber, and power source.
6. Loading dye.
7. Ethidium bromide.
8. DNA size marker.
9. UV light lamp.
10. 0.22-μm pore size filter system for sterilizing solutions
11. YPD (yeast–peptone–dextrose): dissolve 10 g of yeast extract, 20 g of mycological peptone, and 20 g of glucose in 700 mL of H_2O. When dissolved, add H_2O to a final volume of 1 L and autoclave. For solid YPD medium add 20 g of agar before autoclaving.
12. 0.5 M EDTA disodium salt stock solution, pH 8.0: dissolve 93.05 g of EDTA disodium salt in 400 mL of H_2O, and adjust the pH with sodium hydroxide (NaOH) pellets. EDTA will not dissolve completely until the pH is about 8.0. Add H_2O to a final volume of 500 mL and autoclave.
13. 1× TAE buffer: 50-mM EDTA disodium salt, 2-M Tris base and 1-M acetic acid. Preparation of a concentrated 50× stock solution is recommended: dissolve 242 g of Tris base in 750 mL of H_2O, add 57.1 mL of acetic acid and 100 mL of 0.5-M EDTA (pH 8.0), and, finally, adjust the volume to 1 L by adding H_2O. The final pH of the 50× TAE buffer should be ~8.5. Prepare the 1× TAE working buffer by adding 49 parts of H_2O to 1 part of 50× TAE buffer.
14. 1 M Tris–HCl buffer, pH 8.0: dissolve 121.14 g of Tris base in 700 mL of H_2O, adjust the pH using concentrated HCl, add H_2O to a final volume of 1 L, and autoclave.

2.2 Fusion PCR

1. Specific oligonucleotide primers designed for the target locus.
2. High-fidelity DNA polymerase kit (*see* **Note 1**).
3. DNA cleanup kit and DNA gel extraction kit.
4. 1× TE buffer (10-mM Tris-HCl and 1-mM EDTA disodium salt). Preparation of a concentrated 10× stock solution is recommended: add 10 ml of 1-M Tris-HCl buffer (pH 8.0) and 2 mL of 0.5-mM EDTA (pH 8.0) to 88 mL of H_2O for a final volume of 100 mL, and autoclave. Prepare the 1× TE working buffer by adding nine parts of H_2O to one part of 10× TE buffer.

2.3 Transformation

1. 1-M lithium acetate: dissolve 6.6 g of lithium acetate in 70 mL of H_2O, then add H_2O to 100 mL, and sterilize the solution using a 0.22-μm pore size filter system.

2. 50% polyethylene glycol MW 3350 (PEG-3350): mix 50 g of PEG-3350 and 50-mL H_2O, and then autoclave.
3. Transformation Mix: combine 240-μL 50% PEG-3350, 36-μL-1 M lithium acetate, and 10 μL of 10-mg/mL denatured herring sperm DNA (denature the herring sperm DNA by incubating at 95 °C on a heat block for 10 min followed by cooling on ice for 5 min).
4. Solid YPD medium (YPD plates) with selection drug at appropriate concentration.

2.4 Checking Transformants

1. Taq DNA polymerase or high-fidelity DNA polymerase (*see* **Note 1**).
2. 0.02-M NaOH dilution: dissolve 0.08 g of NaOH in 70-mL H_2O, and then add H_2O to 100 mL.
3. 50% glycerol: mix 50-mL glycerol with 50-mL H_2O, and then autoclave.

2.5 Genomic DNA Extraction

1. Acid-washed glass beads.
2. Multi-tube mixer or sample homogenizer.
3. DNA extraction buffer: prepare a bottle with around 100 mL of H_2O, and then add 20 mL of 1-M NaCl, 4 mL of Triton X-100, 2 mL of 1-M Tris–HCl buffer at pH 8.0, 400 μL of 0.5 M EDTA, and 2 g of sodium dodecyl sulfate (SDS). After mixing add H_2O up to 200 mL and autoclave.
4. Phenol/chloroform/isoamyl alcohol (25:24:1).
5. 3 M sodium acetate pH 5.2: dissolve 12.3 g of sodium acetate in 30-mL H_2O, and then add H_2O up to 50 mL and autoclave.
6. 100% ethanol.
7. 70% ethanol: mix 70 mL 100% ethanol with 30-mL H_2O.
8. 1 mg/mL RNase stock solution: dissolve 10 mg of RNase A in 10-mL 10-mM Tris–HCl (pH 7.5), 15-mM NaCl, incubate at 100 °C for 5 min to inactivate contaminating DNases, and cool slowly to room temperature. Finally dispense into aliquots and store at −20 °C.

3 Methods

All procedures are carried out at room temperature unless otherwise specified. The protocol described here is for one transformation and one control.

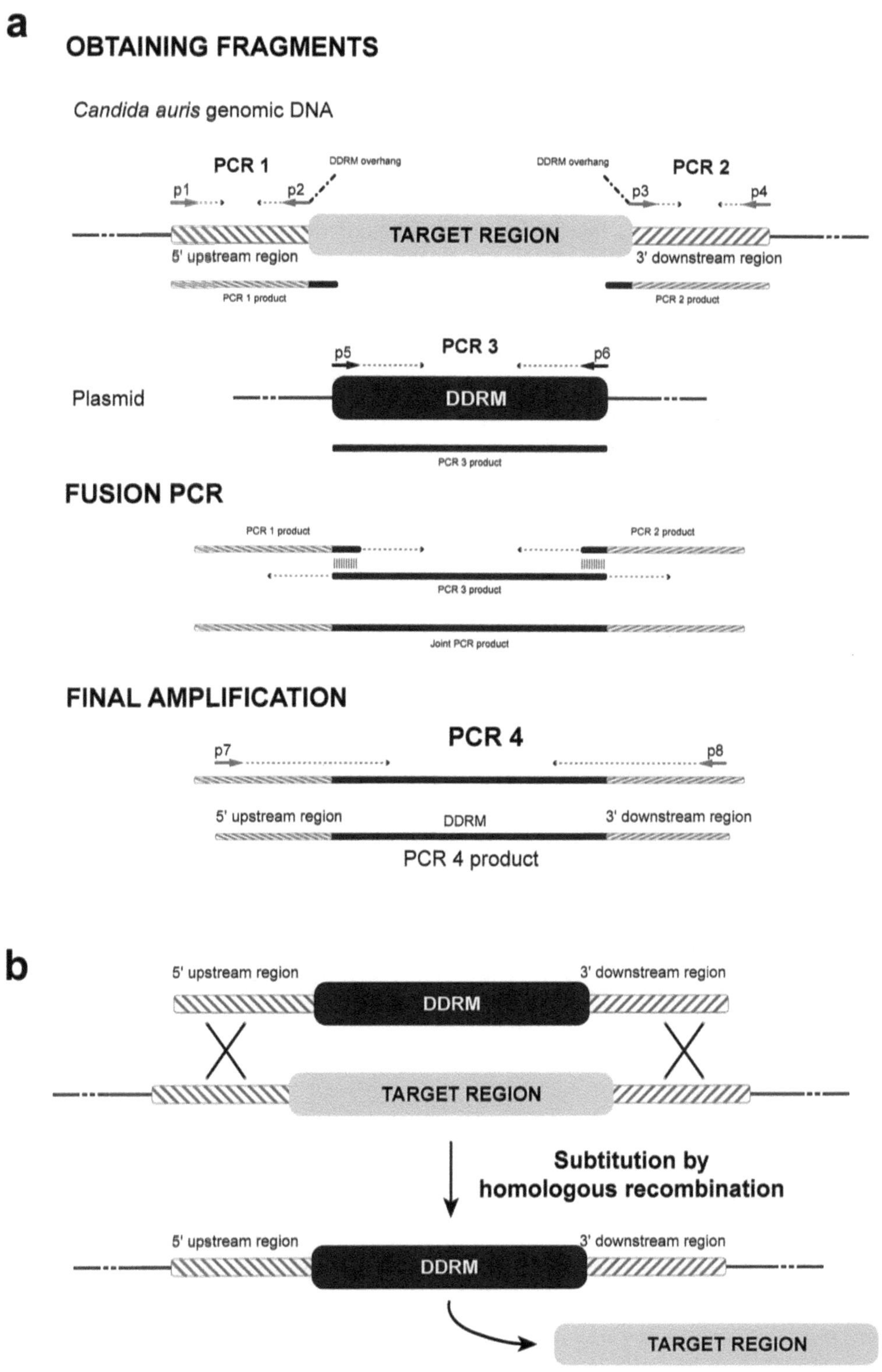

Fig. 1 Schematic representation of the transformation method. (**a**) Marker cassette obtaining by fusion PCR (*see* main text, Subheading 3.1). (**b**) Target region substitution by homologous recombination within the cell after transformation (*see* main text, Subheading 3.2)

3.1 Obtaining the Transformation Cassette by Fusion PCR

The fusion PCR method allows joining of DNA fragments, which share homology with the target region, with a DDRM [16]. Using this method, a cassette consisting of a DDRM gene [17] flanked by 1–2 kb of the up- and downstream sequence of the target DNA region is constructed (Fig. 1a). Afterward, this cassette is directly used for transformation.

Initially, three DNA fragments are obtained by standard PCR (Fig. 1a): PCR1 and PCR2 correspond to the upstream and downstream sequences of the target DNA region including overhang regions (20–30 bp) at the 3′ and 5′ end, respectively (Fig. 1a). The overhang regions are homologous to the promotor and terminator sequences flanking the DDRM gene. PCR3 amplifies the complete DDRM gene including promoter and terminator region (Fig. 1a). These three fragments are joined by fusion PCR, followed by a final standard PCR4 to amplify the final transformation cassette using nested primers (Fig. 1a).

All these PCRs are performed using a high-fidelity DNA polymerase; the manufacturer's instructions need to be followed for primer design, concentration of reaction components, and PCR program. For PCR mixtures combine reagents in small PCR tubes as summarized in Table 1. PCR reagents should be kept on

Table 1
Representative PCR reagents concentrations (*see* Note 32). Add items in the indicated order from top to bottom

Reagent	Final concentration
H_2O	Add to final volume
Buffer	1×
$MgCl_2$	1.5 mM
dNTPs	200 μM (each)
Forward primer (10 pmol μL^{-1})[a]	0.5 μM[b]
Reverse primer (10 pmol μL^{-1})[a]	0.5 μM[b]
Template DNA	0.01–50 ng[c]
DMSO (optional)[d]	5%
Polymerase	1 U

[a]Primer must be diluted to a working concentration of 10 pmol μL^{-1}. Primer dilutions can be made in H_2O, but TE buffer is recommended

[b]Final primer concentrations can vary in a range of 0.2–1.0 μM

[c]For low-complexity DNA (e.g., plasmid), it is recommended to use 0.01–2 ng. For high-complexity genomic DNA, use 10–50 ng

[d]Recommended for GC-rich amplicons, such as the *CaNAT1* and *CaSAT1* cassettes

Table 2
Representative cycling conditions (*see* Note 33)

Cycle step	Temperature	Time	Cycles
Initial denaturation	95–98 °C	30–60 s	1
Denaturation	95–98 °C	10–30 s	
Annealing	T_m[a]	30–60 s	25–35
Extension	72 °C	30–60 s/kb	
Final extension	72 °C	5–10 min	1

[a]A T_m calculator can be found at www.thermofisher.com/tmcalculator

ice while handling. Transfer the PCR tubes containing the reagent mixtures to a thermocycler, and amplify following the conditions detailed in Table 2.

1. Oligonucleotide primer design: Eight primers will be necessary (Fig. 1a) (*see* **Note 2**). Two primers (p1 and p2) for amplifying the upstream flanking region (PCR1), where the reverse primer (p2) includes a homologous overlapping "tail" sequence corresponding to the DDRM promoter region. Two primers (p3 and p4) are used to amplify the downstream flanking region (PCR2); on this occasion the homologous "tail" sequence corresponds to the DDRM terminator region in the forward primer (p3) (Fig. 1a). Note that "tail" sequences always need to be placed at the 5′ end of an oligonucleotide during primer design, as the 3′ end targets it to the desired genomic region. Furthermore, two primers (p5 and p6) are needed to amplify the whole DDRM cassette (PCR3), including its promoter and the terminator, as well as two nested primers (p7 and p8) for the amplification (PCR4) of the transformation cassette (Fig. 1a).
2. PCR1 and PCR2 (Fig. 1a): Use genomic DNA (gDNA) (gDNA extraction is described in Subheading 3.4) from the target strain as template and the primers previously designed to obtain the up- (p1 and p2) and downstream (p3 and p4) fragments. A final PCR reaction volume of 10 μL is sufficient (Table 1).
3. PCR3 (Fig. 1a): Use a plasmid (or DNA) containing the DDRM [6] as template and the primers previously designed to obtain the DDRM fragment (p5 and p6). A final volume of 10 μL is sufficient (Table 1).
4. Purify the PCR products using a DNA cleanup kit following the manufacturer's instructions.
5. Check the PCR results by electrophoresing 1 μL of the samples in an agarose gel. A loading dye must be mixed with the sample

before loading onto the gel, usually, in a proportion of 1 μL of sample, 7 μL of H_2O, and 2 μL of loading dye (check manufacturer's instructions). A DNA ladder marker (5–10 μL) must be included for size comparison. Prepare a 0.7–1% agarose gel in 1× TAE buffer, and add ethidium bromide to a final concentration of approximately 0.2–0.5 μg/mL for DNA visualization. Pour the agarose into a gel tray with the gel comb in place to create wells for loading the DNA into. Place the gel into an electrophoresis unit, cover it with 1× TAE, load the samples in the wells, and run at 100–125 constant V for 45–60 min (*see* **Note 3**). Finally, visualize your DNA fragments using a UV light lamp.

6. Determine DNA concentration of your purified samples using a NanoDrop spectrophotometer (*see* **Note 4**).
7. Use the three previously obtained PCR fragments to perform the fusion PCR (Fig. 1a). Combine the three products in roughly equimolar amounts (~0.15 pmols each; *see* **Note 5**) with the high-fidelity DNA polymerase, its components, and DMSO as indicated by the manufacturer (Table 1) but without oligonucleotide primers. Subject samples to eight PCR cycles (according to the polymerase manufacturer's instructions, Table 2) (*see* **Notes 6, 7**). Do not purify this PCR or run it in gel electrophoresis; proceed directly to the next step.
8. Directly use 1 μL of fusion PCR as template with nested primers p7 and p8 in PCR4 to obtain the final transformation cassette (Fig. 1a) via standard PCR conditions. Before proceeding to the transformation, DNA purification of the transformation cassette is recommended. Check the purified PCR products by electrophoresing 1 μL of the samples in a DNA agarose gel (*see* Subheading 3.1, **step 5**), and determine DNA concentration (*see* Subheading 3.1, **step 6**) (*see* **Note 8**).

3.2 *C. auris* Transformation

The transformation protocol described here is adapted from a *C. albicans* protocol [18] using chemically competent cells and heat-shock for the uptake of the transformation cassette. Within the cell, the transformation cassette is integrated at the target DNA region by homologous recombination guided by the up- and downstream regions flanking the DDRM gene (Fig. 1b).

1. Grow *C. auris* cells in 5-mL YPD in a sterile 50-mL flask or 50-mL tube overnight at 30 °C in an incubator shaking at 200 rpm.
2. Dilute overnight culture 1:100 in 50 mL of fresh YPD in a sterile 250-mL Erlenmeyer flask (*see* **Note 9**), and incubate at 30 °C and shaking at 200 rpm until reaching mid-exponential growth 1–2 $\times 10^7$ cells/ml (corresponding to an OD_{600} reading of 0.5–0.8 on a spectrophotometer) (*see* **Notes 10, 11**).

3. Transfer the culture into a centrifugation tube, and harvest cells by centrifugation at 1000 × *g* for 2 min. Decant supernatant, and resuspend cells gently in 30 ml of sterile H_2O (*see* **Note 12**). Repeat centrifugation at 1000 × *g* for 2 min. Remove H_2O by decanting as before.
4. Resuspend the cell pellet gently in 1-mL 0.1-M lithium acetate, transfer the cell suspension to a 1.5-mL reaction tube, and centrifuge at 1000 × *g* for 60 sec. Then, decant the supernatant and resuspend in 100-μL 0.1-M lithium acetate. Split the sample into two new reaction tubes, 50 μL each, one for the actual transformation and the other one as negative control (*see* **Note 13**).
5. Add transformation mix.
6. Mix the sample, and add around 1 μg of transformation cassette DNA (*see* **Note 14**) or the same volume of sterile H_2O for the negative control.
7. Mix samples by a short pulse on a vortex mixer, and incubate at 30 °C overnight.
8. After overnight (16–18 hrs) incubation, briefly mix the samples, and heat-shock at 44 °C for 15 min in a heat block or water bath.
9. Centrifuge samples at 1000 × *g* for 60 sec. Remove supernatant, resuspend the samples in 1-mL YPD, transfer to a 50-mL centrifugation tube, add 2-mL YPD, and incubate for cell recovery at 30 °C in an incubator shaking at 200 rpm (*see* **Note 15**).
10. Centrifuge samples at 1000 × *g* for 2 min; remove supernatant and resuspend in 100 μL sterile H_2O. Plate transformation sample (with DNA) onto selective media (YPD + selection drug at appropriate concentration) (*see* **Notes 16, 17**). For the negative control sample, use a small aliquot to prepare a 1:10,000 dilution in sterile H_2O, and plate 100 μL (1–5 $\times 10^3$ cells) on a YPD plate as regeneration control; plate the rest of the neat negative control sample on a YPD plate + selection drug (no colonies should grow).
11. Incubate plates at 30 °C. Colonies should be visible 2–4 days after plating (*see* **Note 18**).
12. When visible, patch colonies one by one to a fresh YPD plate + selection drug. Make sure that individual colonies are numbered to keep track of them. Grow overnight at 30 °C.

3.3 Transformant Screening

Homology-directed repair does not seem to work very well in *C. auris*; it tends to integrate DNA randomly into the genome. For this reason, it might be necessary to screen a large number of colonies before finding a transformant in which the marker cassette

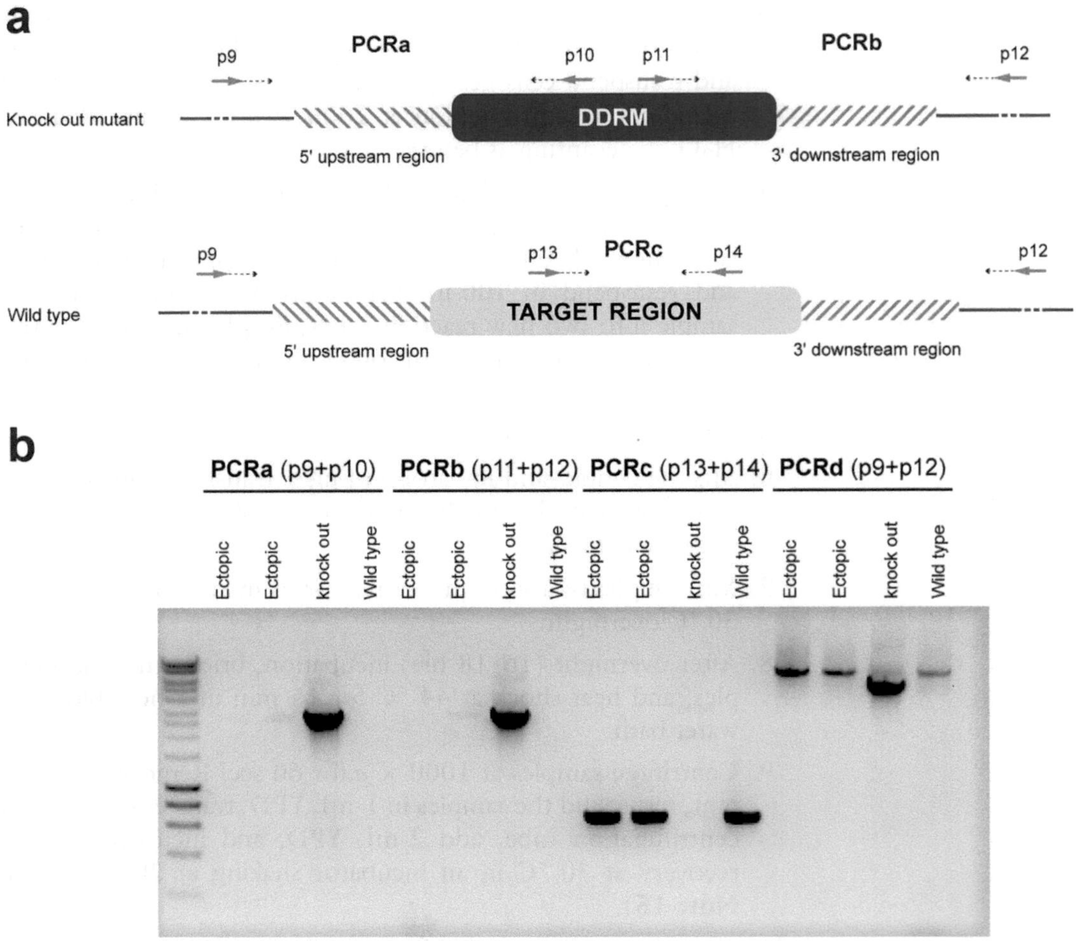

Fig. 2 Transformants screening (*see* main text, Subheading 3.3). (**a**) Schematic representation of PCRs necessary for checking transformants. (**b**) Agarose gel electrophoresis showing a real example of transformants screening including two false positives (ectopic), one positive (knock-out) and the wild-type

is integrated at the desired locus. We recommend a fast screening of the colonies by colony PCR followed by a second check of the positive transformants using high-quality gDNA (gDNA extraction is described in Subheading 3.4) as PCR template (*see* **Note 19**). Colony PCR allows screening hundreds of colonies in one or 2 days.

1. Add 40 μL of a 0.02-M NaOH solution to a 1.5-mL reaction tube.
2. Collect cells with a sterile tip or toothpick taking a small section of the colony. Avoid touching the agar. Transfer the cells in the NaOH solution, and mix until the solution turns slightly cloudy, and then discard the tip or toothpick (*see* **Note 20**).

3. For cell lysis, boil the sample at 100 °C for 10 min using a heat block, and then place samples on ice for 5 min.
4. Centrifuge 2 min at 4000 × *g*. Do not mix or resuspend the pellet after centrifugation.
5. Use 1 μL from the supernatant as template for a standard 10-μl PCR as described in Subheading 3.1 (Tables 1 and 2) (*see* **Note 21**). At least three PCRs are recommended for each sample (Fig. 2) (*see* **Note 22**): PCRa and PCRb using a primer located within the DDRM gene and another placed in the up- (p9 and p10) or downstream (p11 and p12) region of the target site, respectively (ensure that these primers are outside of the homologous flanking sequences of the transformation cassette), and PCRc using primers located within the target DNA (p13 and p14) to check the correct elimination of this region. Genomic DNA from the parental strain should be included as control, negative for the transformation cassette integration PCRs and positive for target DNA PCR (*see* **Notes 23 and 24**).
6. Perform gel electrophoresis using the whole volume of the PCRs (as described in Subheading 3.1, **step 5**) to check the results. The expected result for correct integration of the cassette and deletion of the gene is positive for both PCRs using the primers located within the marker and in the up- and downstream region (PCRa and PCRb) and negative for the PCR using primers of the target gene (PCRc) (Fig. 2b)
7. Double-check the positive transformants using high-quality gDNA (gDNA extraction is described in Subheading 3.4). Repeat PCRs as described above (Fig. 2) using 1 μL of a dilution 1:10 of the high-quality gDNA (*see* **Note 25**).
8. Finally, to store transformants, grow confirmed transformants in 5-mL YPD + selection drug in a 50-mL flask or 50-mL tube overnight at 30 °C shaking at 200 rpm in an incubator. Harvest cells by centrifugation at 1000 × *g* for 2 min and wash with 5 mL of sterile H_2O. Repeat centrifugation at 1000 × *g* for 2 min. Decant the supernatant, resuspend cells in 500 μL of sterile YPD (or H_2O), and mix with 1 ml of 50% of sterile glycerol (final concentration 33%) (*see* **Note 26**).

3.4 Extraction of High-Quality Genomic DNA (gDNA)

The gDNA extraction protocol described here is adapted from a method used in budding yeast [19].

1. Grow *C. auris* cells in 5-mL YPD using a 50-mL flask or 50-mL tube overnight at 30 °C in an incubator shaking at 200 rpm.
2. Harvest the cells by centrifugation at 1000 × *g* for 2 min.
3. Decant supernatant, gently mix cells with 500 μL of H_2O, and transfer cells to a 2-mL reaction tube.

4. Centrifuge sample for 2 min at 12,000 × *g*. Decant the supernatant and resuspend cells in the residual liquid.
5. Add 0.3 g of acid-washed glass beads and 300 μL of DNA extraction buffer. Disrupt cells in a multi-tube mixer or a sample homogenizer for ~3 min.
6. In a fume hood, add 300 μL of phenol/chloroform/isoamyl alcohol (25:24:1) and mix.
7. Centrifuge for 10 min at 12,000 × *g*. In a fume hood, transfer the top aqueous layer to a new 1.5-mL reaction tube (*see* **Note 27**).
8. Determine the volume of the top aqueous layer, either by weighing or using a micropipettor (typically 300–400 μL are recovered), and add 0.1 volumes of 3-M sodium acetate pH 5.2 and 2 volumes of ice-cold 100% ethanol. Mix by inversion four to five times.
9. Incubate at −20 °C for at least 20 min to allow DNA precipitation (*see* **Note 28**).
10. Centrifuge for 10 min at 12,000 × *g*.
11. Decant supernatant and wash pellet in 70% ethanol. Centrifuge for 5 min at 12,000 × *g*. Repeat this step (*see* **Note 29**).
12. Remove ethanol and air-dry the pellet (optional; use a Speed-Vac) (*see* **Note 30**).
13. Dissolve DNA pellet in 50–100 μL of either TE buffer or sterile H_2O (*see* **Note 31**).
14. Add 2.5–5 μL of 1 mg/mL RNase A stock solution (final concentration of 50 μg/mL) and incubate sample 30–60 min at 37 °C.
15. Check gDNA quality by electrophoresing 1 μL of the sample in an agarose gel (as described in Subheading 3.1, **step 5**). High-quality gDNA size should run slower than commercial DNA markers (>10 kb) and form a tight band without a lower-weight smear which would indicate degradation or shearing of the DNA sample.

4 Notes

1. Commercially available DNA polymerase kits contain all necessary buffers and supplements.
2. Good primer design is fundamental for efficient PCRs. Many primer design guidelines are available online, for example, https://www.thermofisher.com/blog/behindthebench/pcr-primer-design-tips/.
3. Voltage and time might need to be adjusted depending on the make of your DNA gel electrophoresis kit; always allow for

enough time to separate the DNA size marker bands without running them off the gel.

4. Alternatively, DNA concentration can be determined by the band intensities on a gel in comparison with DNA ladder marker bands similar in size to the PCR product. The DNA amount included in certain bands of the DNA ladder is usually provided by the manufacturer.
5. To calculate the pmols of a double-stranded DNA, use the following equation: pmol = M/(0.66*N), where M is the mass of the DNA in ng and N is the number of nucleotide pairs; 0.66 accounts for the average molecular weight of a nucleotide pair. 1 μg (1000 ng) of a 1000 base pair DNA fragment thus corresponds to 1.52 pmol. DNA amount conversion calculators can also be found online, e.g., https://www.promega.co.uk/resources/tools/biomath/.
6. A melting temperature (T_m) of 55 °C is recommended, but it could be higher to minimize the formation of unspecific DNA fragments.
7. Fusion PCR is not meant to amplify the product but to join the DNA fragments. Therefore, only a few cycles are needed.
8. Due to the complexity of the process, it is not unusual to obtain unspecific DNA fragments in PCR4. In this case, it is recommended to run the whole PCR reaction on an agarose DNA gel and purify the fragment of correct size from the gel using a DNA gel extraction kit. This extraction process is of lower efficiency than using a DNA purification kit to clean up the PCR reaction itself. Therefore, a large volume of the final PCR product might be required to generate sufficient amounts of the transformation cassette.
9. If several transformations are to be performed with the same strain, or more cells/transformation are required, using a higher volume of YPD is advisable. Please note the indicated volume on the Erlenmeyer flask must be five times bigger than the volume of culture medium used. If in this step a higher volume is used, the volumes in following steps must be adjusted accordingly.
10. Usually, an OD_{600} of 0.5–0.8 is reached after 3–5 h. Cells in late exponential growth or stationary phase can be used, but the transformation efficiency will be considerably reduced.
11. Please note that OD_{600} readings are influenced by making and modeling of spectrophotometer and the strain background.
12. Mix cells gently by pipetting up and down or on a vortex mixer at low-medium velocity, mixing conditions harsher than that

could cause cell death and, thus, reduce transformation efficiency.

13. If the culture volume was changed in Subheading 3.2, **step 2**, ensure that each tube contains between 1 and 5×10^8 cells.
14. A higher amount of DNA will result in a higher total number of transformants. However, the frequency of correctly targeted transformants is not affected. Therefore, the use of a higher concentration of DNA will allow the screening of a higher number of transformants, which might be advantageous when targeting efficiency is low.
15. This incubation step gives cells time to express DDRM gene after the transformation in order to become resistant. The incubation time would depend on the DDRM used, but a few hours tend to be sufficient.
16. Samples are best split over two to three plates to avoid overcrowding of colonies.
17. The antimicrobial drugs shown to work in *C. auris*, at the final concentrations indicated, are nourseothricin (also known as NAT or clonNAT) at 200 mg/L, hygromycin B at 400 mg/L, and mycophenolic acid at 50 mg/L.
18. If the mutant has a growth defect, colonies may need more than 4 days to appear. However, after prolonged incubation small colonies can appear, likely untransformed background colonies tolerant or spontaneously resistant to the antimicrobial. In such circumstances it can be difficult to separate true transformants from background.
19. The DNA yield and quality in colony PCR tend to be low. Therefore, the colony PCR might not work or give ambiguous results. For this reason, we highly recommended confirming possible positive transformants with a standard PCR using high-quality gDNA as template. Performing the initial screen with good-quality gDNA would be too laborious.
20. A very high amount of cells can negatively affect the PCR efficiency.
21. This PCR can be carried out using a high-fidelity or a Taq polymerase; follow manufacturer's instructions to set up reactions.
22. A PCR multi-well plate can be used when a very large numbers of transformants are analyzed.
23. An additional PCR using a pair of primers from a region known to be present in all strains, for example, the actin gene (*ACT1*), can be included as a positive control for DNA quality.
24. When a large number of transformants need to be screened, preparing at least three PCRs for each one will be tedious. A faster and cheaper alternative is to perform PCRa first and

continue only with positive candidates. Then, perform the other PCRs to confirm the correct integration at the downstream end (PCRb) and the elimination of the target gene (PCRc).

25. Alternatively, if the target DNA region and the marker sizes are different, a PCRd using primers located in the up- or downstream region flanking the targeted DNA sequence (p9 and p12) can be done (Fig. 2). Using these primers, the fragment would be amplified in both cases, either the cassette has been correctly integrated or not, but the size of the fragment would be different.
26. Any final glycerol concentration between 25 and 50% works well.
27. After centrifugation three layers are formed. It is important to take only the top layer without touching the middle white layer. Transfer of material from the middle layer would contaminate the gDNA, decreasing its quality and its efficiency as a PCR template.
28. Samples can be kept overnight at −20 °C. However, long incubation times at low temperature will precipitate undesired components such as large amounts of sodium acetate, reducing the quality of the gDNA sample.
29. For a better elimination of salt in the final gDNA, washes with serially diluted ethanol can be performed. Perform three to four washes using ethanol dilutions of descending concentrations between 95% and 70%.
30. Samples are dry when ethanol smell cannot be detected anymore. Do not overdry the DNA pellet as this makes it difficult to dissolve it again.
31. Heating the sample to 50 °C for a few min after adding H_2O or buffer improves solubility of the sample.
32. Concentrations are indicative since different polymerase kits have particular specifications. Always check manufacturer's instruction for specific reaction conditions.
33. Conditions are indicative since different polymerase kits have particular specifications. Always check manufacturer's instruction for specifics.

Acknowledgments

We are grateful for support from the Wellcome Trust [grant numbers 212524/Z/18/Z and 204815/Z/16/Z].

References

1. Rhodes J, Fisher MC (2019) Global epidemiology of emerging *Candida auris*. Curr Opin Microbiol 52:84–89. https://doi.org/10.1016/j.mib.2019.05.008
2. de Jong AW, Hagen F (2019) Attack, defend and persist: how the fungal pathogen *Candida auris* was able to emerge globally in healthcare environments. Mycopathologia 184:353–365. https://doi.org/10.1007/s11046-019-00351-w
3. Bravo Ruiz G, Lorenz A (2021) What do we know about the biology of the emerging fungal pathogen of humans *Candida auris*? Microbiol Res 242:126621. https://doi.org/10.1016/j.micres.2020.126621
4. Grahl N, Demers EG, Crocker AW, Hogan DA (2017) Use of RNA-Protein complexes for genome editing in non-*albicans Candida* species. mSphere 2:e00218–e00217. https://doi.org/10.1128/mSphere.00218-17
5. Day AM, McNiff MM, da Silva DA et al (2018) Hogl regulates stress tolerance and virulence in the emerging fungal pathogen *Candida auris*. mSphere 3:e00506–e00518. https://doi.org/10.1128/mSphere.00506-18
6. Defosse TA, Courdavault V, Coste AT et al (2018) A standardized toolkit for genetic engineering of CTG clade yeasts. J Microbiol Methods 144:152–156. https://doi.org/10.1016/j.mimet.2017.11.015
7. Defosse TA, Le Govic Y, Vandeputte P et al (2018) A synthetic construct for genetic engineering of the emerging pathogenic yeast *Candida auris*. Plasmid 95:7–10. https://doi.org/10.1016/j.plasmid.2017.11.001
8. Kim SH, Iyer KR, Pardeshi L et al (2019) Genetic analysis of *Candida auris* implicates Hsp90 in morphogenesis and azole tolerance and Cdrl in azole resistance. MBio 10:e02529–e02518. https://doi.org/10.1128/mBio.02529-18
9. Rybak JM, Doorley LA, Nishimoto AT et al (2019) Abrogation of triazole resistance upon deletion of *CDR1* in a clinical isolate of *Candida auris*. Antimicrob Agents Chemother 63:e00057–e00019. https://doi.org/10.1128/AAC.00057-19
10. Rybak JM, Muñoz JF, Barker KS et al (2020) Mutations in *TAC1B*: a novel genetic determinant of clinical fluconazole resistance in *Candida auris*. MBio 11:e00365–e00320. https://doi.org/10.1128/mBio.00365-20
11. Bravo Ruiz G, Ross ZK, Gow NAR, Lorenz A (2020) Pseudohyphal growth of the emerging pathogen Candida auris is triggered by genotoxic stress through the S phase checkpoint. mSphere 5:e00151–e00120. https://doi.org/10.1128/msphere.00151-20
12. Mayr E-M, Ramírez-Zavala B, Krüger I, Morschhäuser J (2020) A zinc cluster transcription factor contributes to the intrinsic fluconazole resistance of *Candida auris*. mSphere 5:e00279–e00220. https://doi.org/10.1128/mSphere.00279-20
13. Shivarathri R, Jenull S, Stoiber A et al (2020) The two-component response regulator Sskl and the mitogen-activated protein kinase Hogl control antifungal drug resistance and cell wall achitecture of *Candida auris*. mSphere 5:e00973–e00920. https://doi.org/10.1128/mSphere.00973-20
14. Baudin A, Ozier-Kalogeropoulos O, Denouel A et al (1993) A simple and efficient method for direct gene deletion in *Saccharomyces cerevisiae*. Nucleic Acids Res 21:3329–3330. https://doi.org/10.1093/nar/21.14.3329
15. Wilson RB, Davis D, Mitchell AP (1999) Rapid hypothesis testing with *Candida albicans* through gene disruption with short homology regions. J Bacteriol 181:1868–1874
16. Shevchuk NA, Bryksin AV, Nusinovich YA et al (2004) Construction of long DNA molecules using long PCR-based fusion of several fragments simultaneously. Nucleic Acids Res 32:e19. https://doi.org/10.1093/nar/gnh014
17. Shen J, Guo W, Köhler JR (2005) *CaNAT1*, a heterologous dominant selectable marker for transformation of *Candida albicans* and other pathogenic *Candida* species. Infect Immun 73:1239–1242. https://doi.org/10.1128/IAI.73.2.1239-1242.2005
18. Walther A, Wendland J (2003) An improved transformation protocol for the human fungal pathogen *Candida albicans*. Curr Genet 42:339–343. https://doi.org/10.1007/s00294-002-0349-0
19. Amberg DC, Strathern JN, Burke DJ (2005) Methods in yeast genetics: a cold Spring Harbor Laboratory Course Manual. Cold Spring Harbor Laboratory Press, New York

Chapter 9

Downregulation of Essential Genes in the Fungal Pathogen *Candida auris*

Kali R. Iyer, Sang Hu Kim, Nicole Robbins, and Leah E. Cowen

Abstract

The recent global emergence of the fungal pathogen *Candida auris* has caused significant concern given that this pathogen often exhibits resistance to multiple antifungal drug classes. In order to effectively combat *C. auris* infections, there is a dire need to expand our current antifungal arsenal. Essential proteins often serve as targets for antimicrobial compounds, and thus being able to study essential genes in a pathogen of interest is a critical first step in drug development. To identify and characterize essential genes in microorganisms, researchers must be able to manipulate microbial genomes using a variety of molecular biology techniques. Given the haploid genome of *C. auris*, genetic alterations have largely been achieved by gene deletion through homologous recombination using a drug resistance marker. However, this approach is not feasible to study essential gene function. Here, we describe a method for the study of essential genes using a tetracycline-repressible promoter replacement system, which can be used to genetically repress essential genes in *C. auris* and, thus, study their function. This method provides a powerful approach to assess and characterize essential gene function in an emerging fungal pathogen.

Key words *Candida auris*, Essential genes, Gene repression, Electroporation, Transformation, Hsp90

1 Introduction

Over the past several decades, invasive fungal infections have dramatically increased in prevalence, which is largely attributed to the parallel rise in immunocompromised individuals along with the emergence of novel pathogenic species [1, 2]. In particular, *Candida auris*, which was first discovered in 2009, has since been identified globally and exhibits widespread resistance to our current antifungal arsenal [3, 4]. This prompted the US Centers for Disease Control and Prevention to classify *C. auris* as one of only five pathogens that are the most urgent threats to public health [5],

Kali R. Iyer and Sang Hu Kim contributed equally.

Alexander Lorenz (ed.), *Candida auris: Methods and Protocols*, Methods in Molecular Biology, vol. 2517,
https://doi.org/10.1007/978-1-0716-2417-3_9,

highlighting the need for enhanced efforts to diagnose, prevent, and treat these infections.

The ability to define and characterize essential genes in pathogenic microbes is a critical first step in order to expand the therapeutic target space for drug development. With advancements in molecular biology techniques, reverse genetic analyses enable the systematic discovery of gene function by specifically altering a particular DNA sequence and subsequently defining the phenotypic consequences associated with such mutations. A key facet of this approach is the ability to manipulate the genetic material in the pathogen of interest. Genetic alterations in *Candida* species have been met with some challenges, notably the alternative codon usage, which results in the CUG codon being translated as serine instead of the universal leucine, and the lack of a complete mating cycle [6]. However, optimization of transformation protocols has enabled genome-wide mutant collections to be generated in the related fungal pathogen *Candida albicans* [7–9]. Specifically, a gene replacement and conditional expression (GRACE) collection, which harbors 2357 strains, carries a precise replacement of one of the two copies of the gene with an auxotrophic marker, while the remaining allele is under the control of a repressible promoter system. This functional genomics resource has been used to study the function of *C. albicans* essential genes, as well as genes important for virulence [10–12]. Notably, these investigations stressed the importance of characterizing essential gene function in the organism of interest, as the genetic screens in *C. albicans* that assessed gene essentiality established that the predictive power from *S. cerevisiae* essential genes was limited to 52.4% [10]. Therefore, defining essential genes directly in *C. auris* is required to enhance our understanding of this emerging pathogen.

In this protocol, we describe a technique to genetically manipulate the *C. auris* genome in order to construct strains that enable the titration of a given gene product using a conditional expression promoter system (Fig. 1). We systematically describe how to generate a DNA cassette consisting of a tetracycline-repressible promoter, a selectable resistance marker, and ~1 kbp of upstream and downstream homology that flanks the genomic location of interest using fusion PCR. As an example, we describe the exact sequences used to repress the expression of the essential molecular chaperone Hsp90. We also outline the transformation procedure that eventually enables the successful incorporation of the DNA fragment into the *C. auris* genome and define how to confirm the successful generation of a conditional expression strain. Finally, we provide recommendations for assay optimization and troubleshooting suggestions to help maximize the likelihood of success. Overall, this protocol provides a useful system in order to study essential gene function in *C. auris*.

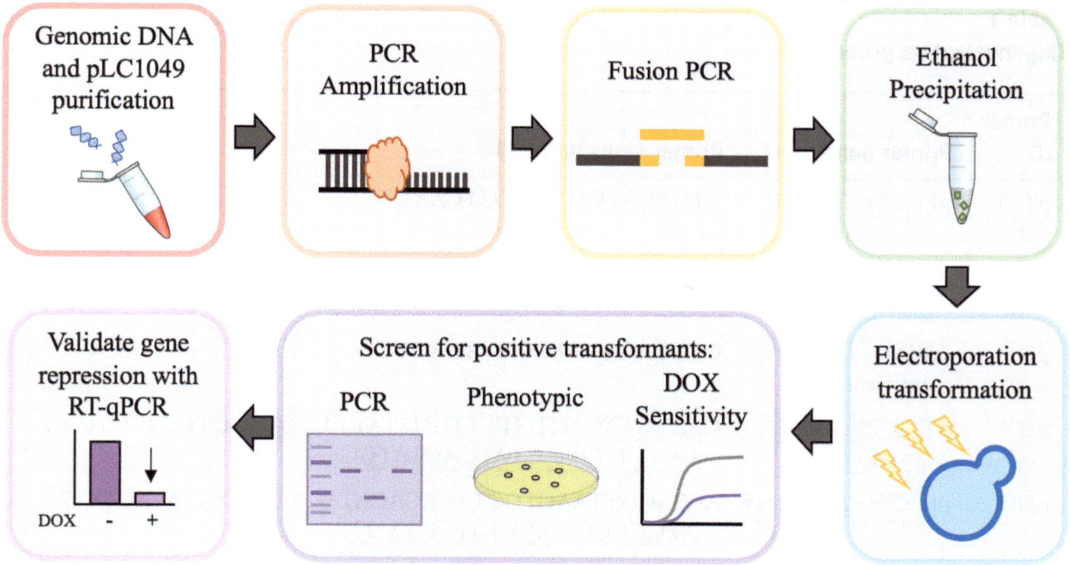

Fig. 1 Workflow for the construction of a conditional expression *C. auris* mutant. In this method, *C. auris* genomic DNA and the vector pLC1049 are used in a PCR reaction to produce three DNA pieces that eventually create the promoter replacement cassette via fusion PCR. This product is ethanol-precipitated and transformed into *C. auris* using an electroporation transformation protocol. Positive colonies can be screened via genotyping PCR or through phenotypic assays, depending on the gene. Finally, reverse transcriptase quantitative PCR (RT–qPCR) is used to confirm successful gene repression in the presence of doxycycline (DOX)

2 Materials

Prepare all solutions using distilled deionized water. Use filter pipette tips, wear gloves, and use sterile technique. Abide by local waste disposal regulations.

1. Genomic DNA from *C. auris* strain of interest stored at −20 °C (*see* **Note 1**).
2. Primers listed in Table 1 stored at −20 °C.
3. Plasmid pLC1049 [13] stored at −20 °C.
4. PCR tubes.
5. PCR Q5 high-fidelity DNA polymerase or suitable high-fidelity PCR polymerase.
6. 5X Q5 reaction buffer.
7. 100 mM dNTP mix.
8. Thermocycler.
9. 1X TAE buffer: 40 mM Tris base, 20 mM acetic acid, 1 mM EDTA.
10. Agarose.

Table 1
Oligonucleotide primers

Primer ID	Primer name	Primer sequence 5′-3′
pTAR-F	pLC1049 TAR-F	TTGAGATGGAGCCGTCAAAT
ptetO-R	pLC1049 tetO-R	CGACTATTTATATTTGTATG
pU1	*HSP90* ~1-kbp upstream F	GTTCACAGTTGGTCAGAGCC
pUT	pTAR-R/HSP90 promoter	GAAAGAAGAGTGTTTTGGATATTTGACGGCTCCATCTCAA GCTGCTTGACTATGAGTATGG
pDO	ptetO-F/HSP90 orf	ACTACGTTAACTCCTACACACATACAAATATAAATAGTCG ATGCCCGAGAAGGTTGAAACG
pD1	*HSP90* ~1-kbp downstream R	TCCTCAGAGTCCACAACACC
pU2	*HSP90* nested primer F	GATGGGATGTTGCAGGTGTG
pD2	*HSP90* nested primer R	GACCCTCGACAGAAAAGTGC
pTAR-R	CaTAR-797 R	CGCAGAAAGTAATATCATGC
ptetO-F	tetOp+488 F	TCGTTTCTGATGGGCTTTTC
ACT1-F	*ACT1*+121 F	ACCCCAAGTCCAACAGAGAG
ACT1-R	*ACT1*+316 R	TCCAGCCAAGTCAAGTCTCA
GPD1-F	*GPD1*+141 F	ATCCTTGCTGAAAACGCTGC
GPD1-R	*GPD1*+318R	TCCTCGGCCACCTTTACAAT
HSP90-F	*HSP90*+287 F	TACCATTGCCAAGTCCGGTA
HSP90-R	*HSP90*+496 R	ACCACGCTTGATTCTCTCGT

Underlined sequence denotes *HSP90*-specific sequences

11. Microwave.
12. SYBR Safe DNA gel stain.
13. DNA size marker.
14. Gel loading dye.

15. DNA gel extraction kit.
16. NanoDrop spectrophotometer.
17. 3 M sodium acetate (pH 5.2 in ddH_2O).
18. 70% ethanol in ddH_2O and stored at −20 °C.
19. Benchtop microcentrifuge cooled to 4 °C.
20. *Candida auris* strain of interest.
21. 15-mL snap cap tubes.
22. YPD medium, sterile: 1% yeast extract, 2% peptone, 2% glucose.
23. Inoculation loops.
24. Spectrophotometer.
25. 15 and 50 mL conical tubes and 1.5 mL tubes.
26. 10X TE: 100 mM Tris-HCl, 10 mM EDTA in ddH_2O.
27. 1 M lithium acetate in ddH_2O.
28. 1 M DTT in ddH_2O.
29. 1 M sorbitol stored at 4 °C.
30. 2 mm electroporation cuvettes.
31. Electroporator.
32. YPD + NAT plates: To YPD medium add 2% agar before autoclaving. Let medium cool to approximately 65 °C before adding nourseothricin (NAT; 250 mg/mL stock filter sterilized in ddH_2O) to a final concentration of 250 μg/mL. Mix well, then pour medium into petri dishes, and let dry. Store at 4 °C.
33. Doxycycline (stock at 50 mg/mL in DMSO).
34. Liquid nitrogen.
35. Yeast RNA extraction kit.
36. ß-mercaptoethanol.
37. 2 mL bead-beating tubes.
38. RNase-free 0.5 mm silica, acid-washed glass beads.
39. Bead beater.
40. DNase I DNA removal kit.
41. cDNA synthesis kit.
42. RT–qPCR reagents (384-well plate and fluorescent master mix).
43. RT–qPCR machine.
44. YPD + doxycycline plates: To autoclaved YPD agar, add doxycycline (50 mg/mL in DMSO) to a final concentration of 50 μg/mL, when cooled to approximately 65 °C. Mix well, then pour medium into petri dishes, and let dry. Store at 4 °C.

3 Methods

3.1 Generation of Gene-Specific Promoter Replacement DNA Cassette

1. To prepare the DNA cassette to be used for the *C. auris* transformation, use primers pTAR-F and ptetO-R (Table 1) to amplify the 4211 bp TAR–NAT–tetO fragment from pLC1049 (Fig. 2a). This contains the tetracycline-repressible transactivator (TAR), nourseothricin resistance marker (NAT), and activator binding sequence (tetO), which contains seven copies of the tet operator sequence [14]. The TAR was adapted from a construct that fused the reverse Tet repressor from *Escherichia coli* with the transcription activation domain of the Gal4 protein from *S. cerevisiae* [14]. Subsequently, five amino

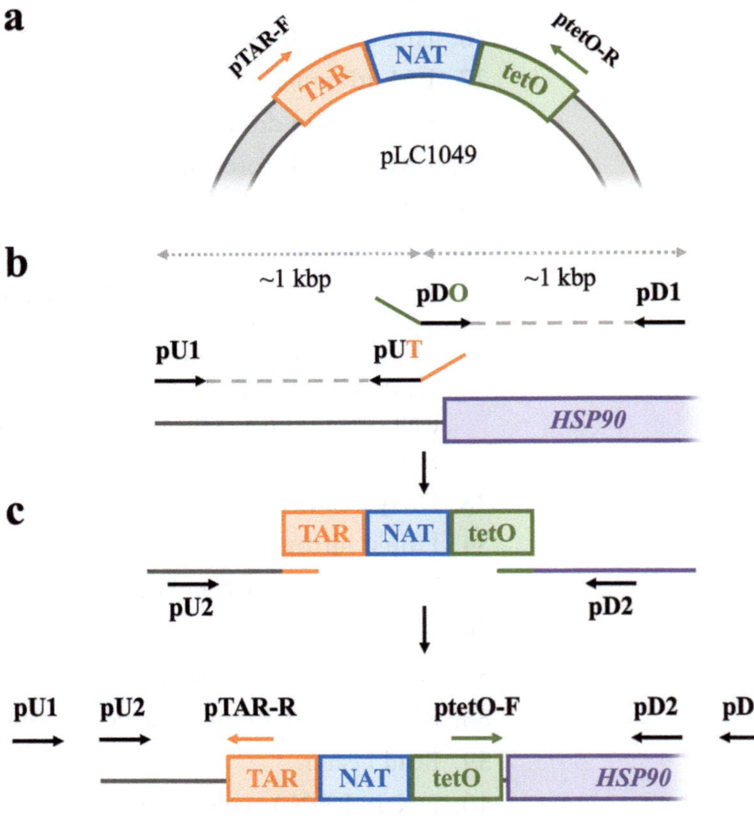

Fig. 2 Primer design strategy to produce deletion construct. (**a**) Using primers pTAR-F and ptetO-R, the tetracycline-repressible transactivator (TAR), nourseothricin resistance marker (NAT), and activator binding sequence (tetO) are amplified from pLC1049. (**b**) Primer pairs pU1 and pUT as well as pDO and pD1 are used to PCR amplify ~1 kbp homology regions upstream and within the gene of interest with a 40 bp overlap to the TAR–NAT–tetO construct. (**c**) Using nested primers pU2 and pD2 and the amplicons created above, fusion PCR is used to generate a gene-specific repressible construct. Successful integration of this construct can be determined via genotyping PCR with primer pairs pU1 and pTAR-R and ptetO-F, and pD1

acids were mutated in the *C. albicans* TAR (G12S, G19E, P56A, E148D, and R1279H) to convert it from a Tet-inducible to Tet-repressible system.

2. Design primers to amplify ~1 kbp of sequence homology upstream of the gene of interest (referred to as pU1/pUT) and ~1 kbp of sequence homology within the open reading frame (referred to as pDO/pD1). Primers pUT and pDO should contain 40 bp overlap to the TAR–NAT–tetO cassette (Table 1) on the 5′ end (Fig. 2b).
3. Design nested primers, pU2 and pD2 (Fig. 2c), that are ~20 bp inset from pU1 and pD1.
4. To amplify the fragments described above, prepare PCR master mixes in 1.5 mL microcentrifuge tubes on ice using the following recipe. Be sure to include a no-template control and set up each reaction in triplicate (*see* **Note 2**).

   Master mix recipe per sample:

   (a) 36.1 μL ddH_2O.

   (b) 10 μL 5X Q5 reaction buffer.

   (c) 0.4 μL 100 mM dNTP mix.

   (d) 0.25 μL of each primer at stock concentration of 100 mM.

   (e) 2 μL template DNA (genomic DNA or plasmid ~10–100-ng/μL stock concentration).

   (f) 1 μL Q5 high-fidelity DNA polymerase.
5. Mix briefly and transfer into PCR tube. Briefly centrifuge to collect all stray liquids. Place PCR tubes in thermocycler.
6. Amplify DNA fragments using the following program:

   (a) 98 °C, 30 s

   (b) 30 cycles of:

      (i) 98 °C, 10 s

      (ii) 55 °C, 30 s

      (iii) 72 °C, 30 s/kb

   (c) 72 °C, 2 min

   (d) 4 °C, hold
7. Run 5 μL of each amplified product on a 1% agarose 1X TAE gel for 30 min at 110 V. To make the gel, combine 100 mL 1X TAE buffer with 1 g agarose and microwave until agarose is dissolved (approximately 1 min at high power). Let the solution cool, then add 10 μL 10,000X SYBR Safe DNA gel stain (or equivalent), and pour into casting tray with comb inserts. Let cool until gel solidifies (approximately 20 min). Add samples mixed with loading dye, and include DNA size marker for band size reference.

8. Gel-purify the fragments by pooling replicate PCR reactions and using a commercial agarose gel extraction kit. Quantify the purified DNA using a NanoDrop spectrophotometer or alternative DNA quantification method.
9. Perform fusion PCR with a high-fidelity polymerase by combining equal molar quantities (calculated here: https://nebiocalculator.neb.com/#!/dsdnaamt) of each of the purified fragments, with ~50 ng of the largest piece.
10. Using the nested primers pU2/pD2, follow the same recipe and cycle as described in **steps 4** and **6** above, with the modification of running 35 cycles rather than 30 (Fig. 2c).
11. Run the samples on 1% agarose gel to verify the amplification of a ~6 kbp product (*see* **Note 3**). See **step** 7 above for how to prepare agarose gel.
12. Ethanol precipitate the PCR product. To the PCR reaction, add 1/10 volume 3 M sodium acetate (pH 5.2 in ddH_2O) and 2X volume 100% ethanol (*see* **Note 4**). Place at −20 °C for at least 30 min (overnight is optimal). Pellet DNA at 4000 × *g* for 30 min at 4 °C. Wash the pellet three times with 70% cold ethanol. Centrifuge again at 4000 × *g* for 5 min at 4 °C, and remove any remaining liquid. Let the pellet dry completely at room temperature, and then resuspend in 50 μL of ddH_2O.

3.2 *C. auris* Transformation

1. Prepare electrocompetent cells fresh for each transformation by inoculating *C. auris* in 50 mL YPD broth and incubate overnight at 30 °C, shaking at 200 rpm. Include several different inoculum densities to ensure the desired OD_{600} the following day.
2. Measure OD_{600} of the overnight culture using a spectrophotometer. Choose the culture with an OD_{600} between 1.6 and 2.2.
3. Transfer the overnight culture into a 50 mL conical tube, and pellet the cells for 5 min at 3000 × *g*.
4. Resuspend in 8 mL ddH_2O, then add 1 mL 10X TE with 1 mL 1 M lithium acetate. Incubate at 30 °C for 1 h, shaking at 200 rpm.
5. Add 250 μL 1 M DTT, and incubate culture at 30 °C for 30 min, shaking at 200 rpm.
6. Transfer the culture into a 10 mL conical tube, and pellet cells for 5 min at 3000 × *g*.
7. Wash cells twice in ice-cold ddH_2O and once in ice-cold 1 M sorbitol.
8. Resuspend pellet in 100 μL ice-cold 1 M sorbitol.

9. Add 40 μL of cell suspension, 200 μL ice-cold 1 M sorbitol, and 3 μg of DNA to a 2 mm electroporation cuvette. Be sure to include a DNA-free control cuvette and perform each electroporation in technical duplicate (*see* **Note 5**).
10. Electroporate cell suspension using the following settings: 1.8 kV, 200 Ω, and 25 μF.
11. Immediately add 800 μL ice-cold 1 M sorbitol to each sample after electroporating.
12. Transfer the ~1 mL electroporated cell mixture to a 1.5 mL microcentrifuge tube, and briefly centrifuge to pellet the cells (pulse for 12 s).
13. Remove supernatant and resuspend the pellet in 1 mL YPD gently by pipetting.
14. Add the 1 mL cell suspension to 9 mL YPD, and culture cells at 30 °C for 4 h, shaking at 200 rpm.
15. Transfer the culture into a 10 mL conical tube, and pellet the cells for 5 min at 3000 × *g*.
16. Resuspend the pellet in 500 μL YPD. Make 1/5, 1/10, and 1/50 cell dilutions in 200 μL YPD.
17. Plate 200 μL of the diluted cultures onto YPD + NAT plates.
18. Incubate at 30 °C for 48 h or until colonies appear (Fig. 3).

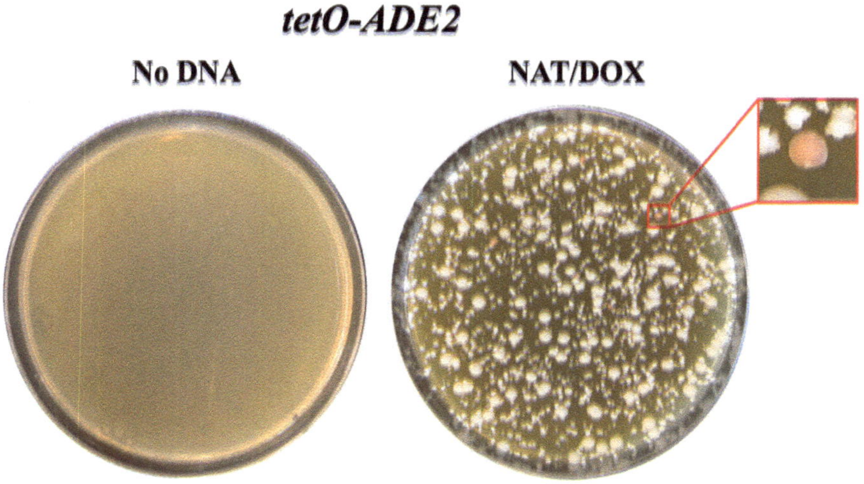

Fig. 3 Example of phenotypic screening for positive transformants. After electroporation, cells can be plated on YPD plates containing nourseothricin (NAT) and doxycycline (DOX), which enables phenotypic screening for positive transformants. Here, positive transformants in which the *ADE2* gene has been successfully repressed in the presence of DOX produce colonies with a visible pink pigmentation. These phenotypic outputs also provide an approximation for the efficiency of the transformation protocol. The "no DNA" control demonstrates what an agar plate looks like after plating *C. auris* cells that underwent the transformation procedure with ddH_2O instead of the DNA cassette

3.3 Genotyping Transformants

1. Choose the transformation plate from dilutions (**step 17** in Subheading 3.2) that produced separated single colonies (Fig. 3).
2. Patch single colonies to a new YPD + NAT (250 μg/mL) plate. In general, correct integration of the TAR–NAT–tetO cassette can be as low as 1%. Thus, patch at minimum ~100 colonies. Incubate at 30 °C for 24 h.
3. Perform colony PCR to check for upstream and downstream integration of the TAR–NAT–tetO cassette to the gene of interest and absence of the native promoter (*see* **Note 6**). Include parent strain as a negative control. Use primers pU1/pTAR-R to test for upstream and ptetO-F/pD1 to test for downstream integration of the DNA cassette. Use pU1/pD1 to verify the absence of the native promoter upstream of the wild-type allele (Fig. 4). This step does not require high-fidelity polymerase.
4. Prepare PCR master mixes in 1.5 mL microcentrifuge tubes on ice using the following recipe:
   (a) 5 μL ddH_2O mixed with a single *C. auris* colony from patched plate.
   (b) 12.2 μL ddH_2O.
   (c) 2 μL reaction buffer.
   (d) 0.2 μL 100 mM dNTP mix.
   (e) 0.2 μL each 100 mM forward and reverse prime.
   (f) 0.2 μL polymerase.
5. Mix briefly and transfer into PCR tube. Briefly centrifuge to collect all stray liquids. Place PCR tubes in thermocycler.
6. Amplify DNA fragments using the following program:
   (a) 98 °C, 30 s
   (b) 30 cycles of:
      (i) 98 °C, 10 s
      (ii) 55 °C, 30 s
      (iii) 72 °C, 30 s/kbp
   (c) 72 °C, 2 min
   (d) 4 °C, hold
7. Run the PCR products on 1% agarose gel (Fig. 4). Sizes of the upstream and downstream integration products can vary depending on the locations of pU1 and pD1, but both should be ~1 kbp in length. In mutants with correctly integrated TAR–NAT–tetO cassette, the native promoter should be absent (*see* **Note 7**).

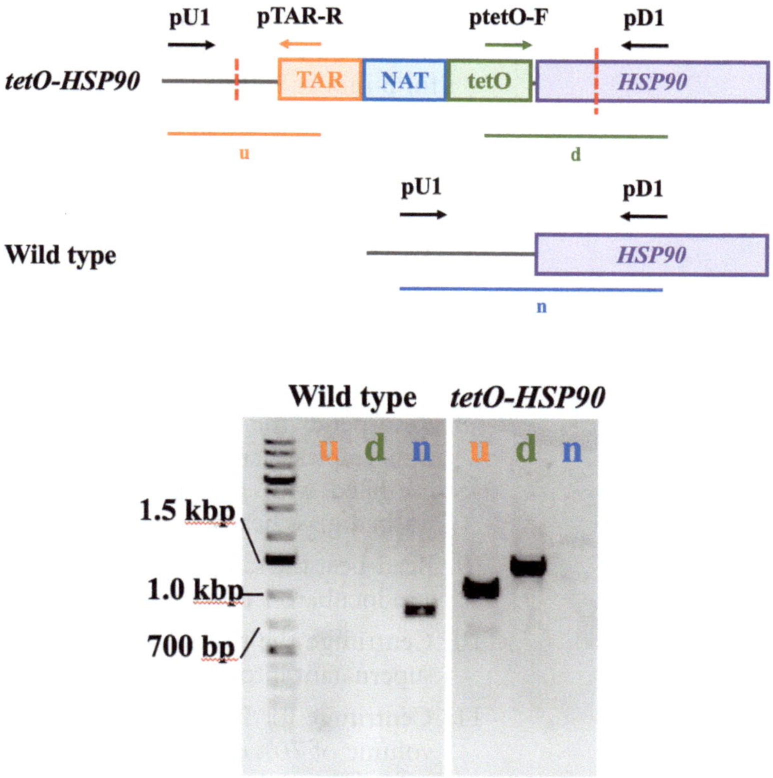

Fig. 4 PCR genotyping of transformants. Using primer pairs pU1 and pTAR-R for upstream integration (u, orange) and ptetO-F and pD1 for downstream integration (d, green), a ~1 kbp band is specifically amplified in those transformants with proper integration of the TAR–NAT–tetO cassette in front of the gene of interest. Dashed red lines indicate the border of the DNA cassette used in the transformation. Additionally, primers pU1 and pD1 are used to amplify the native promoter region (n, blue), which in positive transformants produces a ~6 kbp fragment, often too long for standard colony PCR, whereas in the wild type produces an 800 bp fragment. Gel image adapted from [13]

3.4 Confirm Doxycycline-Repressible Expression

1. To prepare cell pellets for expression analysis, inoculate 3 mL of YPD medium with either the parental strain or any PCR-confirmed positive colonies and grow overnight shaking at 30 °C.
2. The following day, subculture each of these cultures at an OD_{600} 0.1 in 3 mL YPD medium in the absence or presence of 0.05 μg/mL doxycycline overnight shaking at 30 °C to initiate doxycycline-induced gene repression.
3. The following day, set up a second subculture from these overnight cultures at an OD_{600} 0.1 in 10 mL YPD medium in the absence or presence of increasing concentrations of doxycycline (e.g., 0.05, 0.5, 5, and 50 μg/mL) shaking at 30 °C for 4–6 h (*see* **Note 8**).

4. Transfer the cell culture to a 15 mL conical tube (or appropriate size for culture volume), and pellet the cells by centrifugation for 5 min at 3000 × *g*.
5. Remove the supernatant and flash-freeze the cell pellet in liquid nitrogen.
6. Store at −80 °C.
7. To extract RNA from cell pellets, bead-beat samples in lysis buffer (*see* next step) from a standard RNA extraction kit suitable for yeast cells, and follow the manufacturer's instructions. Keep samples cold whenever possible until mixture is added to the spin column.
8. Resuspend the cells in 600 μL lysis buffer with 6 μL ß-mercaptoethanol, and add to 2 mL bead-beating tubes pre-filled with 600 μL of RNase-free 0.5 mm silica, acid-washed glass beads.
9. Bead-beat the samples four times for 45 s in a 4 °C cold room, and incubate 1 min on ice in between.
10. Centrifuge the tubes for 2 min at full speed, and transfer the supernatant into a new 1.5 mL tube.
11. Centrifuge for 2 min at full speed and transfer 350 μL to 1X volume of 70% ethanol. Mix well by gently inverting the tube several times. It is normal to see precipitation in samples from large pellets. Centrifuge on pulse for 5 s and stop.
12. Transfer ~700 μL volume to a spin column provided in the RNA extraction kit, and proceed with RNA extraction as per manufacturer's instructions.
13. Once RNA is extracted, remove potential DNA contamination by treating the samples with DNase I using any standard DNA removal kit.
14. Use any standard cDNA synthesis kit according to manufacturer's instructions to generate cDNA.
15. Dilute cDNA ~1:50, and use this as the DNA for real-time quantitative PCR (RT–qPCR) with standard reaction mixture and gene-specific primers.
16. Include reactions to normalize expression to housekeeping genes such as *ACT1* and *GPD1* by setting up reactions with primers ACT1-F/ACT1-R and GPD1-F/GPD1-R (Table 1) and plot expression relative to the no-doxycycline control treatment (Fig. 5).

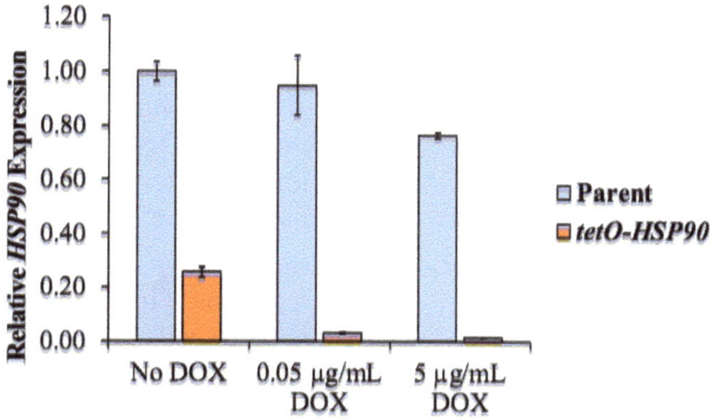

Fig. 5 Doxycycline-repressible expression of *HSP90* in *C. auris*. A *C. auris* strain placed under the control of the *tetO* promoter (orange) exhibits reduced expression of *HSP90* relative to the wild type even in the absence of doxycycline (suggesting that the *tetO* promoter is not as strong as the native *HSP90* promoter), and addition of doxycycline further represses *HSP90* expression in the *tetO-HSP90* strain (blue). Samples are run in technical triplicate, and average expression +/− SEM is plotted relative to the housekeeping genes *ACT1* and *GPD1* and normalized to the no-DOX wild-type strain

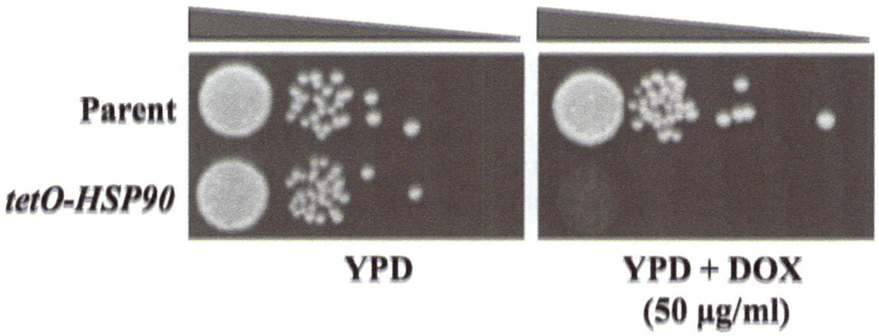

Fig. 6 *HSP90* is essential in *C. auris*. Spotting of the *C. auris* parent and *tetO-HSP90* strain on YPD or YPD + doxycycline (DOX, 50 μg/mL) plates. A 1000-fold dilution of overnight cultures was generated in sterile ddH_2O. Additional 100-fold serial dilutions were generated in sterile ddH_2O. 5 μL of dilutions were spotted onto YPD agar plates. Figure adapted from [13]

3.5 Confirm Essentiality of Gene of Interest

1. Inoculate 3 mL of YPD medium with either the parental strain or any RT–qPCR-confirmed positive colonies, and grow overnight shaking at 30 °C.
2. The following day dilute the overnight culture 1000-fold in sterile ddH_2O.
3. Perform 100-fold serial dilutions in sterile ddH_2O.
4. Spot 5 μL onto YPD agar plates either without or with 50 μg/mL doxycycline supplementation.
5. Incubate at 30 °C for 48 h then image (Fig. 6).

4 Notes

1. Any standard genomic DNA extraction kit used for yeast will suffice. Genomic DNA should be eluted in ddH_2O to avoid any detrimental interactions between buffer components and the Taq used for the PCR.
2. It is advised to set up each master mix, without template DNA, for four reactions. This can then be transferred into PCR tubes, where one tube will receive ddH_2O and the other three will have the appropriate DNA template. Performing each reaction in triplicate ensures sufficient product yield after gel extraction for the transformation procedure.
3. If there are many products in addition to the ~6 kbp DNA fragment in the gel, gel-purify the correct band, and rerun the PCR reaction.
4. Pool multiple (3–5) PCR reactions into the same ethanol precipitation to maximize the concentration of DNA. The concentration should not be <600 ng/μL, to ensure <5 μL of mixture is added to the electroporation cuvette, as adding large amounts can disrupt the salt content and lead to unsuccessful electroporation.
5. Electroporation can be particular. Factors such as condensation on the outside of the electroporation cuvette, high salt concentration in the sample, or bubbles can lead to sample "arcing," where high electrical discharge reduces cell viability. For this reason, it is advised to always set up cuvettes in duplicate to ensure samples are not lost at this stage.
6. Depending on the repressed gene, phenotypic screening can be implemented to streamline the screening for positive transformants. For example, depletion of *ADE2* produces red colonies. Therefore, transformations can be plated on YPD + NAT + doxycycline plates at Subheading 3.2, **step 18**, and colonies with a pink–red phenotype (Fig. 3) can be selected for genotyping by PCR. Alternatively, if repressing a suspected essential gene, patched colonies can be screened for hypersensitivity to doxycycline prior to PCR validation by measuring optical density (OD_{600}) of colonies when grown overnight in 20 μg/mL doxycycline and selecting those that display a growth defect for PCR verification.
7. In theory, the primer combination of pU1/pD1 should produce a 6 kbp product, but in practice, this is hard to obtain. Screening for absence of the native promoter is sufficient. Given the volume of transformants to screen, first screen for upstream or downstream integration; then only proceed with all three PCR reactions with those that pass the first integration test.

8. Every gene is repressed differently; therefore, covering a range of doxycycline concentrations will ensure identification of the correct concentration necessary to collect enough cells for the analysis while sufficiently repressing expression of the target gene. With essential genes with extreme phenotypes, higher doxycycline concentrations will eliminate growth of the subculture. This can be mitigated by growing for longer, increasing the volume of the subculture, or lowering the doxycycline concentration if growth is too low for RNA extraction (generally $< OD_{600}$ 0.2).

Acknowledgments We thank all members of the Cowen lab for helpful discussions. L.E.C. is supported by the Canadian Institutes of Health Research Foundation Grant (FDN-154288) and a National Institutes of Health NIAID R01 Grant (1R01AI127375-01 and 1R01AI120958-01A1). L.E.C. is a Canada Research Chair (Tier 1) in Microbial Genomics and Infectious Disease and co-Director of the CIFAR Fungal Kingdom: Threats and Opportunities program.

Competing Interests L.E.C. is a co-founder and shareholder in Bright Angel Therapeutics, a platform company for development of novel antifungal therapeutics.

References

1. Brown GD, Denning DW, Gow NAR et al (2012) Hidden killers: human fungal infections. Sci Transl Med 4:165rv13. https://doi.org/10.1126/scitranslmed.3004404
2. Fisher MC, Gurr SJ, Cuomo CA et al (2020) Threats posed by the fungal kingdom to humans, wildlife, and agriculture. mBio 11: e00449-20. https://doi.org/10.1128/mBio.00449-20
3. Satoh K, Makimura K, Hasumi Y et al (2009) *Candida auris* sp. nov., a novel ascomycetous yeast isolated from the external ear canal of an inpatient in a Japanese hospital. Microbiol Immunol 53:41–44. https://doi.org/10.1111/j.1348-0421.2008.00083.x
4. Lockhart SR, Etienne KA, Vallabhaneni S et al (2017) Simultaneous emergence of multidrug-resistant *Candida auris* on 3 continents confirmed by whole-genome sequencing and epidemiological analyses. Clin Infect Dis 64: 134–140. https://doi.org/10.1093/cid/ciw691
5. Centers for Disease Control and Prevention (2019) Antibiotic resistance threats in the United States. U.S. Department of Health and Human Services, CDC, Atlanta. https://www.cdc.gov/drugresistance/Biggest-Threats.html
6. Uthayakumar D, Sharma J, Wensing L, Shapiro RS (2021) CRISPR-based genetic manipulation of *Candida* species: historical perspectives and current approaches. Front Genome Ed 2: 606281. https://doi.org/10.3389/fgeed.2020.606281
7. Roemer T, Jiang B, Davison J et al (2003) Large-scale essential gene identification in *Candida albicans* and applications to antifungal drug discovery. Mol Microbiol 50: 167–181. https://doi.org/10.1046/j.1365-2958.2003.03697.x
8. Noble SM, Johnson AD (2005) Strains and strategies for large-scale gene deletion studies of the diploid human fungal pathogen *Candida albicans*. Eukaryot Cell 4:298–309. https://doi.org/10.1128/EC.4.2.298-309.2005
9. Xu D, Jiang B, Ketela T et al (2007) Genome-wide fitness test and mechanism-of-action studies of inhibitory compounds in *Candida*

albicans. PLoS Pathog 3:e92. https://doi.org/10.1371/journal.ppat.0030092

10. O'Meara TR, Veri AO, Ketela T et al (2015) Global analysis of fungal morphology exposes mechanisms of host cell escape. Nat Commun 6:6741. https://doi.org/10.1038/ncomms7741
11. O'Meara TR, Duah K, Guo CX et al (2018) High-throughput screening identifies genes required for *Candida albicans* induction of macrophage pyroptosis. mBio 9:e01581-18. https://doi.org/10.1128/mBio.01581-18
12. Becker JM, Kauffman SJ, Hauser M et al (2010) Pathway analysis of *Candida albicans* survival and virulence determinants in a murine infection model. Proc Natl Acad Sci U S A 107: 22044–22049. https://doi.org/10.1073/pnas.1009845107
13. Kim SH, Iyer KR, Pardeshi L et al (2019) Genetic analysis of *Candida auris* implicates Hsp90 in morphogenesis and azole tolerance and Cdr1 in azole resistance. mBio 10: e02529-18. https://doi.org/10.1128/mBio.02529-18
14. Park YN, Morschhäuser J (2005) Tetracycline-inducible gene expression and gene deletion in *Candida albicans*. Eukaryot Cell 4: 1328–1342. https://doi.org/10.1128/EC.4.8.1328-1342.2005

Part IV

Phenotypic and Biochemical Characterization of *Candida auris*

Chapter 10

Cell Viability Assays for *Candida auris*

Jason L. Brown, Bryn Short, Alicia Ware, Leighann Sherry, Ryan Kean, and Gordon Ramage

Abstract

Cell viability assays are useful for assessing the efficacy of antifungal therapeutics and disinfection strategies in vitro. In recent years these assays have been fundamental for the testing of conventional and novel therapies against the nosocomial fungal pathogen *Candida auris*. Here we provide detailed descriptions of methods for assessing cellular viability of *Candida auris* in vitro, such as metabolic assays (XTT and resazurin), colony-forming unit counting, live/dead quantitative PCR, and fluorescent staining for microscopic analyses.

Key words *Candida auris*, Cell viability, XTT, Resazurin, Colony-forming units, Live/dead qPCR, Fluorescent staining

1 Introduction

Cell viability assays are largely simple yet extremely effective techniques utilized by most microbiology research laboratories. These methodologies can involve several different assays which allow for assessment of cell viability of planktonic or sessile cells of bacterial and fungal species. These include, but are not limited to, the following:

(a) Colorimetric microtiter plate assays using salts (e.g., tetrazolium, XTT) or dyes (e.g., resazurin) are platforms for assessing metabolic activity of microbial biofilms [1–5]. These assays allow for high-throughput screening of various antimicrobial compounds with their "low-input, high-output" applications.

(b) Colony-forming units (CFU) using the Miles and Misra counting technique and/or colony-forming equivalents (CFE) using quantitative polymerase chain reaction (qPCR) in microbial samples. qPCR can be utilized in combination with photoreactive DNA-binding dyes such as propidium

Alexander Lorenz (ed.), *Candida auris: Methods and Protocols*, Methods in Molecular Biology, vol. 2517,
https://doi.org/10.1007/978-1-0716-2417-3_10,

monoazide (PMA) to allow for detection of viable vs. total microorganisms [6, 7].

(c) Fluorescent staining with live/dead viability dyes such as SYTO™ 9 and propidium iodide to allow for microscopical discrimination between viable and dead cells [8]. Fungal-specific dyes have been reported, such as calcofluor white and FUN-1™, with the latter allowing for accurate assessment of viability in yeasts [9].

In the context of medical mycology, these above methods have previously been used to study the response of *Candida albicans* to antifungal therapeutics and disinfection strategies with great success. However, in recent years similar techniques have also been employed for *Candida auris*, a nosocomial pathogen first identified in Japan in 2009 [10]. Since its emergence, *C. auris* has rapidly gained worldwide attention due to its genetic heterogeneity and ability to survive and persist in environments for prolonged periods of time [11]. Furthermore, resistance profiles from *C. auris* clinical isolates vary considerably to the three main classes of antifungals, with a significant number of isolates being multidrug-resistant [12]. At this juncture, a decade of research into *C. auris* has prioritized widespread antifungal susceptibility testing, with more recent investigations into the efficacy of a range of novel therapeutics against the species [13–16]. All such publications have utilized some, if not all, of the techniques highlighted above to assess effectiveness of a given treatment. Herein, we will document in detail the methodologies available for assessing cell viability of *C. auris* in vitro. We hope one or more of these methodologies provide useful assistance for other laboratories testing innovative and/or conventional therapeutics against this fungus.

2 Materials

For all media and working solutions, it is recommended to purchase readily available microbiological powders or tablets to ensure the most accurate batch-to-batch preparations and follow manufacturers' instructions in preparing such solutions. If not possible, ingredients for each media and buffers are described below to allow for preparation in-house. However, concentrations included here for each component may differ slightly from commercially available preparations, which may further vary from company to company.

2.1 Candida auris Storage, Revival, and Cell Standardization

1. Microbank™ beads.
2. Sabouraud's dextrose (SAB) agar (ingredients, 40.0-g/L glucose, 10.0-g/L peptone, and 15.0-g/L agar dissolved in

1000 mL of double distilled H_2O (ddH_2O), pH 5.6 ± 0.2 at 25 °C). The agar should be sterilized via autoclaving at 121 °C.

3. Disposable 10-μL inoculation loops.
4. Yeast extract peptone dextrose (YPD) medium (ingredients, 10.0-g/L yeast extract, 20.0-g/L peptone, and 20.0 g/L dextrose dissolved in 1000 mL of ddH_2O, pH 6.5 ± 0.2 at 25 °C). The medium should be sterilized via autoclaving at 121 °C.
5. Phosphate-buffered saline (PBS) (ingredients, 8.0-g/L NaCl, 0.2-g/L KCl, 1.44-g/L Na_2HPO_4, and 0.24-g/L KH_2PO_4 dissolved in 800 mL of ddH_2O. pH can be adjusted to 7.4 with concentrated HCl, before adding an additional 200 mL of ddH_2O). PBS should be sterilized via autoclaving at 121 °C.
6. Roswell Park Memorial Institute (RPMI) 1640 medium (main ingredients, 2000-mg/L dextrose, 5.0-mg/L phenol red, 5000-mg/L phenol red [pH indicator], 100-mg/L $Ca(NO_3)_2 \cdot 4H_2O$, 48.84-mg/L anhydrous $MgSO_4$, 400.0-mg/L KCl, 2000-mg/L $NaHCO_3$, 6000-mg/L NaCl, and 800.0-mg/L anhydrous Na_2HPO_4 in 1000 mL of ddH_2O. The remaining components include amino acids and vitamins which are listed elsewhere (https://www.thermofisher.com/uk/en/home/technical-resources/media-formulation.114.html). The resulting solution should be pH 8.2 ± 0.3 at 25 °C. Due to the complexity of the media composition, it is highly recommended to purchase RPMI-1640 medium from an appropriate supplier. RPMI media should be sterilized via autoclaving at 121 °C.
7. Dey–Engley Neutralizing Broth (ingredients, 5.0-g/L casein enzymatic hydrolysate, 2.5-g/L yeast extract, 10.0-g/L dextrose, 6.0-g/L sodium thiosulfate, 1.0-g/L sodium thioglycollate, 2.5-g/L sodium bisulfite, 7.0-g/L lecithin, 5.0-g/L polysorbate 80, and 0.02-g/L bromocresol purple in 1000 mL of ddH_2O. pH 7.6 ± 0.2 at 25 °C). Dey–Engley Neutralizing Broth should be sterilized via autoclaving at 121 °C.
8. 6-well, 12-well, 24-well, 48-well, and/or 96-well flat-bottom microtiter plates for biofilm growth (round-bottom 96-well microtiter plates are necessary for planktonic cell preparation).
9. Improved Neubauer Hemocytometer (0.1-mm depth, 0.0025-mm^2 diameter).

2.2 Metabolic Assays and Biomass Assessment

1. XTT [2,3-bis(2-methoxy-4-nitro-5-sulfo-phenyl)-2H-tetrazolium-5-carboxanilide] assay working solution. To prepare the working solution, 0.25-g/L XTT is added to sterile distilled water and fully dissolved by gentle agitation, then filter sterilized through a 0.22-μm-pore-size filter, aliquoted, and stored at −80 °C. All aliquots should be single-use, and not be freeze-

thawed. Upon thawing, XTT solution should be pre-warmed to room temperature prior to addition of menadione which should be prepared to a concentration of 10 mM in 100% acetone in a glass beaker and fully dissolved by gentle agitation. A final concentration of 1 μM should be added to the working solution of XTT (higher concentrations may be required).

2. Resazurin sodium salt (or commercial equivalent, e.g., alamarBlue™ or PrestoBlue™) working solution. For the resazurin assay, a stock solution of 1% is prepared in sterile PBS by vigorous vortexing to completely dissolve the powder. Once dissolved, the solution is filter sterilized through a 0.22-μm pore-size filter, aliquoted, and stored at 4–5 °C for up to 3 months.
3. Crystal violet (CV) powder working solution. To prepare this, CV powder should be dissolved in ddH_2O to give a 1% solution. This is then further diluted in ddH_2O to a working concentration of 0.05%. There is no requirement to filter sterilize or autoclave the CV. The stock solution can be stored at room temperature for up to 3 months.

2.3 Viable Cell Quantification

1. Nunc® Thermanox™ 13-mm coverslips (different sizes available for larger microtiter plates) or alternative substrates for biofilm growth such as 3.14-cm^2 circular stainless steel coupons or cellulose matrix cut to an appropriate size, e.g., 1.25 cm^2.
2. 23-gauge needle (or equivalent).
3. Tweezers.
4. 5-mL bijoux tubes.
5. Ultrasonic bath for biofilm sonication (40-cm handle, 16-mm blade, and sterile cell scrapers can also be used for removing biofilm biomass).
6. 1.5-mL Eppendorf tubes.
7. Propidium monoazide (PMA) is used for qPCR quantification of viable *C. auris* cells. The PMA is supplied as a 1-mg lyophilized solid that requires reconstitution in 200 μL of RNase-free H_2O, dissolved by gentle pipetting—the resulting working solution should be a rich red color. Following dissolution, single-use PMA aliquots should be prepared and stored at −20 °C.
8. 650 W halogen light.
9. QIAamp DNA mini kit.
10. Screw-cap 2-mL BeadBug™ O-Ring tube.
11. Glass beads, acid-washed (425- to 600-μm diameter).
12. BeadBug™ microtube homogenizer or equivalent.

13. Quantitative PCR machine and reagents of choice, e.g., SYBR™ Green PCR Master Mix or TaqMan™ Universal PCR Master Mix with appropriate primers (and probes if applicable).

2.4 Microscopy Imaging

1. FUN-1™ cell stain working solution of 20 μM (the 10-mM stock solution is diluted 1:500 in sterile ddH_2O to get the desired working concentration). Stock solutions of FUN-1™ should be stored at −20 °C in single-use aliquots to limit the amount of freeze–thaw cycles.
2. Glass coverslips (different sizes available).
3. Microscope slides.
4. VECTASHIELD® Vibrance™ Antifade Mounting Medium.
5. Clear nail varnish.
6. Superglue.

3 Methods

3.1 Cell Standardization

The assays described herein can be utilized for planktonic or sessile forms of *C. auris*. To generate such preparations, the following steps can be used for all clinical isolates of *C. auris*. It is noteworthy that all steps should be performed under aseptic conditions wearing appropriate personal protective equipment.

1. Isolates of *C. auris* are stored in vials containing Microbank™ beads at −80 °C, by which stocks are prepared according to manufacturer's instructions (*see* **Note 1**).
2. For revival, vials can be thawed at room temperature, and one bead containing the organism transferred to a Sabouraud's dextrose agar using a disposable 10-μL inoculation loop. The bead is streaked across the agar, and then plates are transferred to a 30 °C incubator and incubated for 24–48 h (Fig. 1a-i).
3. *C. auris* is propagated in liquid media by taking one to two colonies using a disposable 10-μL inoculation loop and then transferred to 10 mL of yeast peptone dextrose (YPD). This suspension is then cultured at 30 °C, shaking at 150–200 rpm for 16–18 h (Fig. 1a-ii).
4. Following growth in liquid media, *C. auris* is pelleted by centrifugation (3000 × *g*), before pellets are washed via resuspension twice in 10 mL of sterile phosphate-buffered saline (PBS, pH 7.2–7.6) (Fig. 1a-iii). Following the final wash step, pellets are then resuspended in 10 mL of PBS prior to counting.
5. *C. auris* cells are standardized via counting on a hemocytometer (0.1 mm depth, 0.0025 mm^2 diameter) using the following

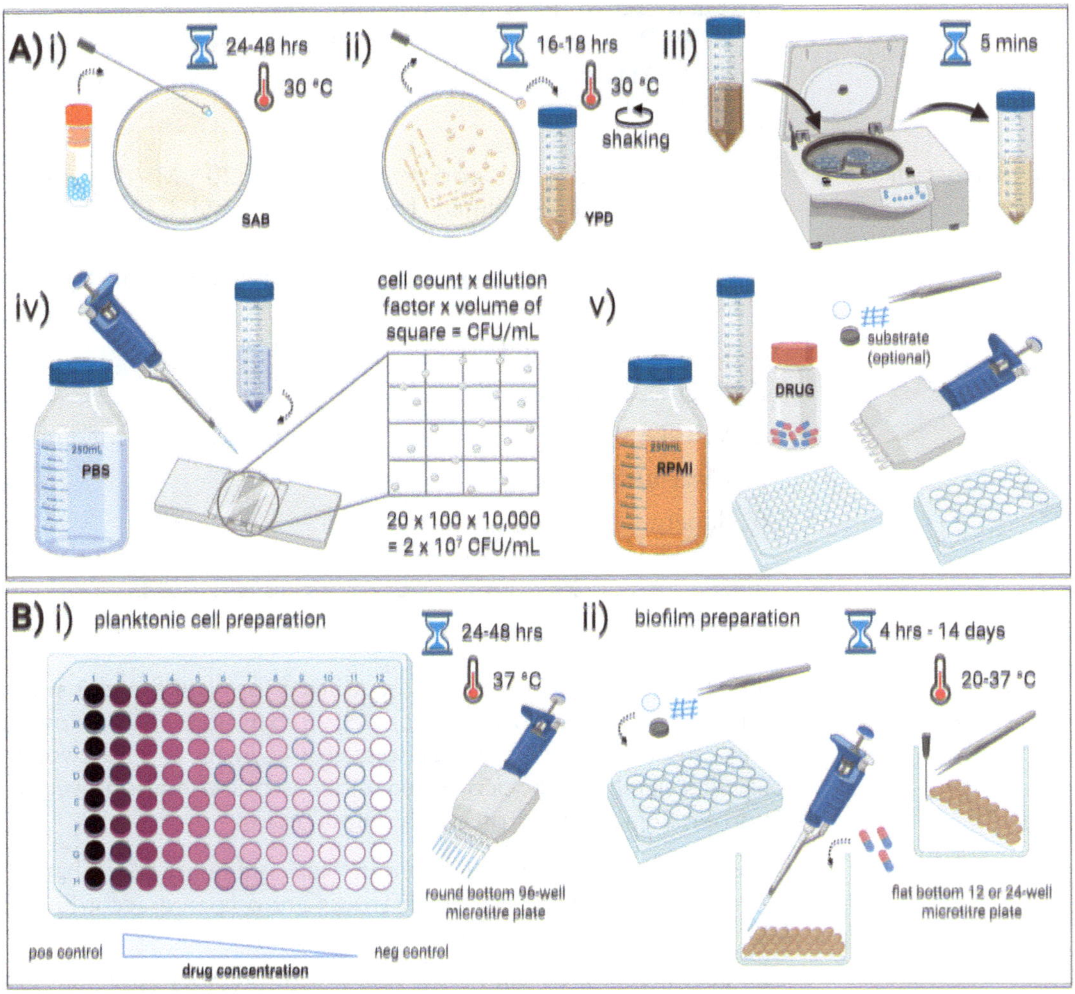

Fig. 1 *Candida auris* cell preparation for planktonic and biofilm growth. Isolates stored in vials are revived by cross-streaking a bead containing the organism on Sabouraud's dextrose agar (SAB). Plates are incubated for 24–48 h at 30 °C to permit growth (**a-i**). Following growth on solid media, one to two colonies are propagated into 10 mL of yeast extract peptone dextrose (YPD) and cultured for 16–18 h at 37 °C, gently shaking at 150–200 rpm (**a-ii**). Following growth in liquid media, suspensions centrifuged at 3000 × *g* for 5 min (**a-iii**). Pelleted cells are washed and resuspended in sterile phosphate-buffered saline (PBS). An aliquot of the pure planktonic *C. auris* culture is diluted 1 in 100 in PBS and then counted using a hemocytometer. The number of colonies counted, the dilution factor, and volume of the square are used to calculate the total CFU/mL (**a-iv**). The standardized cell cultures can then utilized for planktonic or biofilm growth in Roswell Park Memorial Institute 1640 (RPMI) medium in microtiter plates and/or on appropriate substrates (**a-v**). Following standardization, cells can be used for planktonic minimum inhibitory concentration (PMIC) testing. For this, cells are diluted to appropriate volume (usually 2×10^5 CFU/mL) in RPMI, and 100 μL transferred to wells on a round-bottom 96-well microtiter plate containing 100 μL of serially diluted concentrations of drug using a multichannel pipette. Appropriate positive controls (*C. auris* suspensions minus drug) and negative controls (media only) are recommended for all plates (**b-i**). Biofilms can be formed directly in flat-bottom microtiter plates, or on appropriate substrates (e.g., coverslips, steel discs, or cellulose matrix) placed into the microtiter plates with sterile tweezers. For media changes, or washing of biofilms following drug treatment, it is recommended to tilt the pipette at a 45° angle to minimize disruption of the biofilm biomass. A 23-gauge needle with the end formed into a hook and tweezers can be used to remove biofilms grown on substrates from microtiter plates. *C. auris* biofilms can be grown for 4 h to 14 days, at temperatures ranging from 20 °C to 37 °C (**b-ii**). Figure compiled using Biorender.com

formula (cell count × dilution factor × volume of square = colony-forming unit [CFU per mL]) (*see* **Note 2**) (Fig. 1a-iv). Following standardization, *C. auris* cells can then be utilized for planktonic or biofilm growth in Roswell Park Memorial Institute 1640 (RPMI) media in microtiter plates and/or on appropriate substrates (Fig. 1a-v).

3.2 Planktonic and Biofilm Preparation

1. The standardized suspensions of cells are then diluted to appropriate concentrations for experimental procedures. It should be noted that for all antifungal susceptibility testing, appropriate guidelines should be followed using standardized methods such as those of the Clinical and Laboratory Standards Institute (CLSI) and European Committee on Antimicrobial Susceptibility Testing (EUCAST) [2, 17, 18]. Briefly, to prepare plates for planktonic susceptibility testing, all drugs or other therapeutics and cells should be prepared to double the required concentration. Treatments of choice are firstly diluted to working concentrations in RPMI media with 200-μL drug added to the wells (usually the first column) on a round-bottom 96-well microtiter plate using a multichannel pipette.
2. Drugs are then serially diluted twofold in RPMI media across the plate. *C. auris* cells can then be added to the desired wells which will then return the contents of all wells to the desired concentration. Media only and untreated wells should be included for negative and positive controls, respectively. Planktonic minimum inhibitory concentrations (PMICs) are defined as the drug concentration that inhibits cell growth. For PMIC testing, the usual cell concentration initially used is 2×10^4 CFU per well (100 μL of 2×10^5 CFU/mL), thereby when diluted 1:1 with the drug will give a final concentration of 1×10^4 CFU per well. PMIC testing is usually conducted for 24–48 h, with plates incubated at 37 °C (Fig. 1b-i).
3. For biofilm growth, cells should be diluted to appropriate concentrations in RPMI media. Although these concentrations may vary from laboratory to laboratory, we would recommend using a final cell density of 2×10^5 per well (usually in 200 μL, e.g., 200 μL of 1×10^6-CFU/mL suspension) of a flat-bottom 96-well microtiter plate to initiate biofilm formation (concentrations and volumes should be adjusted accordingly for larger microtiter plates).
4. Biofilms can be formed on the bottom of flat-bottom microtiter plates or on appropriate substrates aseptically placed in microtiter plates using sterilized tweezers. Alternative substrates are described in the Materials section and documented elsewhere [19]. From our experience, *C. auris* can be left to form biofilms for a varied amount of time ranging from 4 h to 14 days at different temperatures (20 °C, e.g., room

temperature, to 37 °C) depending on the experimental hypothesis (Fig. 1b-ii). However, usually, biofilms are formed between 4 and 48 h at 37 °C.

3.3 Metabolic Activity Assessment Using Microtiter Plate Assays

To assess the metabolic activity of *C. auris*, two main assays can be employed: XTT and resazurin colorimetric assays. These methods can be used for both planktonic and sessile forms of *C. auris*, although predominantly utilized for the latter. For the methods described below, biofilms are formed in flat-bottom 96-well microtiter plates (larger microtiter plates can be used for biofilm formation, with concentrations and volumes adjusted accordingly). For treatment studies, biofilms can be tested with therapeutics for appropriate times prior to assessment of metabolic activity. Both methodologies are useful techniques for assessing efficacy of treatments in vitro, e.g., a reduction in metabolic activity likely equates to an enhanced antifungal activity, and thereby less of a colorimetric change with the associated assay (*see* **Note 3**). However, it is important to note that careful consideration of assay measurements should be made when comparing the efficacy of antifungals against different isolates of *C. auris* (*see* **Note 4**). Other limitations attached to such assays are discussed elsewhere in detail [20]. It is also recommended that studies using the metabolic assays described below should consider parallel assessment of biofilm biomass using the crystal violet assay. An optional step is described in Subheading 3.3.2, part 4 for this additional assay.

3.3.1 XTT Assay

1. Biofilms are formed by addition of 200 μL of 1×10^6 CFU/mL *C. auris* in RPMI media to flat-bottom 96-well microtiter plates (final concentration of 2×10^5 CFU/well). Biofilms are left to develop for an appropriate amount of time as discussed above (*see* Subheading 3.2, **step 4**).
2. Following growth (and treatment if necessary), media is aspirated, and biofilms are gently washed three times with 200 μL of sterile PBS to remove non-adherent cells. For the wash steps, media is aspirated and replaced with PBS carefully using a multichannel pipette, with tips placed into the wells at a 45° angle at the same point within the plate (bottom left of the well) to limit unnecessary disruption of the biofilm (as seen in Fig. 1b-ii). After the final wash step, XTT (*see* Subheading 3.3.1, **steps 2–6**) or resazurin (*see* Subheading 3.3.2) can be used to assess metabolic activity of the *C. auris* biofilms.
3. For the XTT assay, a total of 100 μL of XTT working solution (0.25-g/L XTT with 1 μM of menadione (*see* **Note 5**)) is added to every well of the 96-well plate containing the pre-washed biofilms using a multichannel pipette. Wells with no biofilms are used as negative controls. For treatment studies, positive control biofilms minus the drug or therapeutic of choice should be included on every plate.

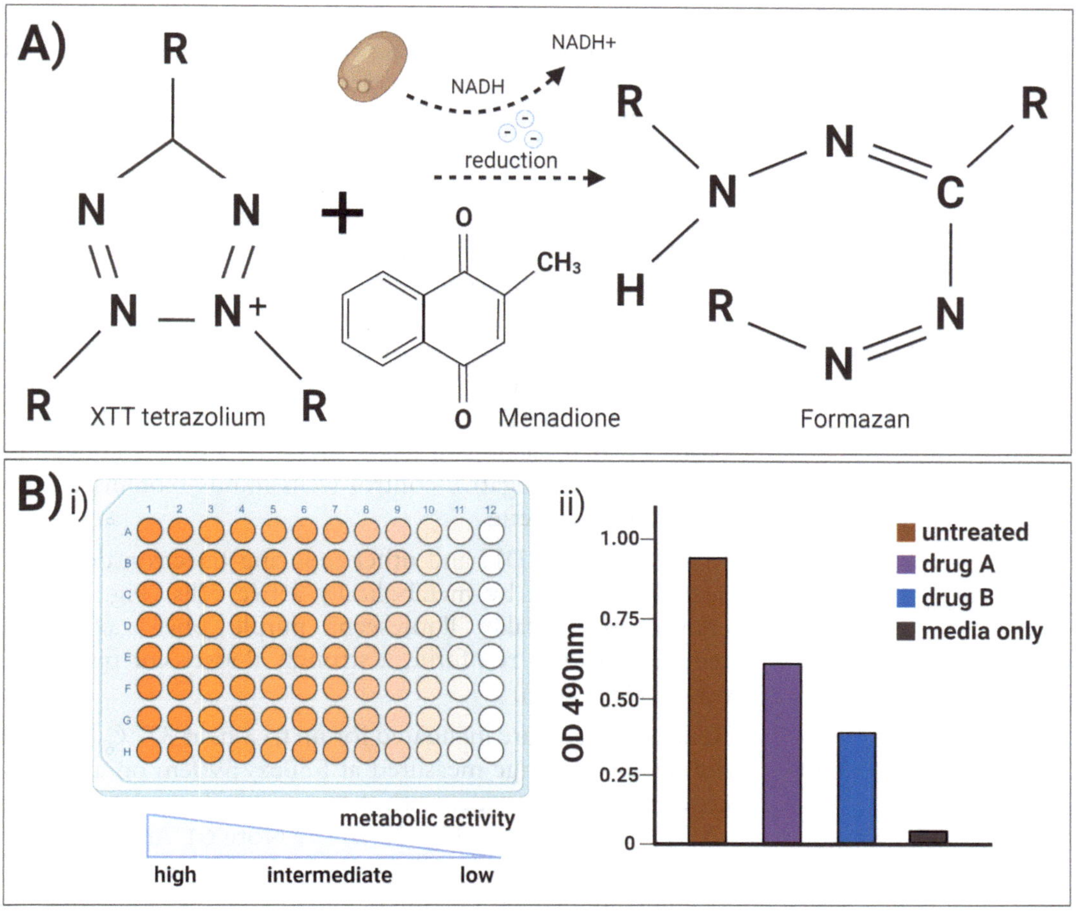

Fig. 2 XTT metabolic assay. The positively charged "oxidized" XTT tetrazolium salt can be used to assess the metabolic activity of *Candida auris*. In the presence of NADH and other reducing (electron donating) agents produced by metabolically active cells, the XTT tetrazole ring structure will be broken down into the water-soluble formazan. An intermediate electron acceptor such as menadione can also be used to accelerate the rate of reaction (**a**). In the reaction, a colorimetric change from a colorless solution (XTT) to a bright orange solution (formazan) is characteristic of increasing levels of XTT conversion and by, thereby, increased level of metabolic activity within the biofilm (**b-i**). The colorimetric change can be quantified at OD_{492nm} (**b-ii**). Figure assembled using Biorender.com

4. During the reaction the positively charged "oxidized" XTT compound will be broken apart in the presence of NADH and other reducing (electron donating) agents produced by metabolically active cells (Fig. 2a). This reaction is accelerated by use of an intermediate electron acceptor such as menadione.
5. XTT plates are incubated in the dark for 2 h at 37 °C, with the colorimetric change measured in a microtiter plate reader at absorbance of 492 nm (*see* **Note 6**). We recommend transferring 75 μL of the XTT reagent following incubation to a new flat-bottom 96-well microtiter plate before reading as the biofilms are likely to interfere with the absorbance reading.

6. Within the assay, a colorimetric change occurs in the presence of metabolically active cells. A colorless solution of XTT changes to a bright orange-colored soluble compound called formazan which can then be quantified at OD_{492nm} (Fig. 2b-i). Absorbance readings should be blank corrected with negative control wells (e.g., media only) to compensate results for background XTT conversion. In treatment studies, results are generally presented with treatment type on x-axis and absorbance values at OD_{492nm} along y-axis (Fig. 2b-ii).

3.3.2 Resazurin Assay

1. Biofilms are formed as described above (*see* Subheading 3.3.1, **steps 1** and **2**). For the resazurin assay, the stock solution of resazurin (1% in PBS) is diluted to 0.001% in RPMI media (*see* **Note 7**). A total of 100 μL is added to all wells containing biofilm and to negative control wells minus biofilms using a multichannel pipette.
2. In this assay, NADH and other reducing (electron donating) agents produced by metabolically active cells (Fig. 3a) will convert the weakly fluorescent purple phenoxazine dye reduced to form resofurin, a strongly fluorescent pink compound (Fig. 3b-i).
3. Resazurin plates are incubated for up to 3 h, at 37 °C, with the fluorescence change measured at 530ex/590em on a fluorescence microplate reader with data presented as arbitrary units (AU) of fluorescence (Fig. 3b-ii) (*see* **Note 6**). Alternatively, if you do not have access to a fluorescence plate reader, the color change can also be measured by reading the absorbance at 570 and 600 nm, respectively. The extent of resazurin reduction can be calculated compared to the negative control wells using the following equation following the manufacturer's instructions:

$$\%\ \text{reduction} = \frac{(\varepsilon_{ox}\lambda 2)(A\lambda 1) - (\varepsilon_{ox}\lambda 1)(A\lambda 2)}{(\varepsilon_{RED}\lambda 1)(A\lambda' 2) - (\varepsilon_{RED}\lambda 2)(A\lambda' 1)} \times 100$$

In the equation above, $\varepsilon\lambda 1$ and $\varepsilon\lambda 2$ are constants that represent the molar extinction coefficient of resazurin (alamarBlue™) at 570 and 600 nm, in their oxidized ($_{ox}$) and reduced ($_{RED}$) forms, respectively. Again, we recommend using blank corrected values in the equation in order to account for background conversion. In the equation, the operands are defined as follows:

$\varepsilon_{ox}\lambda 1 = 80{,}586$

$\varepsilon_{ox}\lambda 2 = 117{,}216$

$\varepsilon_{RED}\lambda 1 = 155{,}677$

$\varepsilon_{RED}\lambda 2 = 14{,}652$

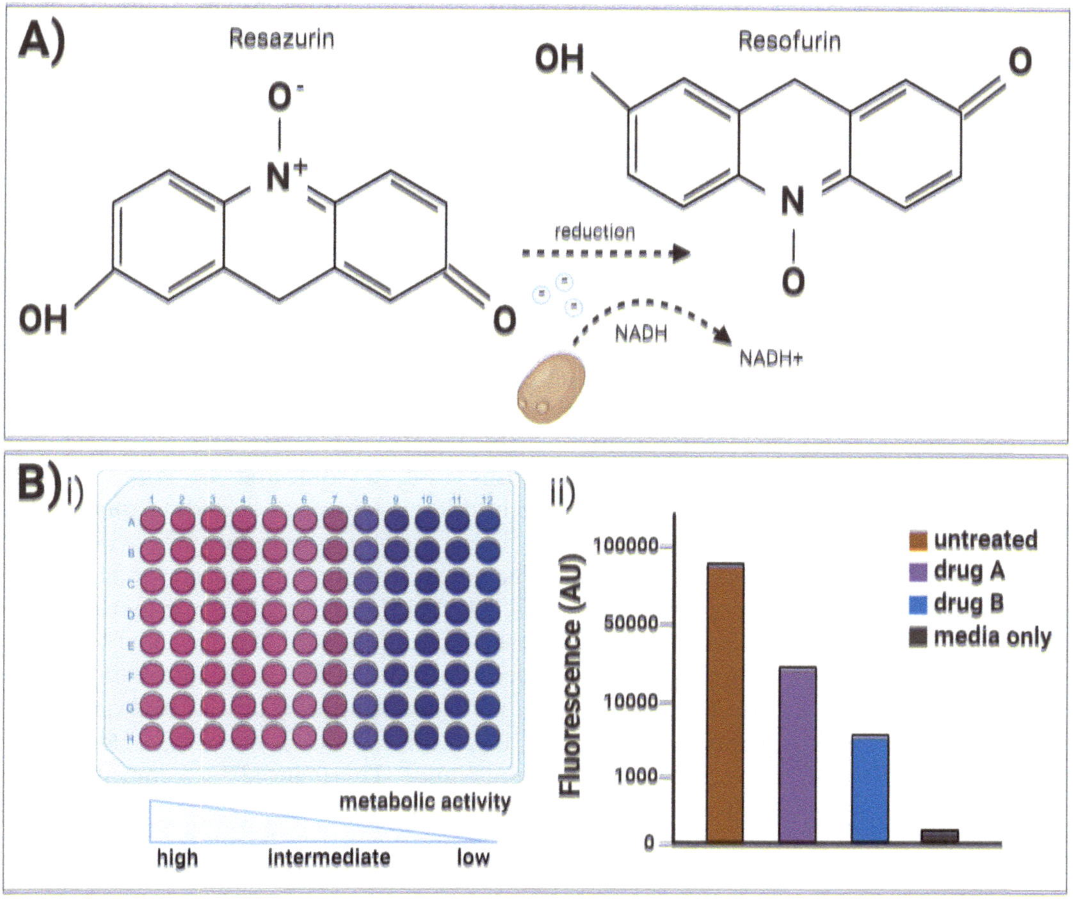

Fig. 3 Resazurin metabolic assay. The phenoxazine dye, resazurin, can be used to assess the metabolic activity of *Candida auris*. In the presence of NADH and other reducing (electron donating) agents produced by metabolically active cells, the positively charged resazurin will be converted into resofurin (**a**). In the reaction, the weakly fluorescent resazurin will be converted into the fluorescent resofurin, visually seen by a change from purple to pink in increasing levels of metabolic activity within the biofilm (**b-i**). The colorimetric change can be quantified on a fluorescence microtiter plate reader at 530 excitation/590 emission, with data presented as arbitrary units (AU) of fluorescence (**b-ii**), or alternatively, at OD_{570nm} and OD_{600nm}, with extent of conversion calculated relative to a negative control. Figure created using Biorender.com

$A\lambda 1$ = Observed absorbance measurement of test well at 570 nm

$A\lambda 2$ = Observed absorbance measurement of test well at 600 nm

$A\lambda' 1$ = Observed absorbance measurement of control well at 570 nm

$A\lambda' 2$ = Observed absorbance measurement of control well at 600 nm

4. **[Optional step]** After quantifying metabolic activity using one of the above assays, biofilm biomass can also be assessed to

reinforce metabolic activity measurements. This can be achieved using the easily obtainable crystal violet (CV) dye. Detailed protocols of CV staining procedures can be found elsewhere [21, 22]. In brief, biofilms are washed, air-dried then stained with 0.05% CV for 20 min at room temperature, washed thoroughly to remove excess stain, and finally destained with 95–100% ethanol. Absorbance of the solubilized dye can be read spectroscopically at 570 nm.

3.4 Cell Counting and Live/Dead Quantitative PCR

Colony-forming unit (CFU) cell counting is still an effective and common methodology used in microbiology research laboratories worldwide. This method employs the Miles and Misra technique [23] for quantifying the number of viable cells in a microbial suspension (e.g., in this case, a sonicated *C. auris* biofilm) (Fig. 4). A second methodology, known as live/dead quantitative PCR, has become a useful tool for assessing the viability of *C. albicans* in planktonic or sessile mono- and mixed species models following treatment [7, 24, 25]. However, the same methodology can be applied to *C. auris* cultures as well (Fig. 5). The following sections provide detailed descriptions for biofilm formation, sonication, and subsequent CFU counting and/or total and viable CFE quantification via qPCR. In the following example, *C. auris* biofilms will be formed on Nunc® Thermanox™ 13-mm coverslips placed at the bottom of 24-well microtiter plates, prior to CFU counting and live/dead assessment (*see* **Note 8**). As discussed above, treatments can be tested on these biofilms for an appropriate time prior to the steps described below (*see* **Note 3**).

3.4.1 Biofilm Preparation and Sonication for Viable Counts

1. Biofilms are formed by addition of 500 μL of 1×10^6 CFU/mL *C. auris* in RPMI media to flat-bottom 24-well microtiter plates (final concentration of 5×10^5 CFU/well) containing Nunc® Thermanox™ 13-mm coverslips. Biofilms can be left to develop for an appropriate amount of time as discussed above (*see* Subheading 3.2, **step 4**).
2. After growth, media is aspirated, and biofilms are gently washed three times with 500 μL of sterile PBS to remove non-adherent cells. As above, these steps were carried out carefully, and pipette tips were placed to the edge of the coverslips to remove media and wash biofilms to ensure limited mechanical disruption of biofilms by pipetting (*see* Fig. 1b-ii).
3. After washing, coverslips containing biofilms are removed from microtiter plates using a 23-gauge needle and tweezers and then transferred to a bijoux tube containing 1 mL of sterile PBS (*see* **Note 9**) (*see* Fig. 1b-ii).
4. Biofilms are then sonicated at 35 kHz for 10 min in an ultrasonic water bath followed by vortexing for 30 s (Fig. 4-i).

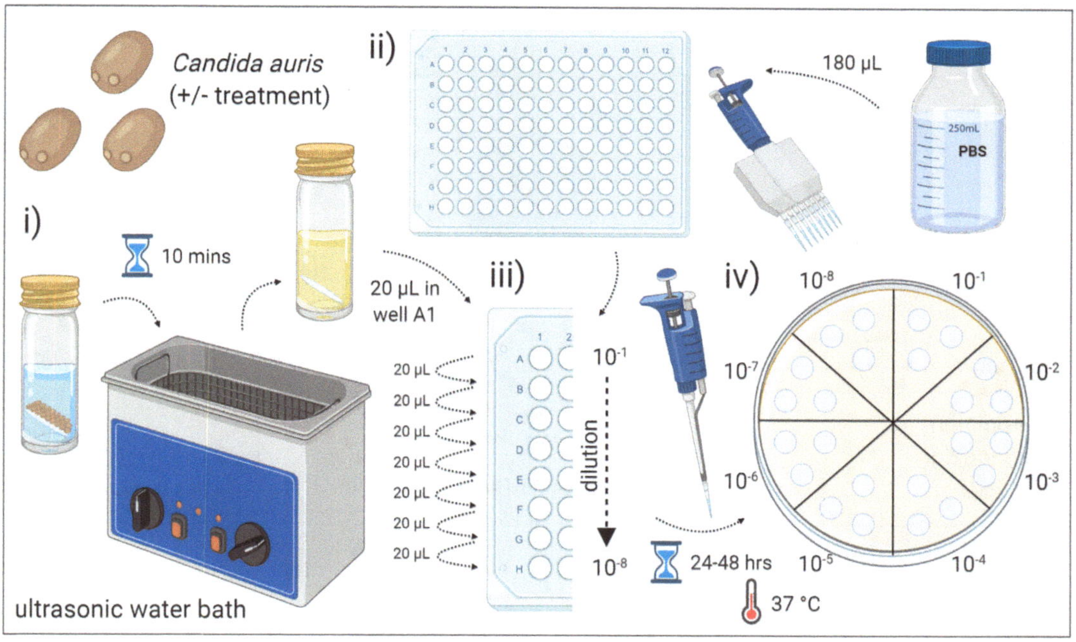

Fig. 4 Cell counting using the Miles and Misra technique. *Candida auris* biofilms (following treatment) are sonicated at 35 kHz for 10 min in an ultrasonic water bath (**i**). In the meantime, a total of 180 μL of sterile PBS is added to every well of a round-bottom 96-well microtiter plate using a multichannel pipette (**ii**), prior to addition of 20 μL of biofilm sonicate into well A1. Each column can be used for individual biofilm sonicates. Suspensions are then serially diluted 1 in 10 down the microtiter plate from row A (10^{-1}) to row H (10^{-8}) (**iii**). Following serial dilutions, starting from the lowest dilution factor (e.g., 10^{-8}), 3 × 10-μL drops from each dilution are transferred to the appropriate sector on a Sabouraud's dextrose agar plate. Once dry, plates are then incubated at 37 °C for up to 48 h until visible growth. Cells can then be counted from sectors containing between 10 and 100 colonies (**iv**). Figure generated using Biorender.com

Alternative methods to sonication exist such as manual removal of the biofilm using cell scrapers. For this, 1 mL of PBS should be pipetted onto the biofilm and biomass mechanically disrupted using the cell scraper. The biofilm cells can then be transferred to a bijoux tube via a pipette in 1 mL of PBS. For this, ensure that no biomass is remaining on the cell scraper following mechanical disruption by visualizing the plates microscopically.

5. The biofilm sonicate can then be used for two purposes: Miles and Misra technique for CFU counting (*see* Subheading 3.4.2) and the live/dead qPCR method for CFE quantification (*see* Subheading 3.4.3). To prepare for these methods, 20 μL of the 1 mL of sonicate is taken for CFU counting, with the remaining 980 μL split equally into 2 × 1.5-mL Eppendorf tubes for live/dead qPCR (*see* **Note 10**).

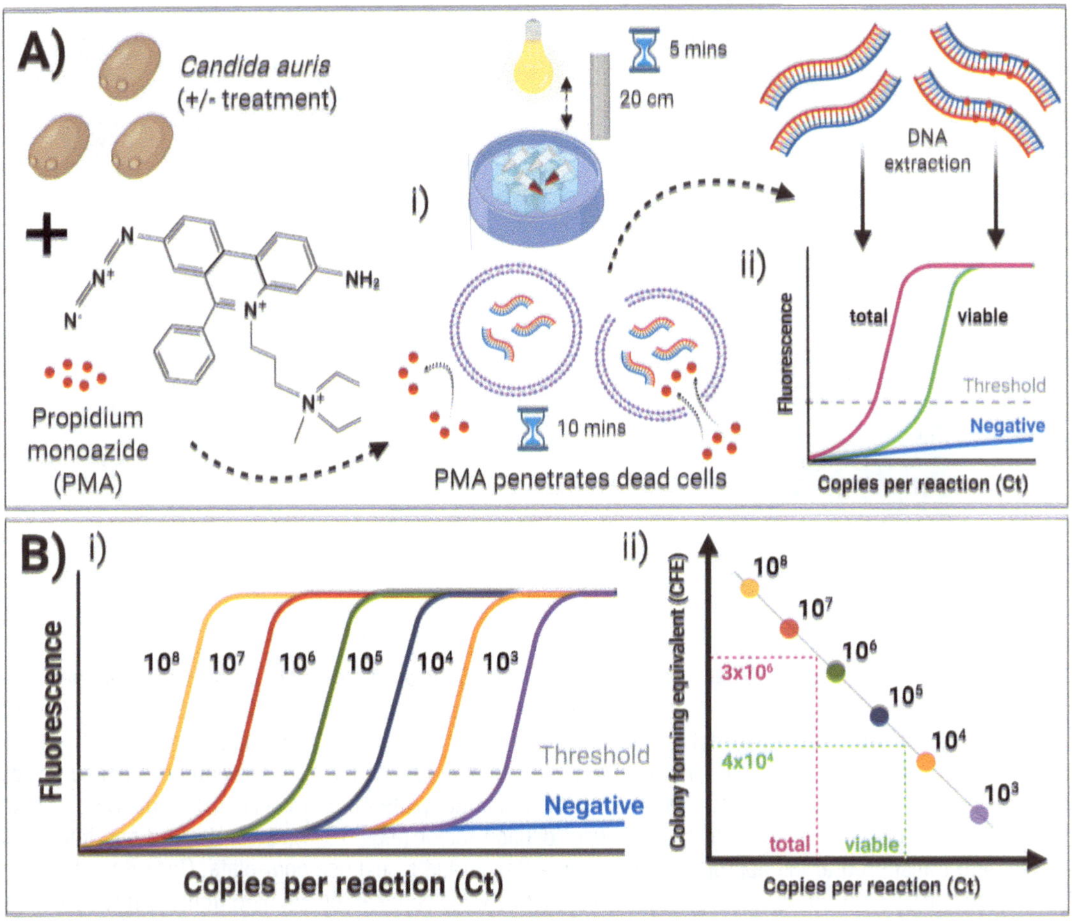

Fig. 5 Live/dead quantitative PCR. The viability of *Candida auris* cells can be assessed using quantitative PCR (qPCR) with the photoreactive DNA-binding dye, propidium monoazide (PMA). In the reaction, the dye can only penetrate cells with compromised membranes (e.g., following antifungal treatment) and not viable cells with an intact membrane structure. Once inside the cell, PMA will intercalate the double-stranded DNA. Samples are incubated with PMA in the dark for 10 min to allow uptake of the dye, and then the samples undergo photo inactivation for 5 min (samples should be laid on a bed of ice approx. 20 cm away from a 650 W halogen light) (**a-i**). Following DNA extraction, the proportion of total and viable cells within a sample can then be determined using qPCR, whereby lower Ct values are observable for untreated (total) cells and higher Ct values for PMA-treated (viable) cells (**a-ii**). To calculate the colony-forming equivalent (CFE)/mL for unknown samples treated with and without PMA, a standard curve of known CFE/mL of *C. auris* is required. For this, DNA is extracted from pure planktonic *C. auris* cultures ranging from 10^8 to 10^3 and used in the qPCR (**b-i**). Unknown viable and total CFE/mL is then extrapolated from the standard curve compiled from the Ct values vs. known CFE/mL for the pure planktonic *C. auris* cultures. In the example provided, the total CFE/mL would be 3×10^6, while the viable CFE/mL would be 4×10^4 (**b-ii**). Figure compiled using Biorender.com

3.4.2 Cell Counting Using the Miles and Misra Technique

1. For CFU counting, 180 μL of sterile PBS is added to every well of a round-bottom 96-well microtiter plate using a multichannel pipette (Fig. 4-ii). Depending on the number of samples, each column should represent a different sample (multiple microtiter plates are required if >12 samples).

2. To row A, add 20 μL of the *C. auris* biofilm sonicate to 180 μL of PBS; this will represent the 10^{-1} dilution. Once all samples are added, the suspensions should be serially diluted 1 in 10 by adding 20 μL of samples from row A to 180 μL of PBS in row B, then 20 μL of diluted sample from row B to 180 μL of PBS in row C, and so on down to row H. At each row, the 20 μL of suspension should be mixed well by pipetting with 180 μL of PBS. It is recommended to change pipette tips between each dilution step. Once complete, the dilution factors for each row should range from 10^{-1} in row A down to 10^{-8} in row H (Fig. 4-iii).
3. One Sabouraud's dextrose agar plate per sample should be split into six or eight equal sectors. Starting from the lowest dilution factor (e.g., 10^{-8}), transfer 3 × 10-μL drops from the microtiter plate in the appropriate sector. This step is repeated for all dilutions, until all sectors of the agar plate are covered (Fig. 4-iv). As above, it is suggested that the pipette tip is changed between each 10-μL drop and/or dilution factor.
4. Plates should be left to completely dry before inverting for incubation. For this, plate lids can be left slightly askew so that they can air-dry (*see* **Note 11**). Once dry, the plates can be incubated overnight at 37 °C for up to 48 h, with monitoring of growth after 24 h.
5. After incubation, visible colonies will appear on the plates at lower dilutions, while higher dilutions will appear as heavily concentrated lawns of growth without discernible colonies. The sector with the dilution containing between 10 and 100 colonies should be selected for cell counting. The colonies should be counted in the 3× spot dilutions per sector. From these triplicate values, take an average, and then multiply by the dilution factor (e.g., 10^4) and also by 100 to account for the scaling factor (e.g., 10 μL to 1 mL). This will now give you the number of colony-forming units in each mL of the original sample (CFU/mL). For example, if an average of 25 colonies were counted in the 10^{-4} sector, the concentration of cells in the original sample would be 2.5×10^7 CFU/mL (e.g., $25 \times 10^4 \times 100$).

3.4.3 Viable and Total Cell Count Quantification Using Live/Dead qPCR

1. Live/dead quantification by qPCR is achieved using a photoreactive DNA-binding dye such as propidium monoazide (PMA) (*see* **Note 12**) prior to qPCR analyses. In its mode of action, the compound will penetrate dead fungal cells or those with compromised membranes, e.g., following antifungal treatment, and preferentially bind double-stranded DNA (dsDNA) within the cell (Fig. 5a-i). The 2 × 1.5-mL Eppendorfs containing the microbial samples (approx. 490 μL in each) are used for live/dead qPCR. One sample is treated

with 2.5 μL of PMA working solution. The other sample remains untreated (no PMA added) and will act as the control for total fungal cells in the initial sample.

2. Both Eppendorf tubes should then be incubated in the dark for 10 min to allow uptake of the dye. All samples are then transferred to ice.
3. Samples are then exposed to a 650-W halogen light for 5 min for photoactivation of the PMA compound; exposure to the bright visible light will induce cross-linking of the PMA with dsDNA. Cross-linking between the compound and DNA structure following exposure to intense visible light will render it unsuitable for qPCR. Thereby, after DNA extraction, only DNA from viable cells will amplify in the qPCR, while untreated samples will allow for distinguishing total fungal cells. This will give rise to lower Ct values for untreated (total) cells and higher Ct values for PMA-treated (viable) cells (Fig. 5a-ii).
4. The Eppendorf tubes should be evenly laid on a "bed of ice," with the halogen light positioned approx. 20 cm away (as seen in Fig. 5a-i). The "bed of ice" should be approx. 5–10-cm-thick and placed in a polystyrene box lid or equivalent heat-resistant container. The purpose of the ice is to prevent the Eppendorf tubes from melting due to the intense heat of the halogen light. The halogen light should be positioned so that all of the samples are exposed to the same degree of light. After 2 min and 30 s, samples should be inverted and placed back on the ice for the remaining duration to ensure light exposure to all the samples. It is essential to take extra caution when handling the Eppendorf tubes following this step as they may be hot.
5. Following photoactivation, samples are centrifuged at 13,000 × *g*, for 10 min, before PBS is carefully aspirated and discarded. Here, pelleted cells treated with PMA may appear red in color.
6. DNA is then extracted from the pellets using the QIAamp® DNA mini kit, as per manufacturer's instructions. A minor modification during the extraction process should be made, so that samples can be mechanically disrupted using glass beads to ensure high DNA yield. This additional step should be done following the initial lysis step using the QIAamp® DNA mini kit. For this, firstly, an equivalent of 100 μL of sterile acid-washed glass beads of 0.5-mm diameter should be added to a screw-cap 2-mL BeadBug™ O-Ring tube.
7. Samples are transferred to the 2-mL O-Ring tubes and then mechanically disrupted using a bead beating machine

(BeadBug™ microtube homogenizer or alternative), for a total of 90 s.

8. Following mechanical disruption of the samples, the homogenized mixture is then used for DNA extraction with the above kit, with the manufacturer's instructions followed accordingly.
9. Following extraction, fungal DNA can be used for quantitative PCR using specific primers for *C. auris* (e.g., 18S or ITS). [**optional**] At this stage, if necessary, the quality and quantity of the extracted DNA can be assessed using a UV spectrophotometer (e.g., NanoDrop or equivalent).
10. To generate a standard curve for quantification, serial dilutions of pure planktonic *C. auris* cultures (e.g., from 10^8 to 10^3 CFU/mL) should be extracted using the same kit and method as described above (*see* **Note 13**). Unknown concentrations of viable and total *C. auris* can then be determined using the standard curves for each isolate. For example, highly concentrated samples of *C. auris* (e.g., 10^8) will give rise to lower Ct values, while higher Ct values are characteristic of lower concentrations of the organism (e.g., 10^3) (Fig. 5b-i). Unknown live and total CFE/mL values can be extrapolated from the Ct values gathered for the pure planktonic *C. auris* cultures of known concentrations (Fig. 5b-ii).

3.5 Live/Dead Fluorescent Imaging

Fluorescent staining and imaging are also useful tools for assessing viability in planktonic or sessile *C. auris*. A combination of different dyes has been reported in the literature for *C. auris* isolates, such as nucleic acid dyes (SYTO™ 9, propidium iodide, and DAPI), cell membrane/wall stains (calcofluor white (CFW) and fluorescently conjugated concanavalin A), and the two-color fluorescent viability probe FUN-1™ [26–30]. In the below protocol, we will describe the fluorescent staining of *C. auris* biofilms using the viability probe FUN-1™ (Fig. 6-i). In the following example, *C. auris* biofilms will be formed on glass coverslips placed at the bottom of 24-well microtiter plates, prior to fluorescent staining with FUN-1™. As above, it is likely that this methodology will be used to assess efficacy of antifungal drugs or therapies against *C. auris*; therefore these treatments can be tested during or following biofilm formation prior to fluorescent staining and microscopy (*see* **Note 3**).

1. Biofilms are formed by adding 500 μL of 1×10^6 CFU/mL *C. auris* in RPMI media to flat-bottom 24-well microtiter plates (final concentration of 5×10^5 CFU/well) containing glass coverslips for an appropriate amount of time (*see* Subheading 3.2, **step 4**). Other substrates may be used for such studies (*see* **Note 14**).

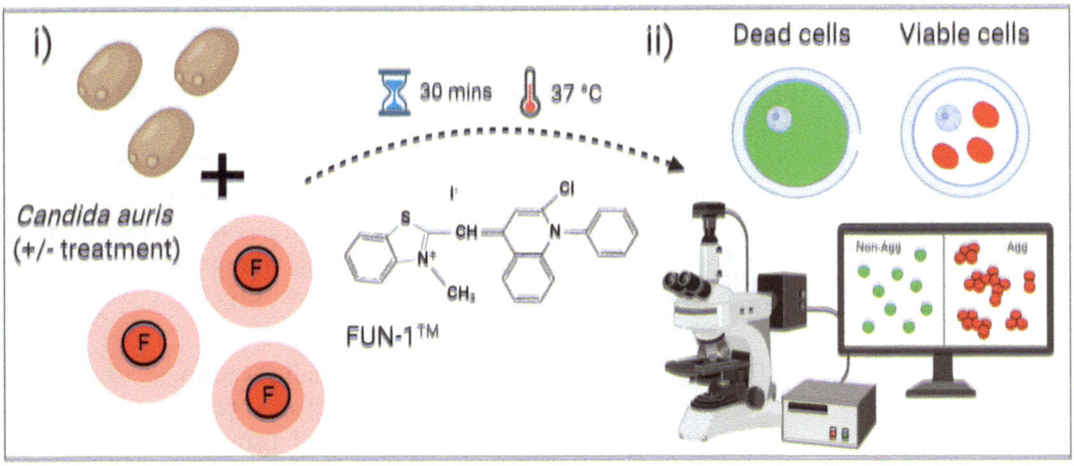

Fig. 6 Fluorescent imaging for assessment of viability. *Candida auris* cells can be stained with the two-color fluorescent viability probe FUN-1™. For such studies, it is likely that cells will be stained following antifungal treatment (**i**). In the mechanism of action, the dye will passively diffuse though all cell membranes staining the cytoplasm green. In viable cells, intracellular processing of the dye in compact, distinctive vacuole structures gives rise to a red fluorescence. This allows for distinguishing between dead and viable cells following confocal imaging (**ii**). In the example provided in (**ii**), images on the computer screen are visual representations of findings presented in Short et al. (2019). In the study, the aggregative phenotype of *C. auris* strain 8978 was found to be more tolerant to antiseptic treatment then the non-aggregative strain 8973, 14-days post-treatment. Image created using Biorender.com

2. Following biofilm maturation, media is aspirated and biofilms washed three times with 500μL of sterile PBS to remove non-adherent cells.

3. After the wash steps, biofilms are then stained with 500-μL working solution of FUN-1™ at a final concentration of 20 μM (*see* **Note 15**). The dye is gently added to the wells as to not disrupt the adhered biofilm and plates incubated for 30 min in the dark at 37 °C.

4. Following incubation, biofilms are washed three times with PBS to remove excess dye. **[optional step]** At this stage, biofilms can be fixed using an appropriate aldehyde or alcohol fixative such as 4% paraformaldehyde (PFA) or ice-cold 100% methanol. It is recommended that 4% PFA should be used for 10 min at room temperature, while alcohol fixation should be used ice-cold (−20 °C) for 10–20 min. Fixation will increase longevity of fluorescently labelled samples (*see* **Note 16**). Following fixation, biofilms should be gently washed three times in PBS to remove excess fixative.

5. Glass coverslips can now be removed using a pair of tweezers with a needle (*see* Fig. 1b-ii) and transferred to a microscope slide (*see* **Note 9**). The coverslips are placed in the center of the microscope slide and 10-μL mounting media added to the

biofilm. An additional larger glass coverslip is gently placed over the biofilm ensuring no bubbles form. The sides of the additional glass coverslip are then kept in place using small amounts of clear nail varnish and allowing it to dry (*see* **Note 17**).

6. Biofilms are then ready to be imaged on a fluorescent microscope or confocal laser scanning microscope. To image biofilms using a conventional fluorescent microscope, the substrate must allow for light penetration. If this is not possible (i.e., biofilms are grown on steel coupons or other nontransparent substrates), it is recommended that images should be captured using a confocal laser scanning microscopy.
7. For the FUN-1™ dye, it is recommended that a fluorescein isothiocyanate (FITC) filter set with excitation ~488 nm and emission ≥530 nm is used.
8. Visually, in viable cells with intact membranes, FUN-1™ will be internalized and the dye then transported to a cytoplasmic vacuole in metabolically active cells. FUN-1™ will subsequently be compacted into distinctive fluorescent red cylindrical intravacuolar structures. Dead cells fluoresce bright yellow–green, with no discernible red structures [31] (Fig. 6-ii). If used in conjunction with CFW (*see* **Note 15**), CFW will bind chitin in the cell wall of fungal cells, staining these structures a fluorescent blue color which can be visualized at ~365-nm excitation and ~435-nm emission (DAPI filter set).

4 Notes

1. Most isolates of *C. auris* can be obtained from the American Type Culture Collection (ATCC), Deutsche Sammlung von Mikroorganismen und Zellkulturen (DSMZ), National Collection of Type Cultures (NCTC), and other microbial culture collections.
2. When using the hemocytometer, fungal cell suspensions should be diluted in PBS by at least 1:100 to enable accurate cell counting. This dilution factor should then be taken into account when calculating the total cell count, as per the following equation (cell count × dilution factor × volume of square = colony-forming unit [CFU per mL]).
3. The methodologies described here document viability assays for untreated *C. auris* biofilms. However, it is likely that such assays will be utilized during treatment studies, testing the efficacy of conventional and/or novel therapeutics. It is recommended for such studies that following treatment, biofilms are washed again three times with sterile PBS to remove any

residual treatment. This is essential as the residues of certain drugs or antiseptics (e.g., chlorhexidine) may interfere with activity of compounds such XTT and resazurin. In some cases, treatments may be neutralized with Dey–Engley Neutralizing Broth for 15 min at 37 °C, prior to wash steps and assessment of viability using metabolic assays.

4. With metabolic assays such as XTT, there still remains a need for standardized testing of fungal biofilms [32]. Therefore, such assays must be used with a certain level of caution in regard to data analysis and interpretation. Here, it is important to note that different isolates of *C. auris* have low, intermediate, and high biofilm-forming capabilities (which can be assessed by a CV assay) each with possible variations in metabolic rates. Therefore, it is not wise to use the XTT or resazurin assay measurements as an absolute assessment of metabolic output.
5. Higher concentrations of menadione can be used to further accelerate the colorimetric change in the XTT assay (e.g., up to 10 μM of menadione). The concentration should be optimized for the cellular density of the tested biofilm, e.g., an increased concentration of metabolically active cells will lead to a faster colorimetric change, and therefore may not require higher concentrations of menadione and vice versa.
6. Regular monitoring of the XTT and resazurin plates is recommended, especially if using new isolates with unknown biofilm-forming capabilities or metabolic profiles. If possible, absorbance/fluorescence of the biofilms should be measured on the plate reader at regular intervals, to prevent saturation of color. For treatment studies, untreated controls should act as "positive controls," which will further allow for determining the optimal incubation period for the metabolic assay.
7. It should be noted that resazurin-containing assays are available commercially as alamarBlue™ or PrestoBlue™ cell viability reagents. As described on the Thermo-Fisher Scientific website, these reagents contain resazurin in a proprietary stabilizing formulation that allows for a convenient "mix, incubate, and read" protocol. Manufacturer's instructions should be followed when using these alternative reagents, although the principle of resazurin conversion and colorimetric change is the same as described in this methodology. Regular monitoring of biofilm conversion of these commercially available resazurin derivatives is again recommended.
8. For the CFU counting and live/dead qPCR methods, biofilms were formed on Nunc® Thermanox™ 13-mm coverslips. However, biofilms can be formed directly on the bottom of flat-bottom 24-, 12-, or 6- well microtiter plates for this assay,

with the resulting biomass mechanically disrupted in 1 mL of PBS using a cell scraper (and transferred to the bijoux tube for sonication or vortexing, where appropriate). It is recommended not to use smaller microtiter plates (e.g., 48- or 96-well plates) as there would be insufficient biomass recovered for DNA extraction. Other substrates can also be used as described previously for growth of *C. auris* on clinically relevant substrates [19]. These alternative substrates are listed in the materials section. It should be noted that volumes of inoculum added may need to be adjusted accordingly to ensure appropriate coverage of the substrate tested.

9. To improve ease of which coverslips can be removed from microtiter plates, the end of the 23-gauge needle should be formed into a hook. This can be achieved by twisting the end of the needle with tweezers. The hook can then be positioned underneath the coverslips to slowly prize it away from the bottom of the plate (as seen in Fig. 1b-ii). The coverslip can then be grasped by the tweezers and transferred to the bijoux tube for live/dead qPCR or to the microscope slide for fluorescent imaging. For imaging, ensure that the tweezers touch the edges of the coverslips only, so that the biofilm is not disrupted.
10. It is highly recommended to utilize both methodologies of cell counting and quantification (e.g., Miles and Misra, and live/dead qPCR) for studies into treatment efficacy against *C. auris* biofilms. This is to overcome the limitations attached to these techniques, such as underestimation of cell counts in CFU counting due to microbial aggregation (particularly problematic if working with aggregative forms of *C. auris*) compared to overestimation of number of cells associated with qPCR, through detection of extracellular DNA. Combining results from the two methodologies should serve to circumvent these limitations. It is also noteworthy that advantages and other limitations for these techniques are described elsewhere [20].
11. Where possible, for Miles and Misra counting, a category II microbiological hood should be used for preparing dilution plates. This will allow for numerous plates to be prepared at the same time, while also minimizing the chances of airborne contamination when leaving plates to dry with lids askew. If not possible, spot dilutions should be done aseptically directly adjacent to a Bunsen burner to reduce chances of contamination.
12. Other photoreactive dyes such as ethidium monoazide (EMA) are available for quantification of viable and dead microorganisms. However, it has been reported that PMA has a higher

charge than EMA and is therefore less likely to penetrate membranes of viable cells [33]. Recently, Biotium have released a PMAxx^{TM} dye, which has been designed to be a superior alternative to PMA. However, the live/dead qPCR methodology is still the same for PMAxx^{TM} as described here for PMA.^{TM} Alongside this new dye, Biotium have also created a PMA-lite^{TM} LED Photolysis Device which will allow for the photoactivation of PMA and PMAxx^{TM} in live/dead qPCR. This device will circumvent the need to utilise a halogen light for photoactivation and cooling of samples with ice.

13. For extractions from pure planktonic *C. auris* cultures, DNA should be extracted from each isolate used to ensure that accurate standard curves are generated. Furthermore, it is suggested that DNA should be extracted from serially diluted cultures of each *C. auris* isolate, ranging from 10^8 to 10^3 CFU/mL to determine the lowest limit of detection. We find that 10^3–10^4 CFU/mL is the lower limit for which the QIAamp® DNA mini kit can extract from *C. auris* isolates efficiently; however, this may vary from lab-to-lab and may be different for other extraction kits. Finally, live/dead qPCR can be utilized with SYBR™ Green and TaqMan™ qPCR technologies.

14. Alternative growth substrates suitable for confocal microscopy are multi-well chamber slides (typically ranging from 2 to 16 wells) and steel coupons. Steel coupons can be placed in 12-well plates with the same protocols for biofilm growth on Nunc® Thermanox™ 13-mm coverslips or glass coverslips and can be followed as described in Subheading 3.5 (concentrations and volumes may need to be adjusted accordingly for thicker substrates to ensure complete coverage of the surface). Alternatively, Nunc® Lab-Tek® 8-well chamber slides (0.8 cm^2/well) or ibidi μ-Slide 8-well chamber slides are specifically manufactured with cell imaging in mind. We recommend those supplied by ibidi GmbH, which work best for clear biofilm imaging. These are specifically made for confocal imaging and can be ordered coated or uncoated, with the latter requiring manual coating with poly-D-lysin which can be dissolved in ddH_2O. Coating with poly-D-lysin aids in fungal cell attachment and biofilm formation. Chamber slides should be incubated with 200 μL of 0.1-mg/mL poly-D-lysine for 30 min at 37 °C. Coated wells are then washed once with 200-μL PBS. *C. auris* cells can now be grown and exposed to desired treatments before staining, imaging, and fixing as described above. Volumes given are for use in 8-well chamber slides and should be adjusted accordingly if using slides with a different number of wells.

15. If using FUN-1™ in conjunction with CFW, it is recommended as per manufacturer's instructions that concentrations of 10 μM and 25 μM are used, respectively. Both dyes can be added simultaneously and CFW visualized at ~365-nm excitation/~435-nm emission. It is recommended for all new fluorescent stains to be serially diluted to optimize dye concentration, given that cellular density of *C. auris* cells may impact staining efficiency. This could be achieved by using standardized suspensions of planktonic cells of *C. auris* diluted in different concentrations of fluorescent dye. Thermo-Fisher supply a useful manual called "probes for yeast viability" (https://www.thermofisher.com/document-connect/document-connect.html?url=https%3A%2F%2Fassets.thermofisher.com%2FTFS-Assets%2FLSG%2Fmanuals%2Fmp07009.pdf&title=UHJvYmVzIGZvciBZZWFzdCBWaWFiaWxpdHk=) for optimization of dye concentration. It should be noted that other viability stains are available for confocal microscopy and can be utilized for such experiments. SYTO™ 9 green fluorescent nucleic acid stain and propidium iodide can be used in conjunction to distinguish between live and dead fungal cells.
16. Any fixation method used should be optimized by each laboratory. The suggestions provided are only examples, and fixation may also vary depending on dye used.
17. For larger and/or thicker substrates, e.g., the steel coupons, it is recommended not to use the additional glass coverslip to cover the biofilm. For these samples, they should be fixed as described above and then the underside of the substrate superglued into place on a microscope slide. From experience, it is difficult to keep the additional microscope slide in place on these larger substrates which can lead to complications during imaging.

References

1. Ramage G, Vande Walle K, Wickes BL, Lopez-Ribot JL (2001) Standardized method for in vitro antifungal susceptibility testing of Candida albicans biofilms. Antimicrob Agents Chemother 45(9):2475–2479
2. Pierce CG, Uppuluri P, Tristan AR, Wormley FL Jr, Mowat E, Ramage G, Lopez-Ribot JL (2008) A simple and reproducible 96-well plate-based method for the formation of fungal biofilms and its application to antifungal susceptibility testing. Nat Protoc 3(9):1494–1500
3. Pierce CG, Uppuluri P, Tummala S, Lopez-Ribot JL (2010) A 96 well microtiter plate-based method for monitoring formation and antifungal susceptibility testing of Candida albicans biofilms. J Vis Exp 44:2287
4. Dalecki AG, Crawford CL, Wolschendorf F (2016) Targeting biofilm associated staphylococcus aureus using resazurin based drug-susceptibility assay. J Vis Exp 111:53925
5. Van den Driessche F, Rigole P, Brackman G, Coenye T (2014) Optimization of resazurin-based viability staining for quantification of microbial biofilms. J Microbiol Methods 98:31–34
6. Loozen G, Boon N, Pauwels M, Quirynen M, Teughels W (2011) Live/dead real-time polymerase chain reaction to assess new therapies

against dental plaque-related pathologies. Mol Oral Microbiol 26(4):253–261

7. Agusti G, Fittipaldi M, Morato J, Codony F (2013) Viable quantitative PCR for assessing the response of Candida albicans to antifungal treatment. Appl Microbiol Biotechnol 97(1): 341–349
8. Boulos L, Prevost M, Barbeau B, Coallier J, Desjardins R (1999) LIVE/DEAD BacLight : application of a new rapid staining method for direct enumeration of viable and total bacteria in drinking water. J Microbiol Methods 37(1): 77–86
9. Millard PJ, Roth BL, Thi HP, Yue ST, Haugland RP (1997) Development of the FUN-1 family of fluorescent probes for vacuole labeling and viability testing of yeasts. Appl Environ Microbiol 63(7):2897–2905
10. Satoh K, Makimura K, Hasumi Y, Nishiyama Y, Uchida K, Yamaguchi H (2009) Candida auris sp. nov., a novel ascomycetous yeast isolated from the external ear canal of an inpatient in a Japanese hospital. Microbiol Immunol 53(1): 41–44
11. Lockhart SR, Etienne KA, Vallabhaneni S, Farooqi J, Chowdhary A, Govender NP, Colombo AL, Calvo B, Cuomo CA, Desjardins CA, Berkow EL, Castanheira M, Magobo RE, Jabeen K, Asghar RJ, Meis JF, Jackson B, Chiller T, Litvintseva AP (2017) Simultaneous emergence of multidrug-resistant Candida auris on 3 continents confirmed by whole-genome sequencing and epidemiological analyses. Clin Infect Dis 64(2):134–140
12. Kean R, Ramage G (2019) Combined antifungal resistance and biofilm tolerance: the global threat of Candida auris. mSphere 4(4): e00458–e00419
13. Hager CL, Larkin EL, Long L, Zohra Abidi F, Shaw KJ, Ghannoum MA (2018) In vitro and in vivo evaluation of the antifungal activity of APX001A/APX001 against Candida auris. Antimicrob Agents Chemother 62(3): e02319–e02317
14. de Oliveira HC, Monteiro MC, Rossi SA, Peman J, Ruiz-Gaitan A, Mendes-Giannini MJS, Mellado E, Zaragoza O (2019) Identification of off-patent compounds that present antifungal activity against the emerging fungal pathogen Candida auris. Front Cell Infect Microbiol 9:83
15. Rudramurthy SM, Colley T, Abdolrasouli A, Ashman J, Dhaliwal M, Kaur H, Armstrong-James D, Strong P, Rapeport G, Schelenz S, Ito K, Chakrabarti A (2019) In vitro antifungal activity of a novel topical triazole PC945 against emerging yeast Candida auris. J Antimicrob Chemother 74(10):2943–2949
16. Arias LS, Butcher MC, Short B, McKloud E, Delaney C, Kean R, Monteiro DR, Williams C, Ramage G, Brown JL (2020) Chitosan Ameliorates Candida auris Virulence in a Galleria mellonella Infection Model. Antimicrob Agents Chemother 64(8):e00476–e00420
17. CLSI (2008) Reference method for broth dilution antifungal susceptibility testing of yeasts, 3rd edn. Clinical and Laboratory Standards Institute (CLSI). Document M27-A3.
18. Allkja J, Bjarnsholt T, Coenye T, Cos P, Fallarero A, Harrison J, Lopes S, Oliver A, Pereira M, Ramage G, Shirtliff M, Stoodley P, Webb J, Zaat S, Goeres D (2020) Minimum information guideline for spectrophotometric and fluorometric methods to assess biofilm formation in microplates. Biofilms 2:100010
19. Kean R, Sherry L, Townsend E, McKloud E, Short B, Akinbobola A, Mackay WG, Williams C, Jones BL, Ramage G (2017) Surface disinfection challenges for Candida auris: an in vitro study. J Hosp Infect 98(4):433–436
20. Azeredo J, Azevedo NF, Briandet R, Cerca N, Coenye T, Costa AR, Desvaux M, Di Bonaventura G, Hebraud M, Jaglic Z, Kacaniova M, Knochel S, Lourenco A, Mergulhao F, Meyer RL, Nychas G, Simoes M, Tresse O, Sternberg C (2017) Critical review on biofilm methods. Crit Rev Microbiol 43(3):313–351
21. Jose A, Coco BJ, Milligan S, Young B, Lappin DF, Bagg J, Murray C, Ramage G (2010) Reducing the incidence of denture stomatitis: are denture cleansers sufficient? J Prosthodont 19(4):252–257
22. Mowat E, Butcher J, Lang S, Williams C, Ramage G (2007) Development of a simple model for studying the effects of antifungal agents on multicellular communities of Aspergillus fumigatus. J Med Microbiol 56(9): 1205–1212
23. Miles AA, Misra SS, Irwin JO (1938) The estimation of the bactericidal power of the blood. J Hyg (Lond) 38(6):732–749
24. Kean R, Rajendran R, Haggarty J, Townsend EM, Short B, Burgess KE, Lang S, Millington O, Mackay WG, Williams C, Ramage G (2017) Candida albicans mycofilms support staphylococcus aureus colonization and enhances miconazole resistance in dual-species interactions. Front Microbiol 8:258
25. Sherry L, Lappin G, O'Donnell LE, Millhouse E, Millington OR, Bradshaw DJ, Axe AS, Williams C, Nile CJ, Ramage G (2016) Viable compositional analysis of an eleven species oral polymicrobial biofilm. Front Microbiol 7:912

26. Yue H, Bing J, Zheng Q, Zhang Y, Hu T, Du H, Wang H, Huang G (2018) Filamentation in Candida auris, an emerging fungal pathogen of humans: passage through the mammalian body induces a heritable phenotypic switch. Emerg Microbes Infect 7(1):188
27. Wang X, Bing J, Zheng Q, Zhang F, Liu J, Yue H, Tao L, Du H, Wang Y, Wang H, Huang G (2018) The first isolate of Candida auris in China: clinical and biological aspects. Emerg Microbes Infect 7(1):93
28. Short B, Brown J, Delaney C, Sherry L, Williams C, Ramage G, Kean R (2019) Candida auris exhibits resilient biofilm characteristics in vitro: implications for environmental persistence. J Hosp Infect 103(1):92–96
29. Cleare LG, Li KL, Abuzeid WM, Nacharaju P, Friedman JM, Nosanchuk JD (2020) NO Candida auris: nitric oxide in nanotherapeutics to combat emerging fungal pathogen Candida auris. J Fungi (Basel) 6(2):85
30. Larkin E, Hager C, Chandra J, Mukherjee PK, Retuerto M, Salem I, Long L, Isham N, Kovanda L, Borroto-Esoda K, Wring S, Angulo D, Ghannoum M (2017) The emerging pathogen Candida auris: growth phenotype, virulence factors, activity of antifungals, and effect of SCY-078, a novel glucan synthesis inhibitor, on growth morphology and biofilm formation. Antimicrob Agents Chemother 61(5):e02396–e02316
31. Essary BD, Marshall PA (2009) Assessment of FUN-1 vital dye staining: yeast with a block in the vacuolar sorting pathway have impaired ability to form CIVS when stained with FUN-1 fluorescent dye. J Microbiol Methods 78(2):208–212
32. Ramage G (2016) Comparing apples and oranges: considerations for quantifying candidal biofilms with XTT [2,3-bis(2-methoxy-4-nitro-5-sulfo-phenyl)-2H-tetrazolium-5-carboxanilide] and the need for standardized testing. J Med Microbiol 65(4):259–260
33. Nocker A, Cheung CY, Camper AK (2006) Comparison of propidium monoazide with ethidium monoazide for differentiation of live vs. dead bacteria by selective removal of DNA from dead cells. J Microbiol Methods 67(2):310–320

Chapter 11

Measuring Gene Expression via mRNA Abundance in *Candida auris*

Mohd Wasi, Remya Nair, Nitesh Kumar Khandelwal, and Rajendra Prasad

Abstract

The recently emerged human pathogenic yeast *Candida auris* has become a major global threat. As compared to other *Candida* species, *C. auris* often displays a high level of resistance to commonly used antifungals and poses additional therapeutic challenges. There is a great need to understand the molecular basis of its success as a drug-resistant human pathogen. The study of condition-specific gene expression can provide good cues of regulatory circuitry governing high drug resistance. Here, we describe the protocol of quantitative reverse transcription PCR (RT–qPCR) which can be conveniently employed as a highly reproducible method for measuring individual transcripts in *C. auris* cells.

Key words *Candida auris*, Drug resistance, RNA, Polymerase chain reaction, Transcriptome, Quantitative reverse transcription

1 Introduction

Gene expression in *C. auris* is highly responsive to various environmental conditions which aids in its rapid adaptation to diverse niches within the host [1, 2]. A variety of in vivo and in vitro techniques have been used to measure gene expression. Among the various methods used for expression studies, quantitative reverse transcription PCR (RT–qPCR) represents a sensitive, reliable, and highly reproducible method for measuring individual transcripts [3]. RT–qPCR meets the need for molecular medicine, biotechnology, microbiology, and diagnostics for quantitative data analysis and has become the tool of choice for mRNA quantification.

In this chapter, we provide the detailed procedure from the isolation of total RNA from *C. auris* cells, complementary DNA (cDNA) preparation, and the setup of RT–qPCR reactions. As an example, we also show relative condition-specific (antifungal trea-

Alexander Lorenz (ed.), *Candida auris: Methods and Protocols*, Methods in Molecular Biology, vol. 2517,
https://doi.org/10.1007/978-1-0716-2417-3_11,

ted vs untreated samples) quantification of transcripts to calculate the fold differences of mRNA levels. Common useful tips and hints are also included to help beginners understanding the technique.

2 Materials

Reagents for RNA and cDNA sample preparation should be made in diethyl pyrocarbonate (DEPC)-treated ultrapure DNase-/RNase-free water. The workbench should be kept clean with RNAseZap.

2.1 Growth

1. Yeast extract peptone dextrose (YEPD) medium: 2% Bacto Peptone, 1% yeast extract, and 2% dextrose in deionized water for broth. YEPD agar plates are prepared by adding 2% agar in the YEPD broth recipe. The broth needs to be autoclaved before it can be used.
2. 50-mL falcon tubes and 200-mL conical flasks.

2.2 RNA Isolation and cDNA Preparation

1. Microcentrifuge tubes.
2. RNAseZAP (RNase decontamination solution).
3. DEPC-treated water.
4. Acid-washed glass beads (0.4–0.6 mm).
5. DNase I 1 U/μL.
6. 10× reaction buffer with $MgCl_2$ (DNase I treatment).
7. 50-mM EDTA: 18.61-g EDTA disodium salt dissolved in 1-L deionized water and adjusted to the pH 8.0.
8. RevertAid First Strand cDNA Synthesis Kit.
9. Agarose.
10. Ethidium bromide.
11. 1× TAE buffer: 40-mM Tris, 20-mM acetic acid, and 1-mM EDTA.
12. Taq polymerase.
13. Standard Taq polymerase reaction buffer.
14. dNTPs.
15. RNA isolation kit.
16. Electrophoresis gel apparatus.
17. DNA marker/DNA ladder.
18. NanoDrop™ Spectrophotometer.
19. UV–Vis Spectrophotometer.
20. Yeast–Cell Disruptor.

2.3 RT–qPCR

1. Oligonucleotide primers.
2. SYBR Green Supermix.
3. 96-well plates.
4. Real-time PCR machine.

3 Methods

3.1 Cell Growth

1. For the primary culture, inoculate a single colony of *C. auris* from a YEPD agar plate in 10-mL YEPD broth, and grow for 24 h at 30 °C in a shaking incubator at 200 rpm.
2. Inoculate $O.D._{600}$ of 0.2 in 50 mL of YEPD media for secondary culture. Grow the cells for another 7 h until they reach log phase, and use these cells for the control sample (*see* **Note 1**).
3. For samples with drug treatment, after 6 h of secondary growth, add the nonlethal concentrations of the desired drug, followed by growth for another 1 h.

3.2 RNA Isolation and Quantification

1. After 7 h of growth, harvest the cells by centrifugation at 3200 × *g* for 5 min at 4 °C. Wash the cell pellet twice with DEPC-treated water (*see* **Note 2**).
2. Measure the $O.D._{600}$ cells after washing. Use around 5×10^6–1×10^7 cells for the disruption step.
3. RNA isolation is carried out by using an RNA isolation kit as per the manufacturer's protocol (*see* **Note 3**).
4. The first step includes lysis of the cells (5×10^6–1×10^7 cells) by the mechanical disruption method. Add 300 μL of acid-washed 0.4- to 0.6-mm glass beads to the cell pellet along with 600 μL of lysis buffer provided with the RNA isolation kit.
5. 3 rounds, each of 20 s pulse are used for cell disruption (*see* **Note 4**). Incubate the cells on ice for 1 min after each pulse to keep the samples cold.
6. After disruption, beads and cell debris are separated by centrifugation for 5 min at 18,000 × *g*. The upper layer of cell lysate is used for RNA preparation as per the manufacturer's protocol.
7. Finally, the total RNA is eluted in RNase-free water.
8. Check the RNA concentration and purity on a NanoDrop™ 2000 Spectrophotometer or any similar measuring device. Different dilutions of RNA such as 1:10 or 1:50 should be prepared in DEPC water to determine the concentration.
9. The spectrophotometric absorption ratio between 260 and 280 nm (A_{260}/A_{280}) is a reliable measure to assess the purity of the extracted RNA. An absorption ratio between 1.8 and 2.0 (a value of A_{260}/A_{280} ratio closer to 2 is best) is an indicator of good quality of the prepared RNA (*see* **Note 5**).

3.3 cDNA Synthesis

1. To start with the cDNA synthesis, the first crucial step is to remove DNA contamination. For this, treat the RNA samples with DNase I using the following reaction mixture 6.25 µL of RNA (1.6-µg/µL stock solution), 1 µL of 10× reaction buffer, and 1-µL (1 U/µL) DNase I enzyme. Make up the final volume to 10 µL by adding the nuclease-free water for each reaction. Incubate the reaction mixture at 37 °C for 30 min in a thermocycler [4].
2. Add 1 µL of 50-mM EDTA into the reaction mixture, and incubate at 65 °C for 10 min to stop the DNase treatment reaction. This DNase I-treated RNA sample is used as a template for the downstream cDNA synthesis process (*see* **Note 6**).
3. For cDNA synthesis use the RevertAid First Strand cDNA Synthesis Kit, and set the reaction with the following components: 1 µg of DNase I treated RNA, 1 µL of the random hexamer and/or 1 µL oligo$(dT)_{18}$, 4-µL reaction buffer from 5× stock, 1 µL of RiboLock RNase inhibitor, 2 µL of 10-mM dNTP mix, and 1 µL of RevertAid M-MuLV RT enzyme. Adjust the volume of the reaction mixture is to 20 µL with the nuclease-free water (*see* **Note 7**).
4. For oligo$(dT)_{18}$-primed cDNA synthesis, incubate the samples at 42 °C for 60 min. For random hexamer primed synthesis, incubate samples for 10 min at 25 °C followed by 60 min incubation at 42 °C. Terminate the reaction by heating at 70 °C for 10 min in both cases (*see* **Notes** 7 and **8**).

3.4 RT–qPCR and Expression Analysis

3.4.1 Primer Design and Optimization

Primer design is a crucial step for the success of the RT–qPCR assay. The quality and reliability of outcome are significantly dependent upon RT–qPCR primers that otherwise would poorly anneal or bind nonspecifically during amplification. There are many tools available to design the primers such as the NCBI primer blast tool (https://www.ncbi.nlm.nih.gov/tools/primer-blast/) or Primer3. We used the Primer3 web-based tool (https://primer3.ut.ee/) for designing the primers on the default setting (for details *see* **Notes 9**, **10**, and **11**).

1. To validate each primer pair for specific size and specific amplification of the product, a normal PCR reaction with genomic DNA as a template and the gene specific primers should be performed. The PCR product should be analyzed via agarose gel electrophoresis to confirm the specific product at the accurate size along with appropriate DNA size marker.
2. The primer's efficiency for equal amplification should be optimized with a different dilution of the genomic DNA as a template. In an earlier study of *C. auris* gene expression analysis, we have used genomic DNA template in dilutions ranging

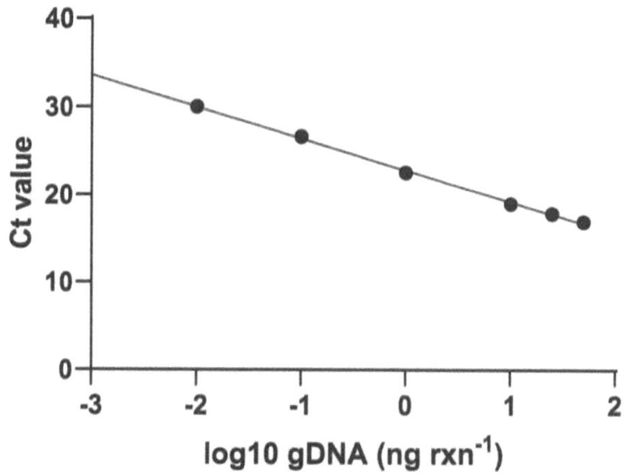

Fig. 1 Primer efficiency curve generated using the Ct values obtained with different dilutions (ranging from 0.01 ng to 50 ng) of genomic DNA as template. The calculated slope value of the plot is −3.456, based on data adapted for *TDH1* gene primers from an earlier study [2]

between 0.01 ng and 50 ng (0.01, 0.1, 1.0, 10, 25, and 50 ng) for setting up the PCR reactions [2].

3. The standard curves are generated by plotting the graph between the log10 values of DNA dilutions against the Ct value as shown in the Fig. 1. A linear relationship (indicated by a straight line) shows that gene primer pairs have equal amplification for a given amount of template. The efficiency of the PCR should be between 90% and 100% having slope values between −3.3 and −3.6.

3.4.2 RT–qPCR Reaction Setup and Selection of Endogenous Controls

To determine the relative expression of different genes, the expression of the target gene of interest must be compared to a housekeeping gene as a control. The expression of the selected housekeeping gene should not be variable under any experimental conditions (*see* **Note 12**). The endogenous control gene not only serves as a control for quality (integrity and to some degree purity) of the RNA in the reaction but also works as a control for the RT step of the RT–PCR quantification.

1. We have used *C. auris TDH1* (*CAUR_02457*), an ortholog of the *Saccharomyces cerevisiae* housekeeping gene *TDH1* as an internal control for normalization of expression.
2. The expression of *TDH1* (*CAUR_02457*) does not change significantly in all the tested conditions (drug treatment) justifying it as a good candidate for internal control.
3. The PCR reaction volume size can vary, but a typical PCR reaction is set up in a 25-μL final volume for the analysis. The

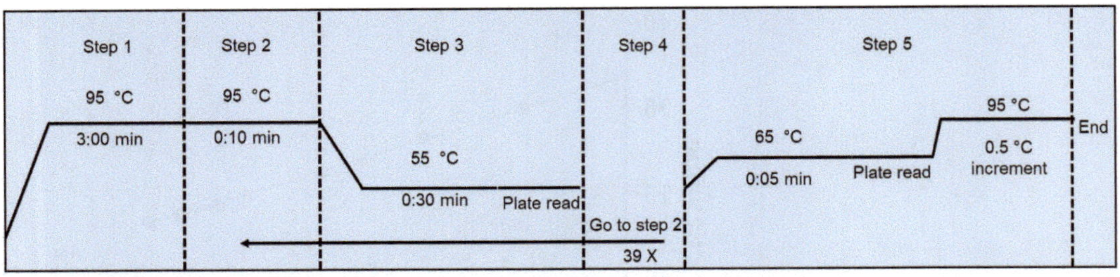

Step 1	95 °C. for 3:00 min
Step 2	95 °C. for 0:10 min
Step 3	55 °C. for 0:30 min
	Plate read
Step 4	Repeat cycle from step 2 (39 more times)
Step 5	Melting curve 65 to 95 °C with increment of 0.5 °C for 0:05 min plate read
	END reaction

Fig. 2 Cycling conditions used for RT–qPCR reactions on LightCycler™ real-time PCR machine

qPCR should be performed in three technical replicates for each sample (*see* **Note 13**).

4. The reaction is set up by using cDNA with SYBR Green Supermix according to the manufacturer's instructions in a real-time PCR machine for 40 cycles in the total volume of 25 μL. The each PCR reaction consists of 12.5 μL of 2× SYBR green super mix, 0.5 μL of each primer (stock concentraion10 μM), and 1 μL of cDNA (1/10 dilution of cDNA obtained after *see* Subheading 3.3) and adjusted with the nuclease-free water to make up the volume to 25 μL. Typical PCR reaction cycling conditions are shown in Fig. 2.

3.4.3 Expression Profile Analysis

The use of fluorescent reporter dyes in RT–qPCR allows us to combine the amplification and detection steps of the PCR reaction in a single reaction [5, 6]. RT–qPCR is based on measuring the variation in fluorescent signal as a function of the amount of DNA produced during each PCR cycle. Each PCR reaction consists of the individual cycles and the number of cycles required to cross the threshold fluorescent signal called threshold cycle (Ct) or crossing point (Cp). The Ct value has an inverse relationship with the amount of starting genetic material; the lower the Ct value, the greater the amount of starting genetic material. The method used to calculate the relative expression changes of two different samples/condition is the comparative Ct method or ΔΔCt method [7]. The comparative Ct method has the advantage of eliminating the need for a standard curve. However, before employing the comparative Ct method, a validation experiment must be carried out to demonstrate that the efficiencies of the target and

Table 1

Table showing the calculation of relative expression of *CAUR_01276* measured by the $2^{-\Delta\Delta CT}$ method in the presence of amphotericin B (0.7 μg/mL) and data calculated as mean ± S.E.

Ct of *TDH1* (Untreated)	Ct of *TDH1* (treated)	Ct of *CAUR_01276* (untreated)	Ct of *CAUR_01276* (treated)	ΔCt *TDH1* (untreated)	ΔCt *CAUR_01276* (treated)	ΔΔCt (ΔCt treated–ΔCt untreated)	Relative fold-change $2^{-\Delta\Delta Ct}$
14.212	14.523	23.335	21.598	9.123	7.075	−2.048	4.135
14.182	14.601	23.343	21.622	9.161	7.021	−2.140	4.407
14.127	14.555	23.304	21.643	9.177	7.088	−2.089	4.254
Mean ± S.E.							
14.174 ± 0.025	14.559 ± 0.023	23.327 ± 0.012	21.621 ± 0.013	9.154 ± 0.016	7.062 ± 0.020	−2.092 ± 0.027	4.266 ± 0.079

Adapted from a previous study [2]

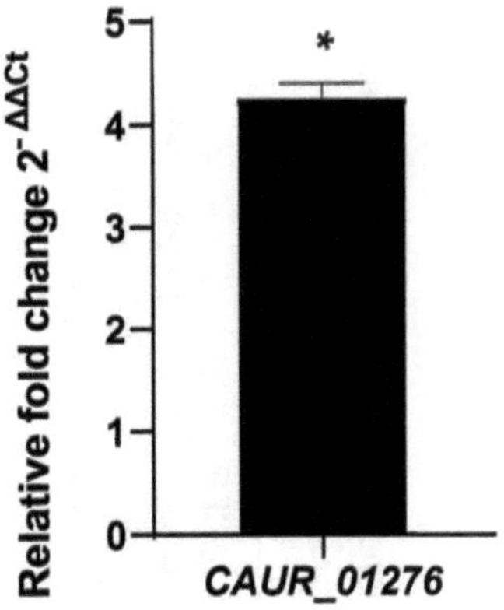

Fig. 3 Drug-induced expression profiles of *CAUR_01276* in *C. auris* checked by RT–qPCR. Data was measured by the $2^{-\Delta\Delta CT}$ method in the presence of amphotericin B (0.7 μg/mL). Data is presented as mean ± S.E. (error bars) from technical triplicates ($n = 3$)

endogenous control amplifications are nearly equal (as described in *see* Subheading 3.4.1).

The first step is to calculate the Ct value by subtracting the Ct value of the internal control (the example here is *TDH1*) from the Ct value of the gene of interest (e.g., *CAUR_01276*) in the treated (e.g., amphotericin B treatment) and untreated conditions, which gives the ΔCt values for the treated and untreated conditions, respectively. The next step is to find the ΔΔCt value by subtracting the ΔCt of the untreated sample from the treated one and finally putting the value of ΔΔCt in the formula $2^{-\Delta\Delta Ct}$ to get the relative fold-change in the expression of the gene (*CAUR_01276*) (Table 1 and Fig. 3). Formulas and details of calculations and steps are given below.

Step 1: ΔCt calculation for treated and untreated samples.

ΔCt treated = Ct (*CAUR_01276*)–Ct (*TDH1*) (amphotericin B-treated sample)

ΔCt untreated = Ct (*CAUR_01276*) – Ct (*TDH1*) (no drug treatment sample)

Step 2: ΔΔCt calculation

ΔΔCt = ΔCt treated–ΔCt untreated

Step 3: Relative fold-change/$2^{-\Delta\Delta Ct}$ calculation

Now relative fold-change can be obtained by 2 to the power of the negative ΔΔCt relative fold-change upon treatment = $2^{-\Delta\Delta Ct}$

4 Notes

1. The total growth time for the secondary culture is 7 h at 30 °C, in which the culture of *C. auris* CBS10913T strain reaches the mid-log phase. The time of secondary culture growth and drug treatment can be decided as per specific experimental requirements and the strain type.
2. It is recommended to prepare a fresh culture for RNA. However, the cell pellet can also be stored at −80 °C and can be used for up to 6–9 months for RNA preparation. RNA should be isolated in triplicates from three independently grown biological samples.
3. RNA isolation can be done via a manual method using TRIzol (available under different names from different manufacturers) or via a kit-based method. We use an isolation kit (RNeasy Mini Kit, Qiagen) because it is simple to use, saves time, and produces high-quality RNA for downstream applications. The kit includes the lysis solution, wash solution, elution buffer, and mini spin columns, as well as collection tubes and elution tubes.
4. Mechanical disruption with glass beads is a low-cost and effective method of yeast cell lysis. For efficient cell lysis, we use 2-mL microcentrifuge tube with flat bottom as the lysis is more efficient as compared to conical tubes. There are many disruptors available in market; we have used Fast Prep-24A Classic cell disrupter, MP Biomedicals for cell disruption. It is recommended to optimize the cycle time as per specific disruptor used.
5. The RNA isolated can be stored at −80 °C and used in the future after a quality check.
6. After DNase I treatment, the success of the reaction can be checked by setting a PCR reaction for amplification of a housekeeping gene using DNase I-treated and DNase-untreated sample as templates. The untreated sample should provide the amplification product which can be used as a PCR reaction control, while the DNase I treated sample should not produce any amplification product.
7. The targets of interest and primer location should be considered when deciding between oligo(dT) and random primers. Using only the random hexamer is not ideal for some RT PCR reaction because of over estimation of mRNA copy number [8]. Similarly, the use of only oligo(dT) also leads to 3′ bias in some targets. In two-step RT–PCR, a mixture of oligo(dT) and random primers is frequently used to maximize the benefits of each primer type. We have used both oligo(dT) and random hexamer primers together in the same reaction for cDNA

synthesis with reaction setup for incubation at 25 °C for 10 min, followed by another 60 min incubation at 42 °C. The reaction is then terminated by heating the reaction the mixture at 70 °C for 10 min.

8. cDNA can be aliquots in tubes and can be stored at −20 °C up to 1 month for further use.
9. Choose primers that will provide a uniform PCR product size of 150–200 bp having a T_m of 55–60 °C for annealing, a GC content of around 40–60% to ensure maximum product stability, and the pair with the least self-complementarity.
10. The region chosen for transcription analysis is normally the mid-region of the gene of interest.
11. To avoid amplification of intronic regions, blast these primers against each gene sequence using the NCBI tool primer blast, and those with binding sites only in the exonic region are chosen for RT–qPCR assay.
12. The selection of internal control housekeeping genes should be evaluated before subjecting samples for RT–qPCR. In the case of *C. auris*, under some conditions, a variation in the expression of housekeeping genes can be noticed.
13. RNA isolation should be performed with at least two biological replicates, and RT–qPCR should be performed in experimental triplicates.

References

1. Muñoz JF, Gade L, Chow NA et al (2018) Genomic insights into multidrug-resistance, mating and virulence in *Candida auris* and related emerging species. Nat Commun 9: 5346. https://doi.org/10.1038/s41467-018-07779-6
2. Wasi M, Khandelwal NK, Moorhouse AJ et al (2019) ABC transporter genes show upregulated expression in drug-resistant clinical isolates of Candida auris: a genome-wide characterization of ATP-binding cassette (ABC) transporter genes. Front Microbiol 10:1445. https://doi.org/10.3389/fmicb.2019.01445
3. Udvardi MK, Czechowski T, Scheible W-R (2008) Eleven golden rules of quantitative RT-PCR. Plant Cell 20:1736–1737
4. Wiame I, Remy S, Swennen R, Sági L (2000) Irreversible heat inactivation of DNase I without RNA degradation. BioTechniques 29:252–254, 256. https://doi.org/10.2144/00292bm11
5. Higuchi R, Dollinger G, Walsh PS, Griffith R (1992) Simultaneous amplification and detection of specific DNA sequences. Biotechnology (N Y) 10:413–417. https://doi.org/10.1038/nbt0492-413
6. Higuchi R, Fockler C, Dollinger G, Watson R (1993) Kinetic PCR analysis: real-time monitoring of DNA amplification reactions. Biotechnology (N Y) 11:1026–1030. https://doi.org/10.1038/nbt0993-1026
7. Livak KJ, Schmittgen TD (2001) Analysis of relative gene expression data using real-time quantitative PCR and the $2^{-\Delta\Delta CT}$ method. Methods 25:402–408. https://doi.org/10.1006/meth.2001.1262
8. Zhang J, Byrne CD (1999) Differential priming of RNA templates during cDNA synthesis markedly affects both accuracy and reproducibility of quantitative competitive reverse-transcriptase PCR. Biochem J 337(Pt 2):231–241

Chapter 12

Gas Chromatography–Mass Spectrometry (GC–MS) Analysis of *Candida auris* Metabolites

Bahgat Fayed, Ghalia Kohder, and Sameh S. M. Soliman

Abstract

Unique metabolic features allow fungi to colonize and persist within the human host. Investigations of unique metabolic fingerprints of a pathogenic fungus can provide a more complete understanding of the infection process and an interpretation of associations between genotype and phenotype. Gas chromatography–mass spectrometry (GC–MS) has proved to be one of the most powerful analytical techniques used for qualitative and quantitative detection of cellular metabolites. This technique has been used for comparative metabolomic analyses of both intracellular and secreted metabolites under variable conditions. This book chapter describes the use of GC–MS in the detection of both intracellular and secreted metabolites from *Candida auris*, a newly emerging fungal pathogen representing a serious global health threat due to its multidrug resistance profile. The identified fungal metabolites are compared using available software in order to assign a correlation between the pattern of accumulation of metabolites and behavior of the organism.

Key words Gas chromatography–mass spectrometry, GC–MS, *Candida auris*, Metabolomics, Metabolic profiling

1 Introduction

Metabolomic analysis is employed for qualitative and quantitative identification of the cellular metabolome including both intracellular and secreted metabolites. Thus, it can be used for comparative analysis and assigning a metabolic network in specific conditions. Metabolites are the end products that result from a cellular chemical reaction following several enzymatic interactions [1]. The recent advances in the use of analytical tools including, in particular, gas chromatography–mass spectrometry (GC–MS) [2, 3] and its integration with chemical libraries allow the detection and identification of metabolites at a large scale. Such data with the help of available software can provide biochemical information that can be used to better understand the cellular metabolic pathways

Alexander Lorenz (ed.), *Candida auris: Methods and Protocols*, Methods in Molecular Biology, vol. 2517,
https://doi.org/10.1007/978-1-0716-2417-3_12,

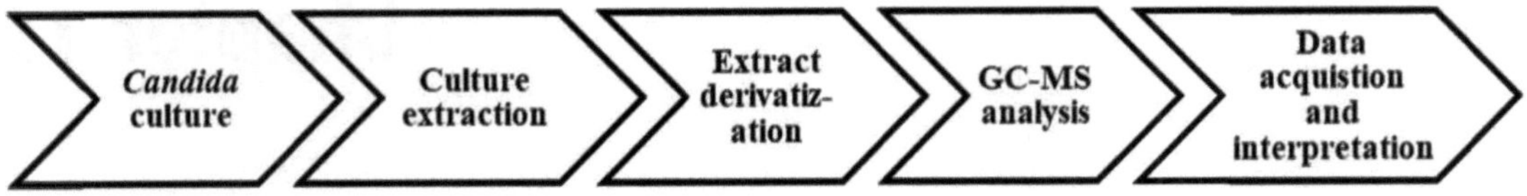

Fig. 1 Schematic representation of the GC–MS analysis procedure of *Candida auris*

enriched under specific conditions along with other biological responses.

GC–MS is a robust, rapid, easy, and very powerful technique used to establish a comprehensive qualitative and quantitative analysis of the complex metabolic network within cells [4]. GC separates the metabolites based on molecular weight, while MS helps in the identification of metabolites via their unique fragmentation patterns [5, 6]. Such mass fragmentation patterns have been accumulated and stored in libraries [7, 8] and made available for better computationally matching with the mass spectra of experimental data [5, 6]. The availability of automated mass spectral deconvolution software (AMDIS) has facilitated the identification process [9]. Procedure of GC–MS analysis of *C. auris* can be summarized as in Fig. 1.

GC–MS analysis has been employed for comparative metabolomics of *C. auris* [10]. Accumulation of intracellular metabolites and secreted metabolites has been compared to those identified from *C. albicans*. The pattern of accumulated metabolites indicated that production of specific metabolites is responsible for the inability of *C. auris* to form hyphae.

2 Material

1. *Obtain C. auris* clinical isolates from Centers for Disease Control and Prevention (CDC, Atlanta, GA, USA) [11]. *C. albicans* strain ATCC10231 can be used as reference.
2. Maintain the *Candida* isolates on yeast nitrogen base (YNB) supplemented with 2% glucose (*see* **Note 1**).
3. RPMI-1640 (*see* **Note 1**).
4. Neubauer-type hemocytometer.
5. Whatman filter paper (0.22 μm) (*see* **Note 2**).
6. 3-(4,5-Dimethylthiazol-2-yl)-2,5-diphenyl tetrazolium bromide (MTT) (*see* **Note 3**).
7. Anhydrous sodium sulfate (*see* **Note 4**).
8. Chloroform (*see* **Note 5**).
9. Methanol (*see* **Note 5**).
10. Ultrasonic water bath (*see* **Note 6**).

11. Rotatory evaporator (*see* **Note 7**).
12. *N*-trimethylsilyl-*N*-methyl trifluoroacetamide and trimethylchlorosilane (MSTFA +1% TMS) (*see* **Notes 8**, **9**, and **10**).
13. Pyridine.
14. Vortex mixer (*see* **Note 10**).
15. GC–MS machine equipped with an auto-injector [12] (*see* **Note 11**).
16. Chromatography column Restek Rtx-5ms (30-m length × 0.25-mm inner diameter × 0.25-μm film thickness; Restek, Bellefonte, PA) (*see* **Note 11**).
17. Helium gas (99.9% purity) in a cylinder (*see* **Note 12**).

3 Methods

3.1 Sample Preparation

1. Maintain the *Candida* isolates on yeast nitrogen base (YNB) supplemented with 2% glucose at 37 °C for 16 h (*see* **Note 1**).
2. Grow *Candida* isolates separately in 10 mL of RPMI-1640 in 125-mL culture flasks at inoculums of 5 × 10^6 yeast/mL (*see* **Note 1**) [13].
3. Incubate the flasks for 16 h at 37 °C.
4. Filter the supernatants prior to extraction using 0.22-μm Whatman filter paper.
5. Analyze the fungal viability using 3-(4,5-dimethylthiazol-2-yl)-2,5- diphenyl tetrazolium bromide (MTT), and measure the final absorbance at 540 nm (*see* **Note 3**).

3.2 Sample Extraction

1. Extract the supernatants with chloroform.
2. Dehydrate the chloroform layer over anhydrous sodium sulfate followed by evaporation using a rotatory evaporator (*see* **Notes 5** and 7).
3. Collect the *Candida* cells, and extract the metabolites by methanol and sonication using an ultrasonic water bath at 25 °C for 30 min (*see* **Notes 5** and **6**).
4. Filter the extract through 0.22-μm filters, and evaporate the solvent using a rotatory evaporator at 45 °C (*see* **Notes 5** and 7).

3.3 Derivatization

Derivatization is employed to enhance the volatility of metabolites and make them compatible with the GC conditions (*see* **Notes** 7, **8**, and **9**) [14].

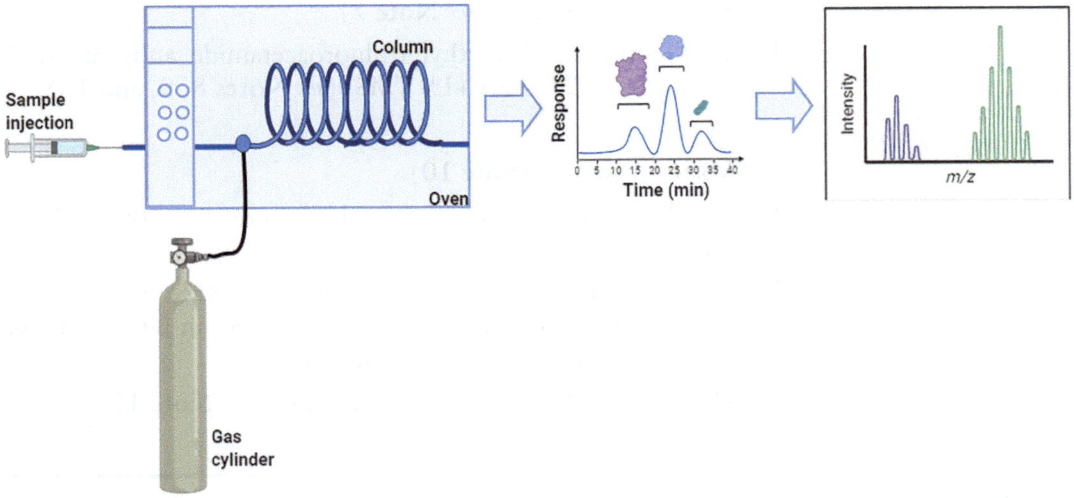

Fig. 2 Schematic drawing of GC procedure

1. Derivatize the dried residue from 100 μL of chloroform extract by silylation using 50 μL of *N*-trimethylsilyl-*N*-methyl trifluoroacetamide and trimethylchlorosilane (MSTFA +1% TMS) (*see* **Notes 7**, **8**, and **9**).
2. Dissolve in 100-μL pyridine by mixing for 2 min on a vortex mixer prior to incubation at 50 °C for 30 min.

3.4 GC–MS

1. Run the GC–MS analysis. Schematic drawing of the GC principle is summarized in Fig. 2 (*see* **Note 11**).
2. Adjust the helium gas to a flow rate of 1 mL/min (*see* **Note 12**).
3. Adjust the column temperature regime initially at 35 °C for 2 min, followed by an increase at a rate of 10 °C/min to reach 250 °C (*see* **Note 12**).
4. Increase the temperature by 20 °C/min until reaching 320 °C and maintain for 23 min.
5. Inject the samples at a 10-μL volume and at an injection temperature of 250 °C using splitless injection mode.
6. Operate the mass spectrometer in electron compact mode with electron energy of 70 eV.
7. Set both the ion source temperature and the interface temperature at 240 °C and 250 °C, respectively.
8. Set the MS mode to scan mode starting from 35 to 450 m/z with a scan speed of 1428.
9. Adjust the run time for each sample to 43.67 min.
10. Extract the data in the form of a chromatogram (Fig. 3).

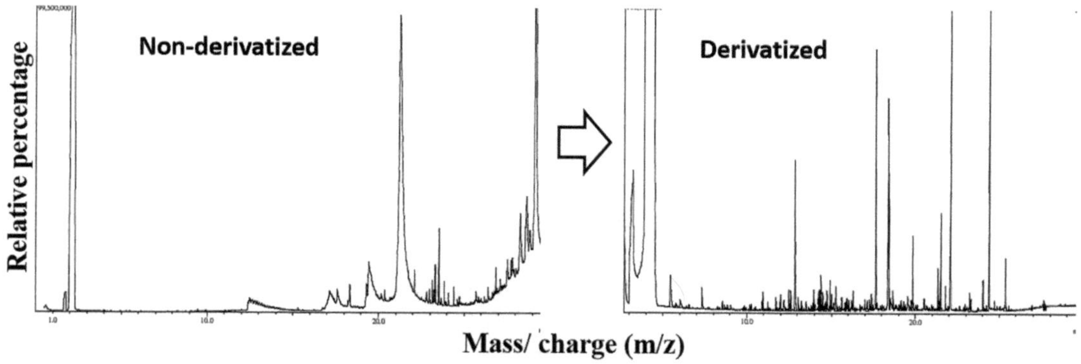

Fig. 3 GC–MS chromatograms of *Candida auris* extract before and after derivatization

3.5 Data Analysis

1. Collect the data and run the data analysis using built-in MSD Enhanced ChemStation software to identify the metabolites.
2. Identify the product spectra by comparison of the measured fragmentation patterns to those present in the NIST 08 Mass Spectral Library.

3.6 Data Interpretation and Software

1. Group the signals that belong to the same category of ions together, and normalize the data based on the total peak area.
2. Analyze the data using various bioinformatics algorithms as indicated below followed by mapping of the results to a biological role.

3.6.1 Excel

1. Quantify each metabolite as the relative percentage by measuring the area under the peak of each metabolite in relation to the total areas of all other metabolites detected in an extract. This is done automatically by the built-in ChemStation software (*see* **Note 13**).
2. Display the average relative percentage of metabolites in replicates in a heatmap ranging in colors based on fold changes.
3. Create a heatmap using conditional formatting and by employing color scales.

3.6.2 MetaboAnalyst

MetaboAnalyst is a comprehensive platform dedicated for metabolomics data analysis. The current version of MetaboAnalyst (V5.0) supports raw MS spectra processing and can provide several types of analysis including (I) analysis to identify the potential biomarkers by providing the receiver operating characteristic, (II) statistical meta-analysis to pinpoint robust biomarkers across multiple studies, (III) metabolite set enrichment analysis, (IV) metabolic pathway analysis, (V) joint pathway analysis by analyzing metabolites or genes within metabolic pathways, and (VI) functional/meta-analysis for untargeted metabolomic data generated from mass spectrometry. Further, MetaboAnalyst can provide a wide range of statistical and machine learning methods for data statistical analysis. The software tutorial can be accessed at https://www.metaboanalyst.ca/docs/Overview.xhtml.

3.6.3 R-package

R is a system used to compute statistics and graphs. It is a computer language, which allows branching, looping, and modular programming using computer programming functions. It can provide a wide variety of statistical analyses including linear and generalized linear models, nonlinear regression models, time series analysis, classical parametric and nonparametric tests, clustering, smoothing, and graphical techniques. Further, it can produce well-designed publication-quality plots. Overall, the R-package can offer the following, (I) effective data handling and the ability to calculate arrays in matrices, (II) a wide range of intermediate tools for data analysis, and (III) graphical facilities for data analysis and an effective programming language which includes conditionals, loops, user-defined recursive functions and input and output facilities.

Use the R-package to:

1. Generate Principal Component Analysis (PCA) by employing R programming language using the *prcomp* library.
2. Carry out unsupervised hierarchical clustering using R programming language to explore possible grouping and patterns in metabolic coverage generated from the extract.
3. Identify the enrichment of metabolites in a metabolic pathway compared to the total of annotated metabolites in the same pathway using KEGG.

3.7 Statistical Analysis

1. Collect the data graph using either Excel to generate the heatmap or GraphPad Prism (5.04, La Jolla, CA, USA) for Windows to generate box and whisker plots and group comparison.
2. Analyze the quantity of metabolites in the extracts by one-way analysis of variance (ANOVA) using Bonferroni's multiple comparison test or two-way analysis of variance.
3. Consider P value <0.05 as significant.

4 Notes

1. Both YNB and RPMI-1640 culture media are commercially available.
2. Filtration of the metabolite extracts is important to avoid any clogging of the machine or the injector.

3. Fungal viability should be correlated with the metabolic activity. Fungal viability can be measured using 3-(4,5-dimethylthiazol-2-yl)-2,5-diphenyl tetrazolium bromide (MTT).
4. Solvent extracts of metabolites should be filtered through anhydrous sodium solvent to get rid of any residual water.
5. HPLC-grade solvents are recommended to avoid any residual plasticizer.
6. The use of an ultrasonic water bath enhances the extraction of intracellular metabolites.
7. Solvents including chloroform and methanol need to be completely evaporated.
8. The derivatizing solution is commercially available.
9. Derivatization is important to modify the structure of nonvolatile metabolites to volatile and thus to become adapted to the GC conditions.
10. Vortexing and heating following derivatization enhance the derivatization and detection of metabolites by GC.
11. These conditions of GC–MS can be employed on a GC–MS-QP2010 Ultra equipped with (AOC-20i+s) auto-injector (Shimadzu, Kyoto, Japan) [12]. However, GC–MS machines from other companies still can be used following minor modifications of the conditions.
12. Adjusting the flow rate of helium gas and temperature of the column is important for the separation of metabolite peaks.
13. Excel software can be used to generate simple statistical analysis and graphs including heatmaps. However, more representative graphs and enriched metabolic pathways require using MetaboAnalyst and/or the R-package.

References

1. Wishart DS, Tzur D, Knox C, Eisner R, Guo AC, Young N, Cheng D, Jewell K, Arndt D, Sawhney S, Fung C, Nikolai L, Lewis M, Coutouly M-A, Forsythe I, Tang P, Shrivastava S, Jeroncic K, Stothard P, Amegbey G, Block D, Hau DD, Wagner J, Miniaci J, Clements M, Gebremedhin M, Guo N, Zhang Y, Duggan GE, MacInnis GD, Weljie AM, Dowlatabadi R, Bamforth F, Clive D, Greiner R, Li L, Marrie T, Sykes BD, Vogel HJ, Querengesser L (2007) HMDB: the human metabolome database. Nucleic Acids Res 35(suppl_1): D521–D526. https://doi.org/10.1093/nar/gkl923
2. Lee MS, Jung BH, Chung BC, Cho SH, Kim KY, Kwon OS, Nugraha B, Lee Y-J (2009) Metabolomics study with gas chromatography–mass spectrometry for predicting valproic acid–induced hepatotoxicity and discovery of novel biomarkers in rat urine. Int J Toxicol 28(5):392–404. https://doi.org/10.1177/1091581809340329
3. Tan B, Qiu Y, Zou X, Chen T, Xie G, Cheng Y, Dong T, Zhao L, Feng B, Hu X, Xu LX, Zhao A, Zhang M, Cai G, Cai S, Zhou Z, Zheng M, Zhang Y, Jia W (2013) Metabonomics identifies serum metabolite markers of colorectal cancer. J Proteome Res 12(6): 3000–3009. https://doi.org/10.1021/pr400337b

4. Faber JH, Malmodin D, Toft H, Maher AD, Crockford D, Holmes E, Nicholson JK, Dumas ME, Baunsgaard D (2007) Metabonomics in diabetes research. J Diabetes Sci Technol 1(4):549–557
5. Ma Y, Kind T, Yang D, Leon C, Fiehn O (2014) MS2Analyzer: a software for small molecule substructure annotations from accurate tandem mass spectra. Anal Chem 86(21): 10724–10731. https://doi.org/10.1021/ac502818e
6. Fiehn O, Kopka J, Trethewey RN, Willmitzer L (2000) Identification of uncommon plant metabolites based on calculation of elemental compositions using gas chromatography and quadrupole mass spectrometry. Anal Chem 72(15):3573–3580. https://doi.org/10.1021/ac991142i
7. Babushok VI, Linstrom PJ, Reed JJ, Zenkevich IG, Brown RL, Mallard WG, Stein SE (2007) Development of a database of gas chromatographic retention properties of organic compounds. J Chromatogr A 1157(1): 414–421. https://doi.org/10.1016/j.chroma.2007.05.044
8. Horai H, Arita M, Kanaya S, Nihei Y, Ikeda T, Suwa K, Ojima Y, Tanaka K, Tanaka S, Aoshima K, Oda Y, Kakazu Y, Kusano M, Tohge T, Matsuda F, Sawada Y, Hirai MY, Nakanishi H, Ikeda K, Akimoto N, Maoka T, Takahashi H, Ara T, Sakurai N, Suzuki H, Shibata D, Neumann S, Iida T, Tanaka K, Funatsu K, Matsuura F, Soga T, Taguchi R, Saito K, Nishioka T (2010) MassBank: a public repository for sharing mass spectral data for life sciences. J Mass Spectrom 45(7):703–714. https://doi.org/10.1002/jms.1777
9. Kind T, Wohlgemuth G, Lee DY, Lu Y, Palazoglu M, Shahbaz S, Fiehn O (2009) FiehnLib: mass spectral and retention index libraries for metabolomics based on quadrupole and time-of-flight gas chromatography/mass spectrometry. Anal Chem 81(24): 10038–10048. https://doi.org/10.1021/ac9019522
10. Semreen MH, Soliman SSM, Saeed BQ, Alqarihi A, Uppuluri P, Ibrahim AS (2019) Metabolic profiling of Candida auris, a newly-emerging multi-drug resistant Candida species, by GC-MS. Molecules 24(3):399
11. Alketbi EH, Hamdy R, El-Kabalawy A, Juric V, Pignitter M, Mosa KA, Almehdi AM, El-Keblawy AA, Soliman SSM (2021) Lipid-based therapies against SARS-CoV-2 infection. Rev Med Virol 31(5):1–13. https://doi.org/10.1002/rmv.2214
12. https://www.shimadzu.com/an/sites/shimadzu.com.an/files/pim/pim_document_file/brochures/10435/jpz19014.pdf
13. Ibarra-Trujillo C, Villar-Vidal M, Gaitán-Cepeda LA, Pozos-Guillen A, Mendoza-de Elias R, Sánchez-Vargas LO (2012) Ensayo de formación y cuantificación de biopelículas mixtas de Candida albicans y Staphylococcus aureus. Rev Iberoam Micol 29(4):214–222
14. DeJongh DC, Radford T, Hribar JD, Hanessian S, Bieber M, Dawson G, Sweeley CC (1969) Analysis of trimethylsilyl derivatives of carbohydrates by gas chromatography and mass spectrometry. J Am Chem Soc 91(7): 1728–1740. https://doi.org/10.1021/ja01035a022

Chapter 13

Isolation of Extracellular Vesicles from *Candida auris*

Daniel Zamith-Miranda, Lysangela R. Alves, Marcio L. Rodrigues, Leonardo Nimrichter, and Joshua D. Nosanchuk

Abstract

Extracellular vesicles (EVs) are structures released by a variety of cells from all kingdoms of life. EVs are typically involved in communication between tissues and organs, between distinct organisms, or inside microbial communities. The plasticity of these structures is reflected in the range of biological effects they are able to induce or inhibit. The study of fungal EVs is relatively new with the first report in 2007, but investigators have already demonstrated in several model systems that fungal EVs significantly modulate the host immune system and that the immunogenic materials in EV can be harnessed as vaccination platforms. This chapter describes the two main procedures used to isolate EVs from an emerging pathogenic fungus, *Candida auris*.

Key words Extracellular vesicles, *Candida auris*, Fungal, Culture

1 Introduction

Extracellular vesicles (EVs) have been isolated from an incredibly diverse range of cell types including bacteria, parasites, fungi, and mammalian cells [1–3]. Their roles, as experimentally addressed in distinct models, reveal that these structures are important players in the communication between tissues and organs, between distinct organisms, or inside microbial communities [1]. Fungal EVs carry a wide array of different classes of molecules capable of promoting the induction or inhibition of the immune response by the host [4, 5]. This document describes validated methods for isolating suspensions enriched in EVs from *Candida auris* utilizing two distinct methodologies (Fig. 1). Fungal EVs have been isolated from fungi grown in liquid media since their first description in 2007 [6]. Recently, the method evolved to an easier, high-yield procedure in which EVs are isolated from cultures grown in solid media [7]. Some steps are common to both methods, which have their pros and cons. Both methods are extremely effective for

Alexander Lorenz (ed.), *Candida auris: Methods and Protocols*, Methods in Molecular Biology, vol. 2517,
https://doi.org/10.1007/978-1-0716-2417-3_13,

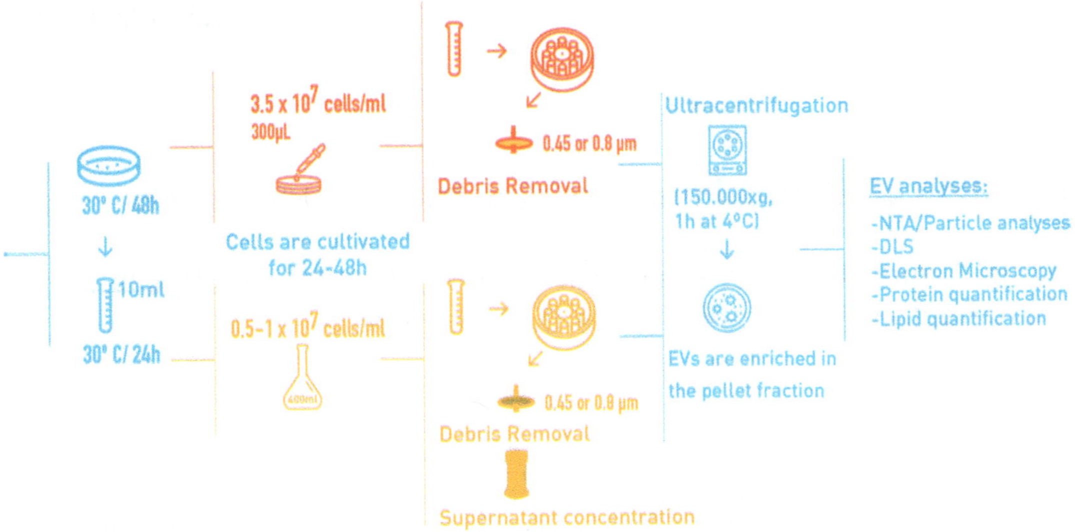

Fig. 1 Methods for isolation of extracellular vesicles from *C. auris*. The figure shows the procedures for the isolation of EVs from *C. auris* cultures in liquid and solid media

isolating fungal EVs, and the procedure of choice depends on the experimental conditions. Whereas the liquid media method requires space in shakers for big flasks and additional equipment (concentration system), it also allows the use of treatments with antifungals or any other compound of choice, including antibodies. On the other hand, the solid media procedure has fewer steps and only requires basic cell culture material. However, the addition of any treatment or conditions can prove troublesome for this method. To date, differences in cargo of EVs from liquid and solid media have not been addressed for *C. auris*, but they are expected as nutritional conditions can alter EV production in other fungi [8, 9].

2 Materials

1. Sabouraud agar, Sabouraud broth, and YPD agar: mix the amount of powdered reagent (Sabouraud or YPD) recommended by the manufacturer in deionized water, and bring it to a boil to ensure solubility. Autoclave and wait until it cools down.
2. PBS: 137-mM NaCl, 2.7-mM KCl, 10-mM Na_2HPO_4, 1.8-mM KH_2PO_4, pH 7.4. After mixing these reagents with deionized water, the solution should be sterilized by filtration to a new flask, using a 0.22-μm pore filter, or autoclaved.
3. Filtration system with membrane pores of 0.22, 0.45, and 0.8 μm.

4. Petri dishes: 15-cm diameter.
5. Inoculation loop: volume – 10 μl.
6. Conical tube: volume – 50 mL.
7. Neubauer chamber/hemocytometer.
8. Sterile Erlenmeyer with foam plug.
9. Incubator and shaker with temperature control.
10. Conical tube centrifuge preferably with temperature control.
11. Concentration system with 100 KDa membrane cutoff.
12. Ultracentrifuge tubes.
13. Ultracentrifuge: preferably with a fixed angle rotor.
14. Microtubes.
15. Nanoparticle tracking analysis (NTA) system or dynamic light scattering (DLS) system.
16. Sterol and/or protein quantification kits.

3 Methods

3.1 Establishing the Cultures and Pre-Inoculum Preparation

1. *C. auris* frozen stock cultures are plated onto Sabouraud agar or YPD agar plates and incubated at 30 °C for 48 h.
2. Once the colonies are grown, the plates can typically be stored at 4 °C for up to 1 month without compromising the viability of the yeast cells.
3. To prepare the pre-inoculum, one colony from the plates is transferred with an inoculation loop to a conical tube containing 10 mL of Sabouraud broth. Cells are incubated at 30 °C for 24 h.

From this point, the protocols for liquid and solid media are different, and follow either Subheadings 3.1.1 or 3.1.2 (*see* **Note 1**).

3.1.1 EV Isolation from Liquid Media

1. After 24 h, the cell suspensions in the pre-inoculum tubes develop into a turbid solution.
2. Cells are centrifuged (1500 × *g* at room temperature) and washed with PBS. After repeating this step twice, the cells are suspended in 10 mL of Sabouraud broth and counted with a hemocytometer.
3. Cells are transferred to an Erlenmeyer flask containing 400 mL of Sabouraud broth (media of choice may vary), reaching a cell density of 0.5–1 × 10^7 cells/mL (*see* **Note 2**).
4. Cultures can be incubated at either 30 °C or 37 °C, for 24 h or 48 h, depending on the application and the growth of the strain used, respectively (*see* **Note 3**).

5. Cells are centrifuged at 5000 × *g* for 15 min at 4 °C, and the yeast pellets are discarded, whereas the supernatant containing the EVs is collected.
6. The supernatant, clear of cells, is filtered through a 0.45- to 0.8-μm membrane to remove cellular debris.
7. The filtered supernatant is concentrated (20–50×) in a concentration system using a membrane with a 100 KDa cutoff (e.g., Amicon stirred cells). This step is important as it assures that the supernatant volume can fit in ultracentrifuge tubes.
8. After reaching the desired volume, the supernatant is loaded into ultracentrifuge tubes.

3.1.2 EV Isolation from Solid Media

1. After 24 h, the cell suspensions in the pre-inoculum tubes develop into a turbid solution.
2. Cells are centrifuged (1500 × *g* at room temperature), washed with PBS, and a cell suspension with a cell density of 3.5×10^7 cells/mL is prepared.
3. Cells are plated onto Sabouraud agar plates by spreading 300 μL of the cell suspension on each plate (prepare at least eight plates).
4. The plates are incubated at 30 °C or 37 °C for 48 h.
5. After lawn formation, cells are gently harvested from the plates using an inoculation loop and transferred to conical tubes containing PBS passed through a 0.22-μm pore size filter.
6. Cell suspensions are centrifuged at 5000 × *g* for 15 min at 4 °C.
7. The supernatant is collected and filtered through a 0.45- to 0.8-μm membrane to remove cellular debris. The filtered samples are then transferred to ultracentrifuge tubes.

3.2 EV Sedimentation Following Collection of EVs from Either Liquid or Solid Media

1. After loading the samples into ultracentrifuge tubes, they are spun at 150,000 × *g* for 1 h at 4 °C.
2. The supernatant is discarded, and the pellet containing EVs is suspended in PBS (washing step) and centrifuged again (150,000 × *g* for 1 h at 4 °C).
3. After two cycles of spin and wash (as in **steps 1** and **2** above), EVs are suspended in PBS (around 200 μL) and stored in microtubes at 4 °C until use (*see* **Notes 4** and **5**).

3.3 Quantification and Visualization of EVs

1. The quantification of EVs can be achieved by both direct and indirect measurements.
2. As a direct method, nanoparticle tracking analysis (NTA) has been most extensively validated, and it requires specialized equipment, such as a particle analyzer [7].

3. Indirect measurements used so far are based on the quantification of structural molecules, known to be present in EVs, such as total protein (content) or total sterol (surface, EV membrane) [10].
4. Transmission electron microscopy (TEM) can be used to visualize EVs, measure their size, and also evaluate the presence of potential pigments [6, 11].
5. As an alternative method for addressing EV size, dynamic light scattering (DLS) can be used. This technique estimates the hydrodynamic radius (or diameter) of particles based on the Brownian motion in solution [12, 13].

4 Notes

1. Depending on the analysis to be performed, blank controls (medium without cells, processed in the same way as the yeast cell cultures) may be necessary. For example, for compositional analysis of EVs by mass spectrometry or RNA sequencing, mock preparations from the sterile medium alone are recommended for appropriate data interpretation and avoidance of the classification of medium contaminants as EV components.
2. The final volume of cell suspension may vary. This protocol suggests 400 mL, but volumes from 100 to 2000 mL can be used depending on the amount of EVs needed.
3. When optimizing yeast cell culture for EV isolation, it is important to monitor the growth to be sure that EVs will be isolated from a culture during logarithmic phase of growth, because strain variations in growth rate are expected.
4. Experimentation should be performed with freshly isolated EVs. EVs can usually be stored at 4 °C for up to a week, but strain-dependent variations can occur. For RNA-seq or mass spectrometry analyses, EVs can be stored (using the recommended buffer for each technique) at −80 °C until the moment of use.
5. After each isolation, a small amount of EV suspension (around 10 μL) should be plated onto a YPD agar plate to check for contaminations.

Acknowledgments

The authors thank Vanessa Koiky for helping with the figure preparation.

References

1. Kalluri R, LeBleu VS (2020) The biology, function, and biomedical applications of exosomes. Science 367(6478):eaau6977. https://doi.org/10.1126/science.aau6977
2. Brown L, Wolf JM, Prados-Rosales R, Casadevall A (2015) Through the wall: extracellular vesicles in Gram-positive bacteria, mycobacteria and fungi. Nat Rev Microbiol 13(10): 620–630. https://doi.org/10.1038/nrmicro3480
3. Mantel PY, Marti M (2014) The role of extracellular vesicles in *Plasmodium* and other protozoan parasites. Cell Microbiol 16(3): 344–354. https://doi.org/10.1111/cmi.12259
4. Zamith-Miranda D, Nimrichter L, Rodrigues ML, Nosanchuk JD (2018) Fungal extracellular vesicles: modulating host-pathogen interactions by both the fungus and the host. Microbes Infect 20(9–10):501–504. https://doi.org/10.1016/j.micinf.2018.01.011
5. Rodrigues ML, Godinho RM, Zamith-Miranda D, Nimrichter L (2015) Traveling into outer space: unanswered questions about fungal extracellular vesicles. PLoS Pathog 11(12):e1005240. https://doi.org/10.1371/journal.ppat.1005240
6. Rodrigues ML, Nimrichter L, Oliveira DL, Frases S, Miranda K, Zaragoza O, Alvarez M, Nakouzi A, Feldmesser M, Casadevall A (2007) Vesicular polysaccharide export in *Cryptococcus neoformans* is a eukaryotic solution to the problem of fungal trans-cell wall transport. Eukaryot Cell 6(1):48–59. https://doi.org/10.1128/ec.00318-06
7. Reis FCG, Borges BS, Jozefowicz LJ, Sena BAG, Garcia AWA, Medeiros LC, Martins ST, Honorato L, Schrank A, Vainstein MH, Kmetzsch L, Nimrichter L, Alves LR, Staats CC, Rodrigues ML (2019) A novel protocol for the isolation of fungal extracellular vesicles reveals the participation of a putative scramblase in polysaccharide export and capsule construction in *Cryptococcus gattii*. mSphere 4(2): e00080–e00019. https://doi.org/10.1128/mSphere.00080-19
8. Marina CL, Bürgel PH, Agostinho DP, Zamith-Miranda D, Las-Casas LO, Tavares AH, Nosanchuk JD, Bocca AL (2020) nutritional conditions modulate *c. neoformans* extracellular vesicles' capacity to elicit host immune response. microorganisms 8(11): 1815. https://doi.org/10.3390/microorganisms8111815
9. Bürgel PH, Marina CL, Saavedra PHV, Albuquerque P, de Oliveira SAM, Veloso Janior PHH, de Castro RA, Heyman HM, Coelho C, Cordero RJB, Casadevall A, Nosanchuk JD, Nakayasu ES, May RC, Tavares AH, Bocca AL (2020) *Cryptococcus neoformans* secretes small molecules that inhibit IL-1β Inflammasome-dependent secretion. Mediat Inflamm 2020:3412763. https://doi.org/10.1155/2020/3412763
10. Cleare LG, Zamith D, Heyman HM, Couvillion SP, Nimrichter L, Rodrigues ML, Nakayasu ES, Nosanchuk JD (2020) Media matters! Alterations in the loading and release of *Histoplasma capsulatum* extracellular vesicles in response to different nutritional milieus. Cell Microbiol 22(9):e13217. https://doi.org/10.1111/cmi.13217
11. Albuquerque PC, Nakayasu ES, Rodrigues ML, Frases S, Casadevall A, Zancope-Oliveira RM, Almeida IC, Nosanchuk JD (2008) Vesicular transport in *Histoplasma capsulatum*: an effective mechanism for trans-cell wall transfer of proteins and lipids in ascomycetes. Cell Microbiol 10(8):1695–1710. https://doi.org/10.1111/j.1462-5822.2008.01160.x
12. Baltazar LM, Zamith-Miranda D, Burnet MC, Choi H, Nimrichter L, Nakayasu ES, Nosanchuk JD (2018) Concentration-dependent protein loading of extracellular vesicles released by *Histoplasma capsulatum* after antibody treatment and its modulatory action upon macrophages. Sci Rep 8(1):8065. https://doi.org/10.1038/s41598-018-25665-5
13. Vargas G, Rocha JD, Oliveira DL, Albuquerque PC, Frases S, Santos SS, Nosanchuk JD, Gomes AM, Medeiros LC, Miranda K, Sobreira TJ, Nakayasu ES, Arigi EA, Casadevall A, Guimaraes AJ, Rodrigues ML, Freire-de-Lima CG, Almeida IC, Nimrichter L (2015) Compositional and immunobiological analyses of extracellular vesicles released by *Candida albicans*. Cell Microbiol 17(3): 389–407. https://doi.org/10.1111/cmi.12374

Chapter 14

Labelling of *Candida auris* Cell Walls to Examine Levels of PAMP Exposure by Flow Cytometry and Fluorescence Microscopy

Chloe Pelletier and Alexander Lorenz

Abstract

Pathogen-associated molecular patterns (PAMPs) of the fungal cell wall are primary targets for the innate immune system of animals. Therefore, characterizing PAMP exposure of fungal pathogens helps to elucidate how they interact with their hosts at a molecular level. Fluorescent labelling can be used to monitor exposure of multiple fungal cell wall PAMPs in a single experiment. Here, we describe a protocol to simultaneously label chitin, mannan, and β-1,3-glucan in *Candida auris* to study these PAMPs by fluorescence microscopy and allow high-throughput examination of their exposure by flow cytometry.

Key words *Candida auris*, Flow cytometry, Fluorescence microscopy, PAMPs, Cell wall, Chitin, Mannan, β-1,3-Glucan

1 Introduction

Candida auris was first isolated from an ear infection in 2009 [1]. Since then, it has been responsible for life-threatening candidemia and outbreaks in intensive care units, with high levels of antifungal resistance (often against multiple drugs) being a hallmark of this pathogen [2]. Yet the infection process is still largely uncharacterized. To help understand the cellular features of infection progression, the fungal cell wall can be examined for the presence of pathogen-associated molecular patterns (PAMPs), which are recognized by pattern recognition receptors (PRR) on host immune cells during infection [3]. The cell wall of *C. auris* has the same basic components as that of many other fungal species, including other *Candida* species [4–6]. To better understand the host–pathogen interaction of *C. auris*, exposure of PAMPs on the surface of the cell wall can be studied.

Alexander Lorenz (ed.), *Candida auris: Methods and Protocols*, Methods in Molecular Biology, vol. 2517, https://doi.org/10.1007/978-1-0716-2417-3_14,

One method to study exposed PAMPs on the cell surface is to use flow cytometry. It is an excellent high-throughput tool for analysis of multiple molecules of interest. The technique uses a fluidic system to analyze cell populations by measuring light scatter and fluorescence of single cells passing a laser and a detector [7]. Generally, measurements of a molecule of interest are produced by conjugating selected fluorophores to agents that bind the molecule of interest, such as lectins or pattern recognition receptors, and detecting this fluorescence. The median fluorescent intensity for a cell population can be calculated and used to compare one experimental condition to another due to fluorescent intensity correlating with level of labelling of the target [8]. Flow cytometry is a technique that has been utilized effectively in immunophenotyping, cell cycle studies, proliferation assays, and many more applications [7]. While flow cytometry allows the cell population to be examined, a complementary technique is to image individual cells with fluorescence microscopy [9]. Using flow cytometry avoids sample size biases that must be considered for microscopy, while fluorescence microscopy allows visualization of exposure at the single-cell level to distinguish between bud scaring and changes in overall PAMP exposure. Fluorescence microscopy has already been used to look at localization of cell wall constituents and used to examine remodelling of the cell wall [10, 11].

For fungal cells, such as *Candida albicans*, flow cytometry has been used to analyze exposed cell wall PAMPs, notably β-1,3-glucan, a key cell wall structure [12, 13]. Other cell wall PAMPs that influence immune cell recognition of fungal pathogens, such as *Candida* species, *Aspergillus fumigatus*, and *Cryptococcus neoformans*, include mannan and chitin [3]. With changes in PAMP exposure and cell wall architecture documented in response to infection conditions, cellular stress induced by the host, and antifungal treatment, it is advantageous to be able to examine and quantify these changes [3, 12–15].

The labelling protocol presented here measures the exposure of β-1,3-glucan, mannan, and chitin on the cell surface of *C. auris*, e.g., this can be used to characterize how differential PAMP exposure could change interactions of the fungal cell with the immune system of a host. β-1,3-glucan exposure is detected using a recombinant protein of the C-terminal extracellular domain of the human pattern recognition receptor Dectin-1 fused to an Fc domain allowing a secondary antibody conjugated to a fluorophore to bind. Exposed chitin is recognized by a lectin called wheat germ agglutinin. In a similar approach, exposed mannans can be detected by the lectin concanavalin A. Labelling of these PAMPs with pattern recognition receptors and lectins enables the study of changes on the cell surface, between populations. *C. auris* cells labelled for these PAMPs can then be analyzed by fluorescence microscopy or flow cytometry.

2 Materials

All solutions should be prepared with sterile, ultrapure, deionized water (ddH_2O) unless stated otherwise and analytical grade reagents. Follow local regulations for disposal of liquid and solid waste.

1. 100-mM thimerosal: Dissolve 2 g in in 50 mL of ddH_2O (*see* **Note 1**).
2. FACS wash: 900 mL of 1× phosphate-buffered saline (PBS), add 100 mL of fetal bovine serum (FBS), and add 0.146-g ethylenediaminetetraacetic acid (EDTA).
3. Fc-fusion protein of the human Dectin-1 pattern recognition receptor to human IgG Fc domain (*see* **Note 2**).
4. Suitable anti-human IgG Fc secondary antibody conjugated to a fluorophore of choice (*see* **Note 3**).
5. Wheat germ agglutinin conjugated to a suitable fluorophore.
6. Concanavalin A conjugated to a suitable fluorophore.
7. Vortex mixer.
8. Hemocytometer (type Neubauer-improved).
9. Fluorescence microscope equipped with filter sets enabling the imaging of the chosen fluorophores and a 100× objective lens.
10. Anti-fade mounting medium.
11. Clear nail polish.
12. Immersion oil for microscopy.
13. Poly-L-Lysine-coated microscopy slides.
14. Cover glasses 22 mm × 50 mm.
15. Flow cytometer capable of detecting the chosen fluorophores.
16. Flow cytometry analysis software.

3 Methods

3.1 Labelling Cells

1. Grow cells in the desired condition(s).
2. Harvest cells by centrifugation 3800 × *g* for 5 min.
3. Retain cell pellet and approximately 200 μL of residual media.
4. Resuspended cells in an equal amounts of 100-mM thimerosal, giving a final thimerosal concentration of 50 mM (*see* **Note 3**).
5. Incubate for at least 16 h in the dark at room temperature (*see* **Note 4**).
6. Centrifuge cells for 5 min at 2400 × *g*.
7. Discard the supernatant containing the thimerosal (follow local waste regulations).

8. Resuspend cells in 500 μL of 1× PBS by using a vortex mixer.
9. Centrifuge cells for 5 min at 2400 × *g*.
10. Retain cell pellet and discard supernatant.
11. Repeat **steps 8–10** above, two more times to ensure that no residual thimerosal is present.
12. Resuspend cell pellet in 500 μL of 1× PBS using a vortex mixer (*see* **Note 6**).
13. Determine cell concentration using a hemocytometer.
14. Add 5×10^6 cells to a 1.5-mL reaction tube.
15. Centrifuge cells at 2400 × *g* for 5 min; remove excess 1× PBS.
16. Resuspend cells in 200 μL of ice-cold FACS wash using a vortex mixer.
17. Pellet cells by centrifugation at 2400 × *g* for 5 min, and remove FACS wash.
18. Add to the pellet 100 μL of ice-cold FACS wash with human Fc–Dectin fusion protein (*see* **Note** 7).
19. Resuspend cells using a vortex mixer.
20. Incubate on ice for 45 min (*see* **Note 8**).
21. After incubation, pellet cells by centrifugation at 2400 × *g* for 5 min (*see* **Note 9**).
22. Discard supernatant, and resuspend pellet in 200 μL of ice-cold FACS wash using vortex mixer.
23. Pellet cells by centrifugation at 2400 × *g* for 5 min; discard supernatant.
24. Repeat **steps 22–23** above twice.
25. Resuspend cells in 100-μL ice-cold FACS wash with 1:200 secondary antibody conjugated to a fluorophore, 50-μg/mL wheat germ agglutinin fluorophore conjugate, 25 μg/mL of concanavalin A fluorophore conjugate (*see* **Note 10**).
26. Incubate on ice in the dark for 30 min.
27. Repeat **steps 21–24** above.
28. Cells are ready to be prepared for microscopy (*see* Subheading 3.2) or flow cytometry (*see* Subheading 3.3).

3.2 Microscopy

1. Cells should be resuspended by pipetting 10-μL FACS wash up and down gently (*see* **Note 11**).
2. Add 10 μL of anti-fade mix by pipetting up and down slowly to mix cells.
3. Add 10 μL to a poly-L-lysine slide.
4. Place cover glass and gently press down to remove air bubbles.
5. Seal cover glass in place with clear nail polish.
6. Image straight away or store at 4 °C in the dark.

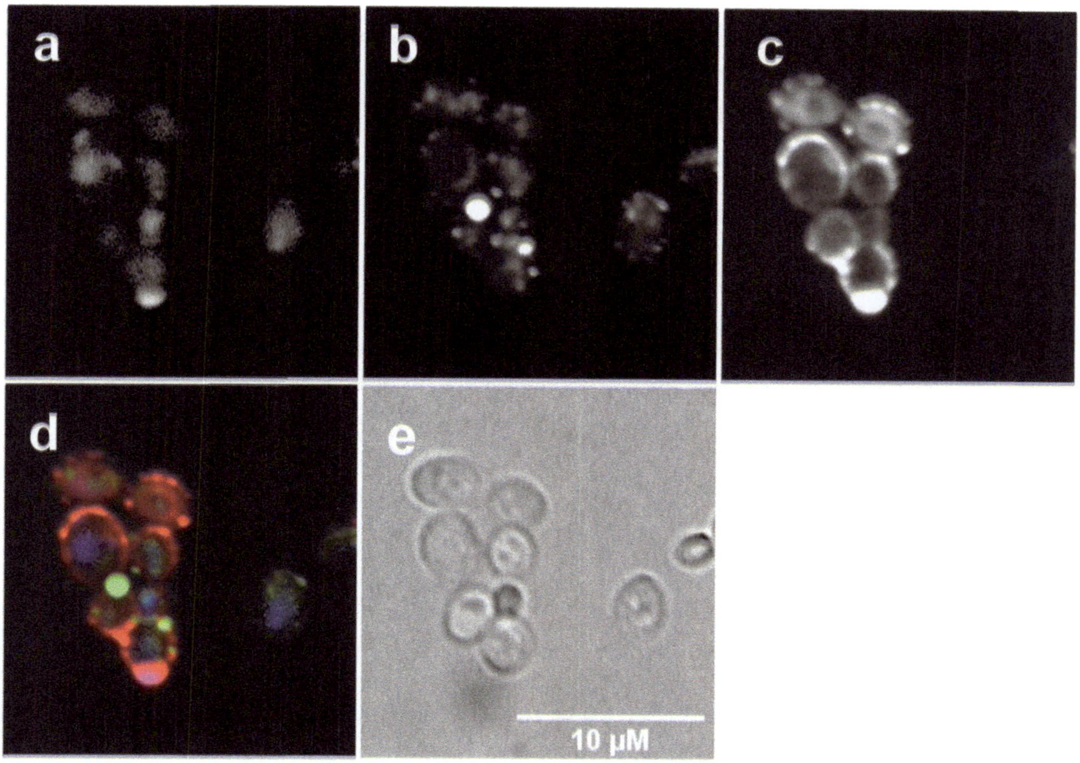

Fig. 1 Labelling of exposed *C. auris* cell wall PAMPs imaged by confocal microscopy. (**a–c**) Individual images of the fluorescent labelling of the given PAMP: (**a**) chitin labelled with wheat germ agglutinin Alexa Fluor 350 conjugate, (**b**) β-1,3-glucan labelled with Fc-hDectin-1a pattern recognition receptor and F(ab')2-goat anti-human IgG Fc secondary antibody conjugated to Alexa Fluor 488, (**c**) mannans labelled with red concanavalin A, Texas Red conjugate. (**d**) Pseudo-colored merged image of (**a–c**) with chitin in blue, β-1,3-glucan in green and mannans in red. (**e**) Differential interference contrast (DIC) image of the cells in (**a–c**)

7. Image cells using a 100× objective lens with oil immersion (*see* **Note 12**).
8. Resulting images show where exposure on the cell wall is occurring (Fig. 1).

3.3 Flow Cytometry

1. Resuspend cells in 300 μL of ice-cold FACS wash, and transfer to flow cytometry tube (*see* **Note 13**).
2. Cells are then ready to be run on a flow cytometer (*see* **Note 14**).
3. Using appropriate controls, set up laser power and threshold for cells (*see* **Note 15**).
4. Measure 50,000 events.
5. Transfer data to preferred software of analysis.
6. Gate for singlets and proceed with this single-cell population; this is done by analysis forward scatter height against forward scatter area (Fig. 2a).

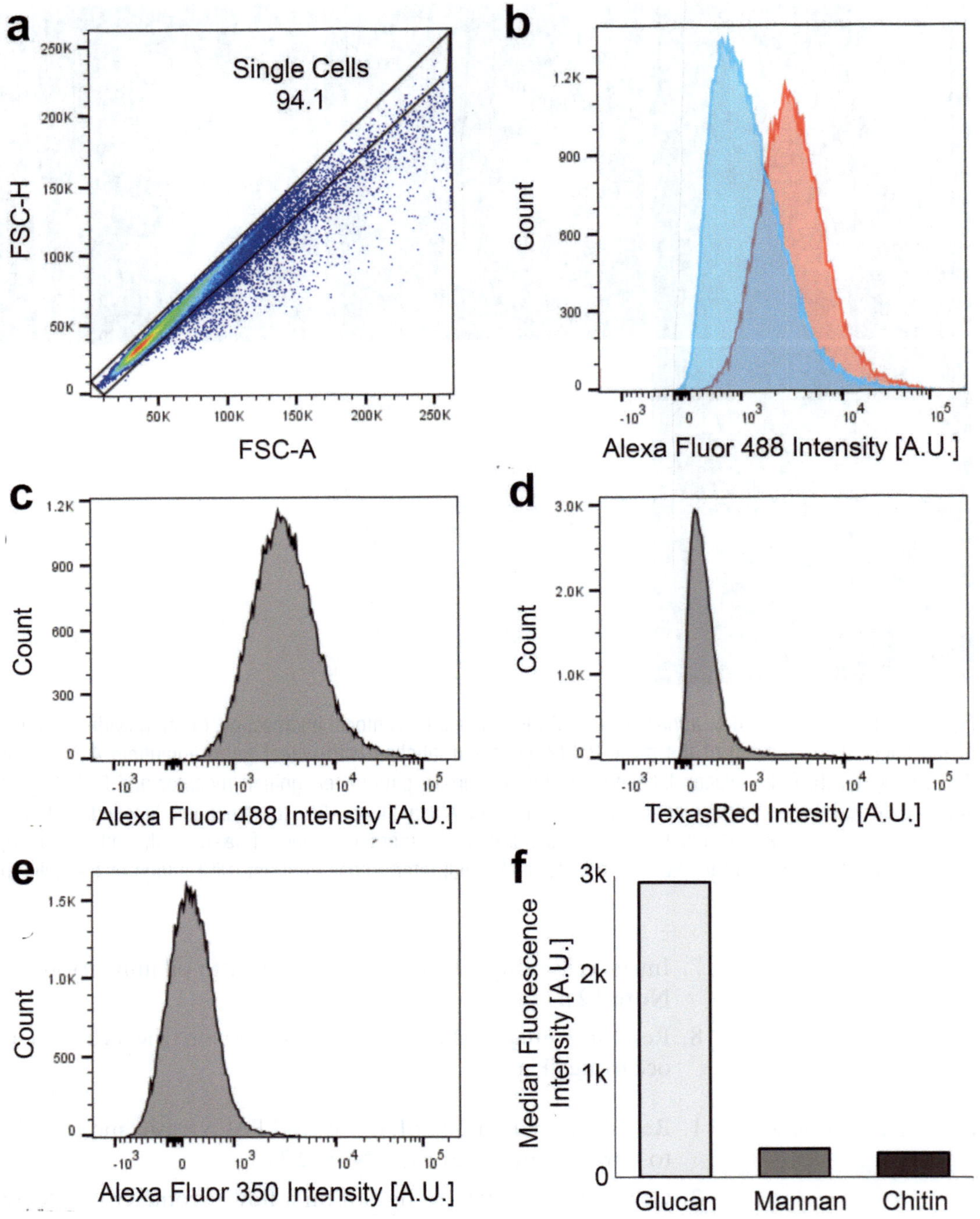

Fig. 2 Flow cytometry results. (**a**) Gating strategy used to gate single cells and exclude events that are budding cells and clumps of cells. (**b**) Histograms can be obtained to show a shift in in PAMP exposure between conditions. Here comparing one fluorescence intensity of β-1,3-glucan labelled with Fc-hDectin-1a pattern recognition receptor and F(ab')2-goat anti-human IgG Fc secondary antibody conjugated to Alexa Fluor 488 between two experimental conditions. (**c–e**) Histograms of individual fluorophores can also be produced: (**c**), (**d**), (**e**). (**f**) Median fluorescence intensity for each PAMP can also be extracted from histograms and shown side by side as a bar chart. (**b–f**) Indicated fluorescent intensities are given in arbitrary units [AU] generated by the flow cytometry software

7. A histogram showing a shift in median fluorescence intensity (Fig. 2b). The median fluorescence intensity of gated single-cell populations can then be directly compared graphically (Fig. 2c–f).

4 Notes

1. Thimerosal powder must be handled in a fume hood with gloves to prevent inhalation and direct contact with the skin; also thimerosal solution should be handled with gloves to prevent skin contact.
2. This protocol was optimized using an Fc-fusion protein from a commercial source. This was constructed by fusing the C-terminal region of human Dectin-1 PRR to a human IgG Fc domain.
3. This antibody must be able to bind the Fc domain of the protein being used to bind β-1,3-glucan. Here, a human IgG Fc domain was used, and thus an anti-human secondary antibody is listed.
4. If cell suspension has a low pH, thimerosal will precipitate. If this happens, add 1 M NaOH (dissolve 0.4-g NaOH pellets in 10-mL water) dropwise until the precipitate has redissolved.
5. Less than 16 h might not completely kill and fix all cells. A test is recommended to ensure no viable cells are left after fixing by streaking cells on appropriate solid growth medium and incubate for 48 h at 37 °C.
6. At this point cells can be stored in 1× PBS at 4 °C for 1 month; do not freeze.
7. A range of concentrations will need to be run prior to labelling final specimens; this is to check that binding is optimal for detection. These concentrations will also be specific to the fusion protein used.
8. The length of incubation is important here. Substantially shorter or longer incubation time can cause less than optimal binding or nonspecific binding.
9. Do not use centrifugation speeds above this, as it can cause shearing of the antibody from the epitope and lead to loss of signal.
10. The concentration of secondary antibody indicated here is the manufacturer's recommendation. For the wheat germ, agglutinin conjugate and concanavalin A conjugate are used at concentrations above the recommended concentration to ensure cells are labelled fully.
11. Try and avoid introducing air bubbles; a short spin in a centrifuge can eliminate most but not all if they do occur.

12. Prepare a non-labelled control. This is used to make sure that the fluorescence signal is real and not autofluorescence, as some samples will have very little exposed chitin and β-1,3-glucan and thus weak fluorescence. Length of exposure will depend on the light source and the sensitivity of the camera used.
13. The liquid volume that the cell pellet is resuspended in and the receptacle might need to be adapted depending on which flow cytometer is used. Some recent machines utilize standard 1.5-mL reaction tubes; then cells can be resuspended in a smaller volume of liquid.
14. Analyze stained cells on the flow cytometer the same day or the next morning at the latest; keep samples at 4 °C in complete darkness. Otherwise fluorophores will lose brightness due to photobleaching, and this might affect results.
15. Settings and filter sets will be determined by the fluorophores and the flow cytometer used but must be kept the same between experimental replicates to ensure the comparability of results. If the threshold of the forward scatter is set too high, this will result in only large cells being recorded, *C. auris* has a small cell size, and events might be lost if not set up properly. Controls needed include an unstained control and a set of single-stained controls. Proper setup with controls prevents false positives caused by autofluorescence (notably seen when using a 488-nm laser) and false negatives by ensuring that the fluorophore is being excited.

Acknowledgments

This work is supported by the Medical Research Council (MRC) Centre for Medical Mycology at the University of Exeter, UK [grant number MR/P501955/1]. The authors are grateful for the support and technical assistance from the Microscopy and Histology Core Facility (Kevin MacKenzie, Lucy Wight) and the Iain Fraser Cytometry Centre (Andrea Holme, Kate Burgoyne) of the University of Aberdeen.

References

1. Satoh K, Makimura K, Hasumi Y et al (2009) *Candida auris* sp. nov., a novel ascomycetous yeast isolated from the external ear canal of an inpatient in a Japanese hospital. Microbiol Immunol 53:41–44. https://doi.org/10.1111/j.1348-0421.2008.00083.x
2. Forsberg K, Woodworth K, Walters M et al (2019) *Candida auris*: the recent emergence of a multidrug-resistant fungal pathogen. Med Mycol 57:1–12. https://doi.org/10.1093/mmy/myy054
3. Hopke A, Brown AJP, Hall RA, Wheeler RT (2018) Dynamic fungal cell wall architecture in stress adaptation and immune evasion. Trends Microbiol 26:284–295. https://doi.org/10.1016/j.tim.2018.01.007

4. Gow NAR, Latge J-P, Munro CA (2017) The fungal cell wall: structure, biosynthesis, and function. Microbiol Spectr 5(3):FUNK-0035-2016. https://doi.org/10.1128/microbiolspec.FUNK-0035-2016
5. Navarro-Arias MJ, Hernández-Chávez MJ, García-Carnero LC et al (2019) Differential recognition of *Candida tropicalis*, *Candida guilliermondii*, *Candida krusei*, and *Candida auris* by human innate immune cells. Infect Drug Resist 12:783–794. https://doi.org/10.2147/IDR.S197531
6. Bruno M, Kersten S, Bain JM et al (2020) Transcriptional and functional insights into the host immune response against the emerging fungal pathogen *Candida auris*. Nat Microbiol 5:1516–1531. https://doi.org/10.1038/s41564-020-0780-3
7. McKinnon KM (2018) Flow cytometry: an overview. Curr Protoc Immunol 120: 5.1.1–5.1.11. https://doi.org/10.1002/cpim.40
8. Macey MG (2007) Flow cytometry: principles and applications. Humana Press, Totowa
9. Pawley JB (2006) Handbook of biological confocal microscopy, 3rd edn. Springer, Boston
10. Walker LA, Munro CA, de Bruijn I et al (2008) Stimulation of chitin synthesis rescues *Candida albicans* from echinocandins. PLoS Pathog 4: e1000040. https://doi.org/10.1371/journal.ppat.1000040
11. Sherrington SL, Sorsby E, Mahtey N et al (2017) Adaptation of *Candida albicans* to environmental pH induces cell wall remodelling and enhances innate immune recognition. PLoS Pathog 13:e1006403. https://doi.org/10.1371/journal.ppat.1006403
12. Ballou ER, Avelar GM, Childers DS et al (2016) Lactate signalling regulates fungal β-glucan masking and immune evasion. Nat Microbiol 2:16238. https://doi.org/10.1038/nmicrobiol.2016.238
13. Pradhan A, Mol Avelar G, Bain JM et al (2019) Non-canonical signalling mediates changes in fungal cell wall PAMPs that drive immune evasion. Nat Commun 10:5315. https://doi.org/10.1038/s41467-019-13298-9
14. Mora-Montes HM, Netea MG, Ferwerda G et al (2011) Recognition and blocking of innate immunity cells by *Candida albicans* chitin. Infect Immun 79:1961–1970. https://doi.org/10.1128/IAI.01282-10
15. Ene IV, Adya AK, Wehmeier S et al (2012) Host carbon sources modulate cell wall architecture, drug resistance and virulence in a fungal pathogen. Cell Microbiol 14:1319–1335. https://doi.org/10.1111/j.1462-5822.2012.01813.x

Chapter 15

High-Pressure Freezing and Transmission Electron Microscopy to Visualize the Ultrastructure of the *C. auris* Cell Wall

Gillian Milne and Louise A. Walker

Abstract

Transmission electron microscopy (TEM) is the main technique used to study the ultrastructure of biological samples. Chemical fixation was considered the main method for preserving samples for TEM; however, it is a relatively slow method of fixation and can result in morphological alterations. Cryofixation using high-pressure freezing (HPF) overcomes the limitations of chemical fixation by preserving samples instantly. Here, we describe our HPF methods optimized for visualizing *Candida auris* at the ultrastructural level.

Key words High-pressure freezing, Freeze substitution, Transmission electron microscopy, Fungal cell wall

1 Introduction

TEM is the principal method to study biological samples at the ultrastructural level because it enables the visualization of interactions of subcellular features. The aim of preparing biological samples for electron microscopy is to preserve fine structures as close to their native state as possible. In the past TEM samples were typically prepared by preservation with a chemical fixative, such as glutaraldehyde [1, 2]. Chemical fixation is a simple and convenient option as a fixative because it does not require specialized equipment; however, it is known to result in alterations to cellular structures within biological samples, which can lead to morphological artifacts [1, 3–7]. Using chemical fixation to preserve samples for ultrastructural analysis can also result in the loss of fine structures within cells which can limit the possibility of using immunological staining [4, 7]. Chemical fixation is considered a slow method for preserving biological samples because it relies on diffusion of the fixative through the cell, which can take several

Alexander Lorenz (ed.), *Candida auris: Methods and Protocols*, Methods in Molecular Biology, vol. 2517,
https://doi.org/10.1007/978-1-0716-2417-3_15,

seconds or even minutes. Once the fixative enters a cell, it takes extra time for the final cross-linking reactions to take place, to complete the fixation process [7]. The time required for chemical fixation of fungal cells is a particularly slow process because the fungal cell wall is relatively impenetrable to many chemicals [8–10]. The limitations of chemical fixation highlight the challenges involved in using this method to observe dynamically occurring events within biological samples, at the ultrastructural level.

Cryofixation is now commonly used to overcome the disadvantages of chemical fixation. It is considered the best way to rapidly preserve biological specimens by freezing samples within milliseconds, allowing dynamically occurring events within cells to be preserved. HPF followed by freeze substitution (FS) is the favored approach for cryofixation. Performing HPF with liquid nitrogen (−196 °C) at a pressure of 2100 bars reduces the freezing temperature and preserves samples without the formation of visible ice crystals [11–14]. The addition of high pressure enables samples up to a thickness of ~300 nm to be preserved, compared to a maximum thickness of 10 nm with chemical fixation. Typically, HPF is followed by FS which replaces any water in the samples with an organic solvent [3, 6]. Acetone is the solvent most commonly used for FS because it remains as a liquid at low temperatures and diffuses over time to replace the water in HPF samples. The introduction of HPF has substantially improved the level of ultrastructural preservation achieved by electron microscopy. Using HPF it is possible to take live observations on the fluorescence microscope and visualize these changes at the ultrastructural level [15–18]. HPF has considerably enhanced the ability to study fungal cells at the ultrastructural level [19, 20]. We have developed HPF methods for fungal pathogens of the *Candida* genus, which have allowed the cell wall to be observed at a level that was previously unachievable using chemical fixation [21–24] (Fig. 1).

2 Materials

All resins and chemicals should be stored according to Material Safety Data Sheets/safety recommendations. All equipment for HPF and processing are supplied by Leica Microsystems Ltd. All freeze substitution solutions and resins are prepared in a fume hood with full protection (safety glasses, lab coat, and gloves) and disposed of in accordance with local rules. *C. auris* was cultured and disposed of according to the local rules for containment level 2 microorganisms.

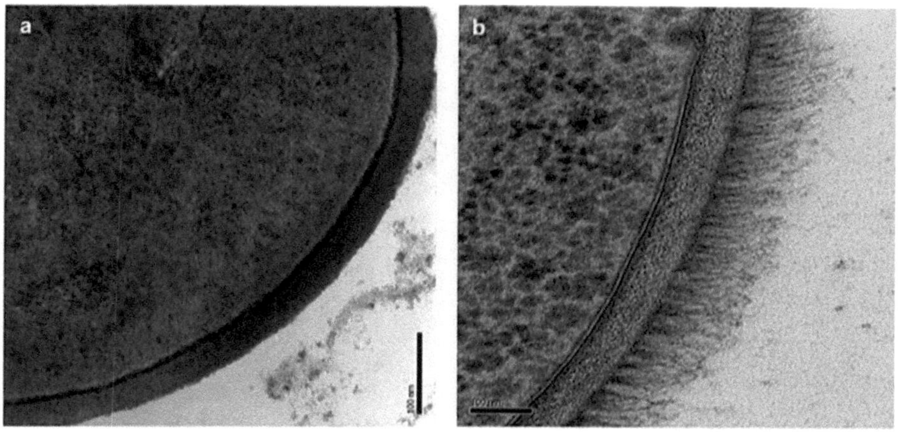

Fig. 1 HPF allows detailed visualization of the ultrastructure of the *Candida* spp. cell wall. TEMs of the cell wall of *Candida albicans* after (**a**) chemical fixation with 2.5% glutaraldehyde and (**b**) after HPF. Scale bars are 100 nm

2.1 Culturing of *C. auris*

1. Sabouraud dextrose liquid: 1% (w/v) mycological peptone, 4% (w/v) glucose, 2% (w/v) technical agar, dissolved in deionized water. Media was sterilized by autoclaving at 121 °C for 15 min at 103.4 kPa (15 p.s.i).
2. Shaking incubator.
3. Petri dishes.
4. Centrifuge.
5. Plastic 50 mL tubes.
6. 1.5 mL microcentrifuge tubes.
7. Microcentrifuge.
8. Benchtop autoclave (121 °C for 15 min at 103.4 kPa (15 p.s. i)).
9. Deionized water.

2.2 High-Pressure Freezing (HPF)

1. Leica HPF machine and Leica EMPACT 2.
2. Liquid nitrogen (~40 L).
3. Methylcyclohexane (MCH).
4. Flat specimen system for HPF (Fig. 2): Torque wrench with 2-mm Allen key insert (Fig. 2a); forceps with insulation coating (Fig. 2b); bayonet pod (Fig. 2c); manual bayonet loading device (Fig. 2d); membrane carrier, 1.5-mm diameter × 0.1-mm thick (Fig. 2e); specimen loader (Fig. 2f), specimen loading station (Fig. 2g), and specimen transfer carrier (Fig. 2h).
5. Cryo transfer box for specimen carrier.
6. P10 micropipette.

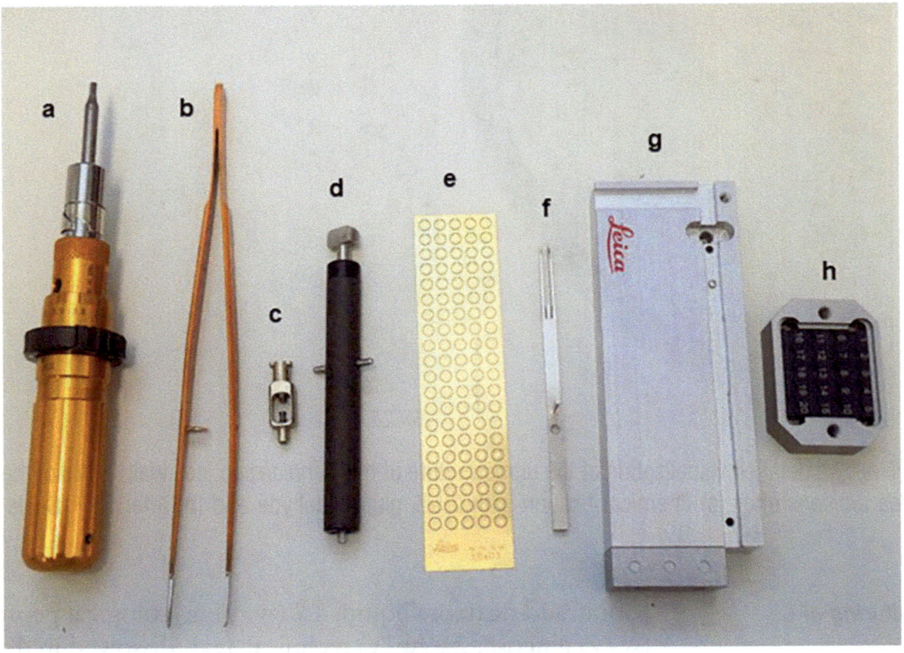

Fig. 2 Leica EM PACT2 Flat Specimen System tools required to prepare samples for HPF. (**a**) Torque wrench with 2 mm Allen key insert, (**b**) forceps with insulation coating, (**c**) bayonet pod, (**d**) manual bayonet loading device, (**e**) flat membrane carrier, (**f**) flat specimen loader, (**g**) flat specimen loading station, (**h**) specimen transfer carrier

2.3 Freeze Substitution (FS)

1. Leica AFS2 automatic freeze substitution processor.
2. Standard FS cocktail: 1% osmium tetroxide (OsO4) prepared using pure acetone.
3. Spurr's resin: 12.5-g nonenyl succinic anhydride (NSA), 5-g epoxycyclohexylmethyl-3,4-epoxycyclohexylcarboxylate (ERL 4221D), 4-g diglycidyl ether of polypropylene glycol (DER 736), 188-μL 2-dimethylaminoethanol (S1).
4. 10%, 30%, 50%, 70%, and 90% Spurr's resin diluted with acetone.
5. Flow-through rings.
6. Specimen processing block with two positions for specimen container.
7. Reagent baths.
8. Specimen carrier detaching tool.
9. Razor blade.
10. Hotplate.

2.4 Ultrathin Sectioning and Staining of TEM Grids

1. Ultramicrotome.
2. Diamond knife.

3. 1% Toluidine Blue in 1% borax.
4. Copper grids.
5. Leica EM AC20 automatic contrasting system.
6. Loading jig for autostainer.
7. Grid holder for autostainer.
8. UranyLess.
9. 3% lead citrate, stabilized solution.
10. Distilled H_2O.
11. Fine forceps.

3 Methods

All recommended safety precautions are followed through this protocol. All safety information must be followed with the utmost attention.

3.1 Preparation of C. auris Cells

1. *C. auris* strains are grown in 10 mL of Sabouraud dextrose liquid overnight at 30°C with shaking at 200 rpm.
2. Cells are collected by centrifugation at 1250 × *g* for 5 min and the supernatant decanted.
3. Pellets are resuspended in 1 mL of deionized water and centrifuged at 9450 × *g* for 5 min. This washing step is repeated twice more.
4. Resuspend final pellet in a small volume of residual liquid (*see* **Note 1**).

3.2 HPF

The protocol described for HPF is carried out using a Leica EM PACT 2 with rapid transfer system (RTS) in manual mode; other HPF machines are available where the manufacturer's instructions should be followed.

1. Switch on compressor and HPF.
2. Replenish hydraulic fluid at the start of each session by filling with MCH.
3. Close drain valve and fill liquid nitrogen tank to the top. After filling close the cover of the tank and maintain a high level of liquid nitrogen as fungal samples are loaded.
4. Fill the liquid nitrogen bath, where the specimens will be stored, by switching the valve to open.
5. Using forceps remove the required number of flat membrane carriers (100 μm) from the packaging (*see* **Note 2**).
6. Insert a membrane carrier into the flat specimen loader.

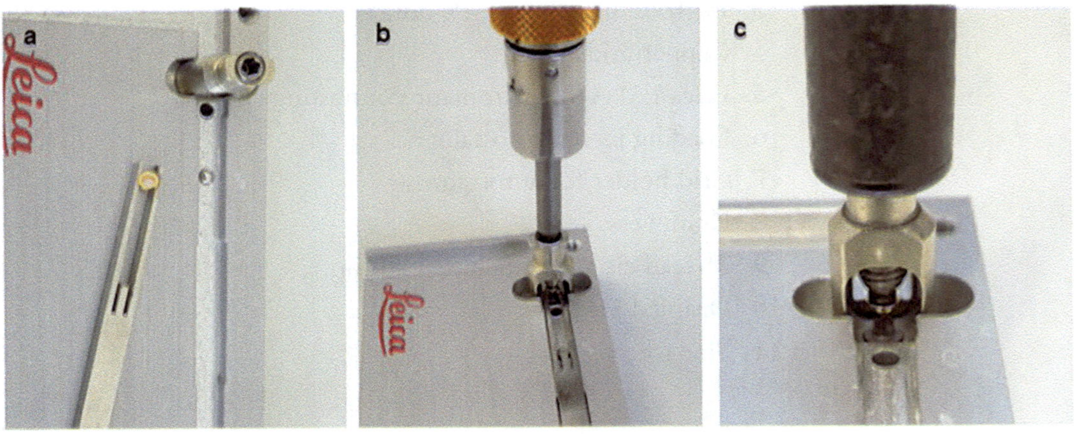

Fig. 3 Leica EM PACT2 Flat Specimen System tools required to prepare samples for HPF using RTS in manual mode. (**a**) *C. auris* cells loaded into a membrane carrier which has been inserted into the flat specimen loader, (**b**) flat specimen loader containing the membrane carrier which has been inserted into the bayonet pod. The torque wrench is then used to tighten the screw in the bayonet pod which will hold the membrane carrier in place. (**c**) The flat specimen loader is removed to leave the membrane carrier in the bayonet pod. The bayonet loading device is then locked onto the bayonet pod by pushing downward and turning anticlockwise

7. Using a P10 micropipette, fill the inner circle of the membrane carrier with *C. auris* cells (approx. 5 μl) (Fig. 3a) (*see* **Note 3**).
8. Place the flat specimen loader into the loading station.
9. Insert the bayonet pod into the loading station, and slide the specimen loader into the bayonet pod (Fig. 3b).
10. Insert the torque wrench into the top of the bayonet pod, and turn to tighten (Fig. 3b) (*see* **Note 4**).
11. The specimen loader can now be pulled backward leaving the membrane carrier (containing the *C. auris* sample) in the bayonet pod (Fig. 3c).
12. The bayonet loading device can then be connected to the bayonet pod by pushing the loading device on top of bayonet pod and turning anticlockwise to lock it into place (Fig. 3c).
13. The assembled bayonet pod (containing membrane carrier with *C. auris* sample) and attached loading device can then be inserted into the rapid transfer system (RTS) of the HPF.
14. The RTS will be used in manual mode because the membrane carrier has already been loaded into the bayonet pod.
15. Ensure both ends of the assembled bayonet complex are clicked into place and close the safety lid of the RTS.
16. Press the START button on the touch screen of the HPF to activate the machine and freeze the sample. The sample will now have been transferred from the RTS to the EM PACT2 (*see* **Note 5**).

Table 1
Standard Leica protocol used for FS

Step	Start temp	End temp	Time (h/min)	Reagent
1	−95	−90	30	Standard FS cocktail
2	−90	−90	10:00	Standard FS cocktail
3	−90	−30	08:00	Standard FS cocktail
4	−30	−10	01:00	Acetone
5	−10	4	01:00	Acetone
6	4	20	01:00	Acetone

17. The sample is then ejected into the liquid nitrogen bath of the EM PACT2.
18. Transfer bayonet pod to the unloading station in the liquid nitrogen bath.
19. Release the bayonet loading device by pushing down and turning clockwise. This will leave the bayonet pod in the unloading station.
20. Insert the torque wrench into the top of the bayonet pod, and turn anticlockwise. This will open the screw inside which is holding the sample on the specimen carrier in place.
21. Cool the forceps with insulation coating in liquid nitrogen, and use them to release the membrane carrier from the bayonet pod.
22. Transfer the membrane carrier containing *C. auris* sample to the cryo transfer carrier (*see* **Note 6**).

3.3 Freeze Substitution (FS)

The protocol described for FS is carried out using a Leica AFS2; other automatic FS processors are available where the manufacturer's instructions should be followed.

1. The cryo transfer carrier containing the membrane carriers is transferred from the HPF inside a cryo transfer box to a Leica AFS 2 (*see* **Note 7**). Here the samples are placed in flow-through chambers containing pre-cooled standard FS cocktail.
2. Using forceps with insulation coating, transfer the membrane carriers to flow rings in the FS (*see* **Note 8**).
3. Freeze substitution is carried out following the standard Leica protocol (Table 1) (*see* **Note 9**). Manual changes are made at each step (*see* **Note 10**).
4. Samples are then removed and placed in 10% Spurr's resin in acetone for 72 h (*see* **Notes 11** and **12**).

5. Processing continues using the following steps: 30% Spurr's resin in acetone overnight, 50% Spurr's resin in acetone for 8 h, 70% Spurr's resin in acetone overnight, and 90% Spurr's resin in acetone for 8 h (*see* **Notes 13**, **14** and **15**).
6. For the final processing step, samples are embedded in Spurr's resin at 60°C for at least 24 h (*see* **Note 16**).
7. The membrane carrier can then be removed from the sample using the Leica detaching tool that has been preheated on a hotplate.

3.4 Sectioning of TEM Samples

The protocol described for sectioning of TEM samples is carried out using a Leica EM UC6 Ultramicrotome (Leica, UK); other ultramicrotomes are available where the manufacturer's instructions should be followed.

1. Prior to initial sectioning, excess resin can be removed from the block using a razorblade.
2. Semi-thin sections can be taken initially to check presence/quality of sample (*see* **Note 17**).
3. Semi-thin sections are cut using a histology diamond knife, or a lower-quality Ultra knife (*see* **Note 18**).
4. 500 nm–1 μm sections are cut and dried onto a slide prior to staining. Sections can be carefully picked up from the diamond knife water bath using a wire loop or small paintbrush.
5. Sections are dried on the hotplate (60°C) and then stained while on the hotplate with 1% Toluidine Blue in 1% borax (*see* **Note 19**).
6. After rinsing with H_2O, dry slide on hotplate prior to viewing under microscope.
7. A specific area of sample can then be selected prior to further trimming.
8. Trim selected area of block into a trapezium for ultrathin sectioning (*see* **Note 20**).
9. Use backlight to view alignment on the block.
10. Move block steadily toward knife edge manually leaving a small gap. Further movement to be carried out using ultramicrotome controls.
11. Ensure blockface is lined up both horizontally and vertically by adjusting alignment controls before block is in contact with knife edge. It may take a few repeats of aligning, advancing, and then realigning before it is safe to begin sectioning.
12. Once as close to knife edge without touching and fully aligned, fine advancement can be made until contact with knife edge, constantly rotating block checking at every stage for contact (*see* **Note 21**).

13. Once contact is made, set cutting window and motor can be started. Recommended speed 1 mm/s (*see* **Notes 22** and **23**).
14. Sections can be manipulated with an eyelash brush (*see* **Note 24**) and flattened out using a heat pen to reduce wrinkling.
15. Sections are picked up onto copper grids and left to dry at room temperature (*see* **Notes 25** and **26**).

3.5 Staining of TEM Samples

The protocol described for staining TEM grids is carried out using Leica AC20 EM Autostainer for Contrast; other automatic contrasting machines are available where the manufacturer's instructions should be followed.

1. Switch on the machine and empty waste bottles.
2. Fill water bottle with distilled H_2O.
3. Turn valve to wash (W), insert grid pad and run wash programme.
4. Position the first slot in the grid holder over the loading jig, and push down to open.
5. Using fine forceps pick up the first grid on the outer edge. Insert the edge of the grid into the first slot of the grid holder.
6. Transfer all grids containing sections to the grid holder for staining (maximum of 20).
7. Insert grid holder into the contrasting chamber. The loaded grids should be pointing toward the machine.
8. Close the contrasting chamber by turning the bayonet lock.
9. Turn valve to stain (S).
10. Start staining program—UranyLess for 20 min and 3% lead citrate for 4 min (*see* **Note 27**).
11. Once completed, remove grids and air-dry. Excess water can be removed by with filter paper.
12. Turn valve to wash (W).
13. Insert grid pad and select W1.
14. Start wash cycle to rinse and switch off the machine.
15. Grids can then be stored for future use, or imaged directly on the TEM (*see* **Note 28**) (Fig. 4).

4 Notes

1. *C. auris* cells must be a thick paste for loading onto the specimen carrier. To achieve this, a cell pellet of *C. auris* yeast cells should be resuspended in the residual liquid that is left after pouring off the supernatant from the wash steps. For strains that are clumpy and do not form a paste well, hexadecane can be added to the sample.

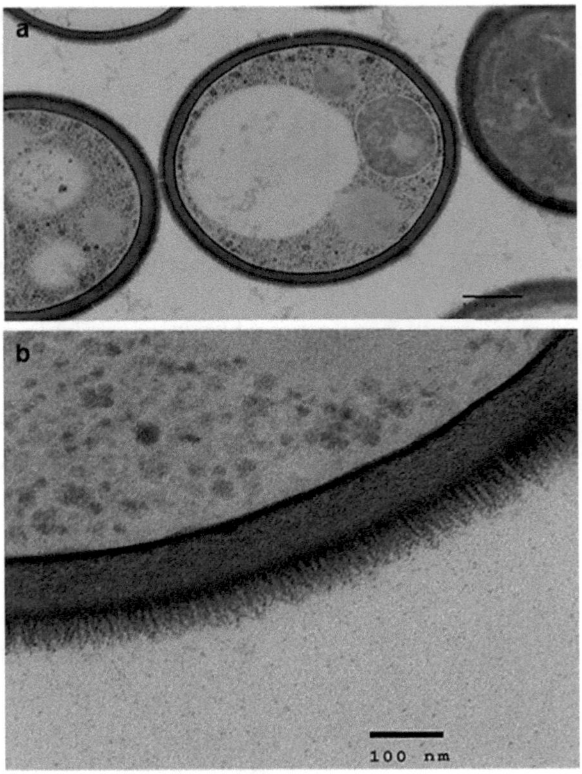

Fig. 4 The ultrastructure of the cell wall of *C. auris*. (**a**) TEM of whole cells of *C. auris*. Scale bar is 500 nm., (**b**) TEM image of the cell wall of *C. auris*, captured using a JEOL 1400 plus TEM with and an AMT UltraVUE camera. Scale bar is 100 nm

2. After membrane carriers have been removed from their original packaging, use a razor blade to remove any excess material that is still attached to the membrane carrier.
3. Do not overflow the specimen carrier with fungal cells when loading the sample, or the specimen carrier will not load into the machine properly.
4. Tighten the screw in the bayonet pod with the torque wrench until one click is heard, overtightening may damage the equipment.
5. Before starting HPF of samples load one to two empty membrane carriers into the system first to check machine is operating correctly.
6. After HPF has been carried out, the Leica EMPACT 2 allows up to 20 individual samples to be stored under liquid nitrogen until transferring to AFS.
7. Set up AFS prior to freezing session:

- Ensure extraction tube is secured in fume cupboard.
- Turn on AFS and start selected program.
- Pause program then fill with liquid nitrogen—this will ensure AFS is at correct temperature before use.
- Place vial of freeze sub to allow to thaw before use.
- For the standard program, ¼ fill AFS with liquid N_2 can then be topped up with leftover liquid N_2 from HPF.

8. Once samples are transferred to Leica flow rings, ensure they have sunk to the bottom and that sample is correctly facing upward. Membrane carriers can be turned over using a dissecting needle or syringe needle.
9. Freeze substitution reagents can be pre-prepared and stored under liquid nitrogen prior to use.
10. When transferring reagents ensure transfer function is set to ON.
11. Remove samples from AFS once it is at room temperature to enable control of evaporation. The samples can then be removed, placed in a low Spurr's dilution, and stored in a sealed container.
12. We prepare our own resin fresh prior to processing and make a fresh batch for embedding. Kits are also available. Resin can also be pre-prepared and frozen:
    - Spurr's resin is a lower viscosity resin and improves infiltration of the sample. More viscous resins do not infiltrate as well.
    - The long initial infiltration step is beneficial for infiltration. Some *C. auris* strains with thicker cell walls may require additional infiltration.
13. All processing steps are carried out in Leica flow-through chambers.
14. PELCO Biowave® Pro+ microwave tissue processor is an alternative to manual processing. Use of microwave and vacuum function improves infiltration decreasing processing times.
15. Continue to carry out polymerization in oven. It is also possible to polymerize in PELCO Biowave® Pro+ microwave tissue processor (quicker option but not generally required).
16. Block reference code/number is inserted prior to polymerization. After polymerization blocks are removed using a razor blade.
17. Initial trimming of the blockface can be carried out with a glass knife; however, with HPF blocks samples are close to blockface, and minimal trimming is required.

18. Diamond knife water bath must be fully filled with water level up to edge of the knife. It can then be levelled off using a syringe. If the water level is too high, water will then be dragged down back of the knife by the block. If level is too low, you risk dry areas and damage to knife edge. The best way to get the correct water level is to overfill bath then remove excess water with the syringe. If water catches on block, this can be dried off using filter paper.
19. Once drying is visible around edge of stain, (around 20 s) remove and rinse well with tap water.
20. Trimming of resin can be done using a razor blade, or for more precise trimming a glass or diamond trimming knife. The recommended size for the diameter of the block after trimming is 1–2 mm.
21. For ultrathin sectioning, it is essential for the preservation of the knife, and blockface, that the block is precisely lined up to the edge of the knife.
22. Initial section thickness can be started ~250 nm. Once full sections are coming off into water bath of the diamond knife, thickness can be gradually decreased to ~70–90 nm.
23. When sectioning, it is essential that the blade edge does not come into contact with any hard objects, e.g., forceps.
24. Eyelash brushes are made by attaching an eyelash to a cocktail stick using nail varnish.
25. 200 mesh copper grids are normally the grid of choice, although 300 can also be used.
26. Where there is an issue with holes in a sample, formvar-coated grids can be used to aid support of the sample.
27. Automated contrast staining is beneficial for staining a large number of grids. This can also be carried out manually.
28. TEM is set at 80 kV for standard applications.

Acknowledgement

The authors acknowledge the Microscopy and Histology Core Facility at the University of Aberdeen for expert assistance.

References

1. Sabatini DD, Bensch K, Barrnett RJ (1963) Cytochemistry and electron microscopy. The preservation of cellular ultrastructure and enzymatic activity by aldehyde fixation. J Cell Biol 17:19–58. https://doi.org/10.1083/jcb.17.1.19
2. Sabatini DD, Miller F, Barrnett RJ (1964) Aldehyde fixation for morphological and enzyme histochemical studies with the electron microscope. J Histochem Cytochem 12:57–71. https://doi.org/10.1177/12.2.57

3. Giddings TH, O'Toole ET, Morphew M et al (2001) Using rapid freeze and freeze-substitution for the preparation of yeast cells for electron microscopy and three-dimensional analysis. Methods Cell Biol 67:27–42. https://doi.org/10.1016/s0091-679x(01)67003-1
4. Sawaguchi A (2013) Advantages of high-pressure freezing technique for fine structural electron microscopy. Plant Morphol 25:7–10
5. Korogod N, Petersen CCH, Knott GW (2015) Ultrastructural analysis of adult mouse neocortex comparing aldehyde perfusion with cryo fixation. eLife 4:e05793. https://doi.org/10.7554/eLife.05793
6. Steinbrecht RA, Müller M (1987) Freeze-substitution and freeze-drying. In: Cryotechniques in biological electron microscopy. Springer, Berlin, Heidelberg, pp 149–172
7. Kellenberger E, Johansen R, Maeder M et al (1992) Artefacts and morphological changes during chemical fixation. J Microsc 168(Pt 2): 181–201
8. O'Donnell KL, McLaughlin DJ (1984) Ultrastructure of meiosis in ustilago maydis. Mycologia 76:468–485
9. Hammill TM (1974) Septal pore structure in trichoderma saturnisporum. Am J Bot 61(7): 767–771. https://doi.org/10.1002/j.1537-2197.1974.tb12299.x
10. Collinge AJ, Markham P (1982) Hyphal tip ultrastructure of Aspergillus nidulans and Aspergillus giganteus and possible implications of woronin bodies close to the hyphal apex of the latter species. Protoplasma 113:209–213. https://doi.org/10.1007/BF01280909
11. Dahl R, Staehelin LA (1989) High pressure freezing for the preservation of biological structure: theory and practice. J Electron Microsc Tech 13:165–174. https://doi.org/10.1002/jemt.1060130305
12. McDonald K (1999) High-pressure freezing for preservation of high resolution fine structure and antigenicity for immunolabelling. In: Nasser Hajibagheri MA (ed) Electron microscopy methods and protocols. Humana Press, Totowa, pp 77–97
13. Moor H (1987) Theory and practice of high pressure freezing. In: Cryotechniques in biological electron microscopy. Springer-Verlag, Heidelberg, pp 175–191
14. Shimoni E, Muller M (2008) On optimizing high-pressure freezing: from heat transfer theory to a new microbiopsy device. J Microsc 192:236–247
15. Murray S (2008) Chapter 1 high pressure freezing and freeze substitution of Schizosaccharomyces pombe and Saccharomyces cerevisiae for TEM. In: Allen T (ed) Introduction to electron microscopy for biologists. Elsevier, Manchester, pp 3–17
16. Nicolas M-T, Bassot J-M (1993) Freeze substitution after fast-freeze fixation in preparation for immunocytochemistry. Microsc Res Tech 24:474–487
17. Villiger W (1991) Lowicryl resins. In: Hayat M (ed) Colloidal gold: principles, methods and applications. Academic Press, New York, pp 59–71
18. Newman G, Hobot J (1993) Handling resin blocks. In: Resin microscopy and on-section immunocytochemistry. Springer, Berlin, Heidelberg, pp 101–105
19. Hoch H (1986) Freeze-substitution of fungi. In: Aldrich H, Todd W (eds) Ultrastructure techniques for microorganisms. Springer, Boston, pp 183–212
20. Mims CW, Celio GJ, Richardson EA (2003) The use of high pressure freezing and freeze substitution to study host-pathogen interactions in fungal diseases of plants. Microsc Microanal 9:522–531. https://doi.org/10.1017/S1431927603030587
21. Walker LA, Gow NA, Munro CA (2013) Elevated chitin content reduces the susceptibility of Candida species to caspofungin. Antimicrob Agents Chemother 57:146–154
22. Walker L, Sood P, Lenardon MD et al (2018) The viscoelastic properties of the fungal cell wall allow traffic of ambisome as intact liposome vesicles. mBio 9:1–15. https://doi.org/10.1128/mBio.02383-17
23. Walker CA, Gomez BL, Mora-Montes HM et al (2010) Melanin externalization in Candida albicans depends on cell wall chitin structures. Eukaryot Cell 9:1329–1342
24. Gow NA, Netea MG, Munro CA et al (2007) Immune recognition of Candida albicans beta-glucan by dectin-1. J Infect Dis 196:1565–1571

Part V

Bioinformatic Methods for *Candida auris*

Chapter 16

Generating Complete Genome Assemblies of *Candida auris*

Terrance P. Shea and Christina A. Cuomo

Abstract

Genomic studies of *Candida auris* are underpinned by the generation of high-quality genome assemblies. These reference genomes have been essential for investigations of the evolution and epidemiology of this emerging fungal pathogen. In addition to genomic epidemiology studies of local outbreaks and analysis of the global emergence of this species, comparisons of genomes of isolates from the five major clades have revealed differences in gene content and genomic structure. Here, we provide a detailed protocol for generating complete genome assemblies for *C. auris*.

Key words Whole genome assembly, Hybrid assembly methods, *Candida auris*, Fungal genome

1 Introduction

Genomic studies have played a primary role in understanding the global emergence and local outbreaks of the human fungal pathogen *Candida auris*. The generation of whole genome assemblies is an essential component of both population and comparative genomic studies. Genome assemblies of selected isolates have been used to characterize the population structure using short-read sequencing of large collections of isolates [1–5]. De novo assembly of isolates from each of the five described clades of *C. auris* [1, 6] has described conservation or expansion of gene content [7, 8], identified the loss of subtelomeric regions, and secreted gene families in clade II isolates [9] and the presence of a rhamnose utilization gene cluster found only in clade III isolates [10]. The generation of multiple high-quality reference genomes therefore provides clade-specific references for genomic analyses.

A hybrid assembly strategy including long and short reads is recommended for generating complete telomere-to-telomere fungal genome assemblies with high base accuracy. While short-read assemblies have been generated, they are typically more fragmented than assemblies that incorporate long reads (e.g., PacBio or

Alexander Lorenz (ed.), *Candida auris: Methods and Protocols*, Methods in Molecular Biology, vol. 2517,
https://doi.org/10.1007/978-1-0716-2417-3_16,

Oxford Nanopore Technology), which are needed to assemble across repetitive regions that may break contigs in short-read assemblies. Polishing steps incorporating both long and short reads ensure high base-level accuracy in the assembly, which is important for genomes used as references for short-read alignment and variant identification.

This protocol covers the bioinformatic steps we use for whole genome assembly of *C. auris*, as well as other haploid fungi. These steps recapitulate our published methods used to generate complete genome assemblies for isolates from each of the five clades [8, 9].

2 Materials

2.1 Sequence Data

This protocol requires two types of sequence data for isolates or strains of interest. Long reads either from Oxford Nanopore or PacBio technology are typically generated to a sequence depth of at least 50× and should be in FASTQ format. We have the most experience with Oxford Nanopore sequencing and recommend the use of the ligation sequencing kit, taking care to reduce pipetting and other handling steps that can shear DNA. For *C. auris*, we have seen good assembly results from reads with an N50 length of 6–29 kb. Oxford Nanopore files should be in FASTQ format for assembly and polishing; the FAST5 format is required for basecalling and for training a new model for polishing. We recommend polishing with Illumina-paired reads of 100 bases or longer and a sequence depth of at least 30×. Illumina sequence files should be in FASTQ format.

2.2 Software

The software required for the methods in this chapter and links to installation instructions are listed in Table 1. We install and run the commands on our local computing cluster. Although we have listed version numbers for software, it is recommended to check for and incorporate new versions that have major changes in performance.

3 Methods

3.1 Computational Environment

All of the steps outlined in this workflow can be run locally as a series of commands in a terminal session on a standard Unix workstation, such as a server or a local machine (Mac or Linux computer).

3.2 Quality Control and Evaluation of Sequence Data

Prior to analysis, we recommend running quality control (QC) checks of sequence data. The sequencing yield should be confirmed; as described above, typically for genome assembly, long reads (Oxford Nanopore or PacBio) of a coverage of 50X or

Table 1
Software used for genome assembly and analysis

Package [version]	Purpose in workflow	Download/installation link
FastQC [v0.11.9]	Quality control	https://www.bioinformatics.babraham.ac.uk/projects/fastqc/
NanoPlot [v1.20.0] [11]	Quality control	https://github.com/wdecoster/NanoPlot
bwa [v0.7.12] [21]	Short-read alignment	http://bio-bwa.sourceforge.net
Samtools [v1.10] [14]	Reference formatting and BAM file manipulation	https://github.com/samtools/samtools
Picard [GATK v4.1.4.1]	Quality control and BAM manipulation	http://broadinstitute.github.io/picard
Guppy [v4.4]	Basecalling for Oxford Nanopore reads	https://community.nanoporetech.com/downloads
Canu [v1.5 or v1.6] [12]	Long-read assembly	https://github.com/marbl/canu
Flye [v2.4.2] [13]	Long-read assembly	https://github.com/fenderglass/Flye
Minimap2 [15]	Long-read alignment	https://github.com/lh3/minimap2
Medaka [v0.8.1]	Nanopore polishing	https://github.com/nanoporetech/medaka
Pilon [v1.23] [16]	Illumina polishing	https://github.com/broadinstitute/pilon
MUMmer [v3.1] [17]	Genome alignment	http://mummer.sourceforge.net/
IGV [v2.3.72]	Alignment visualization	https://software.broadinstitute.org/software/igv/
GAEMR [v1.0.1]	Assembly QC	https://github.com/broadinstitute/GAEMR

more are generated. For polishing, Illumina-paired reads of 100 bases or longer and a coverage depth of greater than 30X are used. For long reads, we run NanoPlot [11] (Oxford Nanopore) or view SMRT Link output (PacBio) to evaluate the quality, N50, and read length distribution. For Illumina data, we run FastQC (https://www.bioinformatics.babraham.ac.uk/projects/fastqc/) and evaluate the sequence quality across the read length, GC content, and level of duplication or over-represented sequences in each run. For Illumina reads, we also recommend using the GATK (Picard) tool CollectGcBiasMetrics (http://broadinstitute.github.io/picard) to determine if there is evidence of GC bias in the sequence data, which can result in higher variation of sequence depth across the assembly. With high-quality sequence runs, we do not typically trim the reads for adapter; however this may be needed for Illumina libraries with a fraction of insert sizes smaller than the read length. We also do not trim such runs for quality, as regions that are more difficult to sequence may be frequently represented by lower-quality bases for which a consensus can still be useful.

1. Oxford Nanopore QC and read metrics

   Software for basecalling of Oxford Nanopore reads has been regularly updated, and we recommend using the most recent version. If native FAST5 files are saved, and then bases can be recalled with newer versions of software. Typically, we run guppy in the high accuracy mode on an Oxford Nanopore GridION; guppy can also be installed locally and run on a computing cluster:

```
guppy_basecaller -c dna_r9.4.1_450bps_hac.cfg -i run_path -s
run_name -x auto -r
```

   If multiplexing was used, next run guppy_barcoder to demultiplex the samples in the same run. Next, run nanoplot to evaluate read quality and N50.

```
NanoPlot -t 4 --fastq_rich nanopore_reads.fastq --outdir /
output/path --raw
```

2. Run FastQC to evaluate Illumina read quality

   FastQC performs a series of quality control evaluations and produces a file of summary statistics for a given FASTQ file. After installing FastQC (*see* Table 1 for installation link), run the following command in a java environment.

```
fastqc -f fastq -o /output/path -d /tmp/directory/path illu-
mina_fq1.fastq illumina_fq2.fastq
```

   In the above example, `/output/path` is the path to the directory where the output reports should be stored, `/tmp/directory/path` is the path to a directory for temporary files, and `illumina_fq1.fastq illumina_fq2.fastq` are the forward and reverse read FASTQ files.

3. To further evaluate the Illumina reads, first align them to a reference genome.

```
bwa mem -M -p reference.fasta illumina_fq1.fastq illumina_fq2.
fastq
```

4. Next, run the Picard GATK command CollectGcBiasMetrics on the BAM file of Illumina reads and inspect the resulting plot for evidence of GC bias.

```
gatk CollectGcBiasMetrics -I input.bam -O gc_bias_metrics.txt
-CHART gc_bias_metrics.pdf -S summary_metrics.txt
```

3.3 Long-Read Assembly

First generate an initial assembly of the long reads using either Canu [12], Flye [13], or other long-read assembly software. In our experience, it can be helpful to generate multiple assemblies with different software to evaluate how the results compare. Both Canu and Flye have a parameter to set the expected genome size. Canu has additional parameters for the expected error rate (for PacBio, change read input flag to "`-pacbio-raw`"). For *C. auris*, we typically specify four cores (can be set maxThreads parameter) with a request of 12Gb per core; lower memory can also be specified, and we have found that 12Gb total memory is sufficient for *C. auris* assembly (*see* **Note 1**).

```
canu -p canu java=/path/to/java-1.8 genomeSize=12000000 -na-
nopore-raw nanopore_reads.fastq stopOnReadQuality=false cor-
rectedErrorRate=0.075
```

```
flye --nano-raw nanopore_reads.fastq genome-size=12000000
```

3.4 Genome Structure Evaluation and Manual Finishing

By incorporating long reads, we expect to be able to resolve fungal genome assemblies as one contig per chromosome that includes both telomeres at both contig ends. For *C. auris*, seven chromosomes are expected with a total assembly size of 12.1–12.7 Mb (*see* **Note 2**). A circular mitochondrial chromosome should also be assembled. In cases where the number of contigs exceeds the number of chromosomes, we look for bridging reads (from the unassembled reads) and seek to support the join based on read alignments and synteny with other assemblies. The preliminary assembly also may contain short contigs that are redundant in the genome assembly, resulting from sequencing errors in a subset of reads. These are confirmed based on an end-to-end alignment of a small contig to a larger contig and total sequence depth across both regions roughly equal to the genome average. Additionally, the assembly may contain one or more small contigs containing only rDNA repeats, when unable to completely assemble across this repetitive genomic region. Chromosome ends should also be evaluated for the presence of tandem short repeats characteristic of telomere. For *C. auris*, a tandem motif (AGACACCACCTA{1,2}GAAA{1,2}CC{1,2}) should be identified at contig ends [9]; contig ends missing this motif may be aligned to the unassembled contigs and manually extended, confirming that the unassembled contig (containing the telomere motif) is aligned uniquely to the assembly contig end. In the example below, the NUCmer output files are given the prefix "vs_unassembled."

```
nucmer -p vs_unassembled canu.contigs.fasta canu.unassembled.
fasta
```

3.5 Long-Read Polishing

Next, the assembly should be polished with Medaka (https://github.com/nanoporetech/medaka) using Nanopore reads. Medaka requires both samtools [14] and minimap2 [15]. In this example, the model "r941_min_high" is used; for more details about setting models, see (https://github.com/nanoporetech/medaka#models).

```
medaka_consensus -i nanopore_reads.fastq -d contigs.fasta -m
r941_min_high -t 4 -o /output/path
```

3.6 Short-Read Polishing

All assemblies should have roughly four or five iterations of Illumina read error correction using Pilon [16]. We recommend running Pilon until very few SNP and indel changes are introduced.

1. Starting with a FASTQ file, use bwa mem to align the reads to the draft assembly file.

```
bwa mem -M -p medaka_consensus.fasta illumina_fq1.fastq illu-
mina_fq2.fastq
```

2. Run pilon polishing in a java environment with the "-fix-all" flag specified.

```
java -Xmx16G -jar pilon-1.23.jar -genome medaka_consensus.
fasta -bam read_alignment.bam -fix-all
```

3. Finally, evaluate the Pilon output to decide if iteration is needed. Count the number of lines in the pilon.changes file, and compare to the number of changes in the previous round. Expect to run at least three iterations of Pilon polishing, with fewer changes detected in each round; this is needed as some larger changes made in early rounds may then be further polished in the following rounds. When few to no differences are detected between rounds, iteration is complete.

3.7 Quality Control Checks

It is essential to perform quality control checks on the resulting assembly to determine if there are any structural issues. This involves evaluating the sequence depth across the assembly and then using whole genome alignment to other genomes of the same species to identify and then validate sites of rearrangement.

1. First, align Illumina reads to polished assembly with bwa and sort the resulting BAM file with samtools.

```
bwa mem -M -p polished.fasta fq1.fastq fq2.fastq | samtools
view -h -b -S -o illumina.bam

samtools sort illumina.sorted
```

2. Align the nanopore reads to the polished assembly with minimap2 and sort the resulting BAM file with samtools.

```
minimap2 -ax map-ont polished.fasta nanopore_reads.fastq.gz |
samtools view -h -b -S -o nanopore_reads.bam -

samtools sort nanopore_reads.bam nanopore_reads.sorted
```

3. Use the GAEMR package (https://github.com/broadinstitute/GAEMR) tool generate_bam_plots.py, which will produce a plot of the depth of each read type (each BAM file) across the assembly.

```
generate_bam_plots.py -g polished.fasta -d cvg_data.txt -o
polished nanopore_reads.sorted.bam illumina.sorted.bam
```

4. To identify potential misassemblies, we also recommend aligning independent assemblies (e.g., the Canu and Flye assemblies of the same data) to each other and to well-curated reference genomes [8, 9]. We use NUCmer (MUMmer v3.22) [17] for whole genome alignment and plot the results with mummerplot (Fig. 1a).

```
nucmer -p ref_vs_asm -mum -o ref.fasta polished.fasta

mummerplot -f -l -p ref_vs_asm ref_vs_asm.delta -png
```

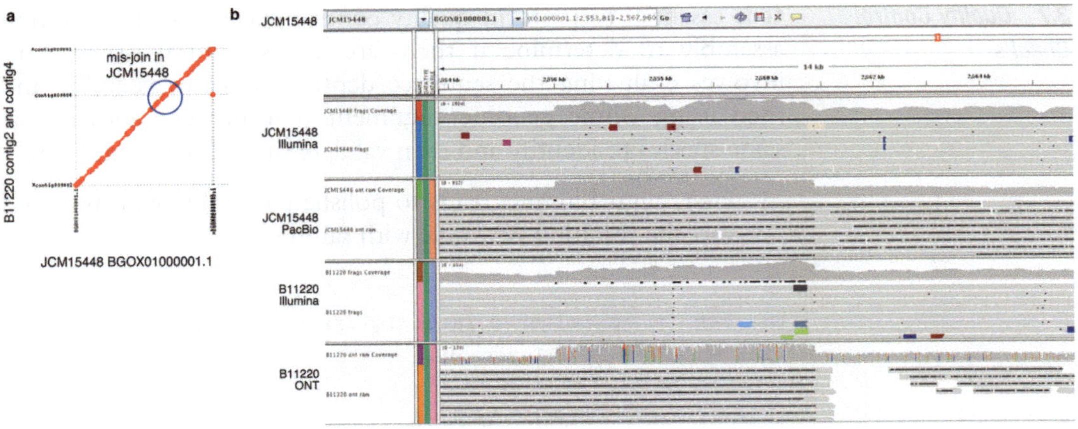

Fig. 1 Analysis of a mis-join in the assembly of *C. auris* JCM15448. (**a**) Mummerplot view of NUCmer alignment of contig1 of *C. auris* JCM15448 to contig2 and contig4 of *C. auris* B11220. Circle indicates a join found only in JCM15448. (**b**) Integrated Genomics Viewer (IGV) snapshot of short- and long-read alignments to the JCM15448 contig1; reads from JCM15448 (first two panels) and B11220 (bottom two panels) are displayed as coverage tracks (top row in each panel) and stacked alignments of individual reads. The join region shows twice the average genome coverage, and long-read alignments include a high rate of mismatches and poorly aligning reads. Even evaluating only the JCM15448 data, this suggests the region is not correctly assembled; B11220 is a clonally related isolate, so major differences are not expected. This region is present at two contig ends of B11220 that contain telomeric repeats, suggesting that repetitive sequence resulted in a mis-join in the JCM15448 assembly. (This figure was reproduced from [9] under terms of the Creative Commons CC BY license)

5. For sites of rearrangements such as translocations or inversions, we recommend aligning long and short reads back to the final polished assembly (as outlined above) and visually inspecting alignments using Integrative Genomics Viewer (IGV) [18] (Fig. 1b).

The final assembly can then be used for short-read alignment or other genomic comparisons. For a detailed protocol for short-read alignment and variant calling for *C. auris*, see the chapter in this issue by Bagal et al. [19]. To predict gene structures on the final assembly, we have typically run BRAKER [20] using RNA-seq [8, 9].

4 Notes

1. The Canu documentation specifies that 8Gb is sufficient to assemble a bacterial genome in a few CPU hours or as little as 1 h with eight cores.
2. Genome size estimate is based on nine *C. auris* assemblies (B8441, B11220, B11221, B11243, B11245, B11809, B12043, B13463, and B18474).

Acknowledgments

The authors would like to thank Sean Sykes, Bruce Walker, Sarah Young, and the previous members of the genome assembly and analysis group for the development of methods and best practices used in this chapter and José Muñoz and the members of the fungal genomics group for evaluation of genome assemblies.

References

1. Lockhart SR, Etienne KA, Vallabhaneni S, Farooqi J, Chowdhary A, Govender NP, Colombo AL, Calvo B, Cuomo CA, Desjardins CA, Berkow EL, Castanheira M, Magobo RE, Jabeen K, Asghar RJ, Meis JF, Jackson B, Chiller T, Litvintseva AP (2017) Simultaneous emergence of multidrug-resistant *Candida auris* on 3 continents confirmed by whole-genome sequencing and epidemiological analyses. Clin Infect Dis 64:134–140
2. Chow NA, Muñoz JF, Gade L, Berkow EL, Li X, Welsh RM, Forsberg K, Lockhart SR, Adam R, Alanio A, Alastruey-Izquierdo A, Althawadi S, Araúz AB, Ben-Ami R, Bharat A, Calvo B, Desnos-Ollivier M, Escandón P, Gardam D, Gunturu R, Heath CH, Kurzai O, Martin R, Litvintseva AP, Cuomo CA (2020) Tracing the evolutionary history and global expansion of *Candida auris* using population genomic analyses. mBio 11:e03364–19
3. Chow NA, Gade L, Tsay SV, Forsberg K, Greenko JA, Southwick KL, Barrett PM, Kerins JL, Lockhart SR, Chiller TM, Litvintseva AP, US Candida auris Investigation Team (2018) Multiple introductions and subsequent transmission of multidrug-resistant *Candida auris* in the USA: a molecular epidemiological survey. Lancet Infect Dis 18:1377–1384
4. Eyre DW, Sheppard AE, Madder H, Moir I, Moroney R, Quan TP, Griffiths D, George S, Butcher L, Morgan M, Newnham R, Sunderland M, Clarke T, Foster D, Hoffman P, Borman AM, Johnson EM, Moore G, Brown CS, Walker AS, Peto TEA, Crook DW, Jeffery KJM (2018) A *Candida auris* outbreak and its control in an intensive care setting. N Engl J Med 379:1322–1331
5. Rhodes J, Abdolrasouli A, Farrer RA, Cuomo CA, Aanensen DM, Armstrong-James D, Fisher MC, Schelenz S (2018) Genomic epidemiology of the UK outbreak of the emerging human fungal pathogen *Candida auris*. Emerg Microbes Infect 7:43
6. Chow NA, de Groot T, Badali H, Abastabar M, Chiller TM, Meis JF (2019) Potential fifth clade of *Candida auris*, Iran, 2018. Emerg Infect Dis 25:1780–1781
7. Chatterjee S, Alampalli SV, Nageshan RK, Chettiar ST, Joshi S, Tatu US (2015) Draft genome of a commonly misdiagnosed multidrug resistant pathogen *Candida auris*. BMC Genomics 16:686
8. Muñoz JF, Gade L, Chow NA, Loparev VN, Juieng P, Berkow EL, Farrer RA, Litvintseva AP, Cuomo CA (2018) Genomic insights into multidrug-resistance, mating and virulence in *Candida auris* and related emerging species. Nat Commun 9:5346
9. Muñoz JF, Welsh RM, Shea T, Batra D, Gade L, Howard D, Rowe LA, Meis JF, Litvintseva AP, Cuomo CA (2021) Clade-specific chromosomal rearrangements and loss of subtelomeric adhesins in *Candida auris*. Genetics 218:iyab029
10. Ambaraghassi G, Dufresne PJ, Dufresne SF, Vallières É, Muñoz JF, Cuomo CA, Berkow EL, Lockhart SR, Luong M-L (2019) Identification of *Candida auris* by use of the updated Vitek 2 yeast identification system, version 8.01: a multilaboratory evaluation study. J Clin Microbiol 57:e00884–e00819
11. De Coster W, D'Hert S, Schultz DT, Cruts M, Van Broeckhoven C (2018) NanoPack: visualizing and processing long-read sequencing data. Bioinformatics 34:2666–2669
12. Koren S, Walenz BP, Berlin K, Miller JR, Bergman NH, Phillippy AM (2017) Canu: scalable and accurate long-read assembly via adaptive k-mer weighting and repeat separation. Genome Res 27:722–736
13. Kolmogorov M, Yuan J, Lin Y, Pevzner PA (2019) Assembly of long, error-prone reads using repeat graphs. Nat Biotechnol 37: 540–546
14. Li H, Handsaker B, Wysoker A, Fennell T, Ruan J, Homer N, Marth G, Abecasis G, Durbin R (2009) The sequence alignment/map format and SAMtools. Bioinformatics 25: 2078–2079

15. Li H (2018) Minimap2: pairwise alignment for nucleotide sequences. Bioinformatics 34: 3094–3100
16. Walker BJ, Abeel T, Shea T, Priest M, Abouelliel A, Sakthikumar S, Cuomo CA, Zeng Q, Wortman J, Young SK, Earl AM (2014) Pilon: an integrated tool for comprehensive microbial variant detection and genome assembly improvement. PLoS One 9: e112963
17. Kurtz S, Phillippy A, Delcher AL, Smoot M, Shumway M, Antonescu C, Salzberg SL (2004) Versatile and open software for comparing large genomes. Genome Biol 5:R12
18. Robinson JT, Thorvaldsdóttir H, Winckler W, Guttman M, Lander ES, Getz G, Mesirov JP (2011) Integrative genomics viewer. Nat Biotechnol 29:24–26
19. Bagal UR, Phan J, Welsh RM, Misas E, Gade L, Litvintseva AP, Cuomo CA, Chow NA (2021) MycoSNP: a portable workflow for performing whole-genome sequencing analysis of Candida auris. Methods Mol Biol
20. Hoff KJ, Lange S, Lomsadze A, Borodovsky M, Stanke M (2016) BRAKER1: unsupervised RNA-Seq-based genome annotation with GeneMark-ET and AUGUSTUS. Bioinformatics Oxf Engl 32:767–769
21. Li H (2013) Aligning sequence reads, clone sequences and assembly contigs with BWA-MEM. ArXiv13033997 Q-Bio.

Chapter 17

MycoSNP: A Portable Workflow for Performing Whole-Genome Sequencing Analysis of *Candida auris*

Ujwal R. Bagal, John Phan, Rory M. Welsh, Elizabeth Misas, Darlene Wagner, Lalitha Gade, Anastasia P. Litvintseva, Christina A. Cuomo, and Nancy A. Chow

Abstract

Candida auris is an urgent public health threat characterized by high drug-resistant rates and rapid spread in healthcare settings worldwide. As part of the *C. auris* response, molecular surveillance has helped public health officials track the global spread and investigate local outbreaks. Here, we describe whole-genome sequencing analysis methods used for routine *C. auris* molecular surveillance in the United States; methods include reference selection, reference preparation, quality assessment and control of sequencing reads, read alignment, and single-nucleotide polymorphism calling and filtration. We also describe the newly developed pipeline MycoSNP, a portable workflow for performing whole-genome sequencing analysis of fungal organisms including *C. auris*.

Key words Fungal, *Candida auris*, Genomic, Single-nucleotide polymorphisms (SNPs), Whole-genome sequencing (WGS), MycoSNP

1 Introduction

Listed as an Urgent Threat in the most recent US Centers for Disease Control and Prevention (CDC) AR Threats Report, *Candida auris* is a serious public health concern [1, 2]. This fungal pathogen is a multidrug-resistant organism that has emerged in healthcare settings worldwide with several countries observing rapid spread once *C. auris* enters a healthcare setting [3]. Factors contributing to rapid spread include the following: (i) *C. auris* is transmissible with person-to-person spread [4], (ii) *C. auris* can indefinitely colonize a patient's skin and shed onto surfaces [5], and (iii) *C. auris* can persist on surfaces for weeks [6] and is resistant to several commonly used disinfectants [7].

As part of the public health response to prevent and control the spread of *C. auris*, molecular epidemiology using whole-genome

Alexander Lorenz (ed.), *Candida auris: Methods and Protocols*, Methods in Molecular Biology, vol. 2517, https://doi.org/10.1007/978-1-0716-2417-3_17,

sequencing (WGS) data has helped epidemiologists perform strain surveillance and investigate local outbreaks [8, 9]. Specifically, the genomic data has identified new introductions into healthcare settings, detected unrecognized clusters of transmission, and described transmission dynamics within a public health jurisdiction. Additionally, genomic data has helped map the global spread of *C. auris* [10].

Phylogenetic analyses of *C. auris* reveal five major clades (Clades I–V) with all but one reported case clustering to Clades I–IV [11]. Clade I, often referred to as the South Asian clade, was originally characterized with *C. auris* cases from India and Pakistan, Clade II (East Asian clade) with a case from Japan, Clade III (African clade) with cases from South Africa, Clade IV (South American clade) with cases from Venezuela, and Clade V with a case from Iran. Additional analyses with cases from over 20 countries have been sequenced and found to cluster with Clades I–IV, usually adhering to a strong phylogeographic structure [10].

Regarding single-nucleotide polymorphisms (SNPs), tens to hundreds of thousands of SNPs separate the five major clades with only hundreds to a few thousands of SNPs separating *C. auris* cases within Clades I–IV. Interpreting SNP numbers in the context of transmission can be complex as the number of SNPs separating any two samples can vary depending on the methods used to call SNPs. In this chapter, we describe a WGS analysis workflow for *C. auris*. This workflow is currently used by the CDC for routine molecular surveillance in the United States and for outbreak investigations conducted with various global partners. It is a multi-step analysis that includes standard bioinformatics tools. We start with a detailed description of each step and move forward to demonstrate its integration into a portable workflow called MycoSNP.

2 WGS Analysis

For WGS analysis, we will describe the reference genome, quality control of sequencing reads, read alignment to the reference, and variant calling and filtration.

2.1 Selecting a Reference Genome

For reference-based SNP calling approaches, SNPs are detected by aligning sequence reads to a closely related reference genome and identifying individual nucleotide differences [12]. The selection of an optimal reference genome is critical to the fidelity of the SNP analysis [13]. Selecting high-quality closed or nearly complete genome assemblies that, in the case of *C. auris,* consist of seven nuclear contigs corresponding to nearly complete chromosomes will allow for SNP calling at a higher accuracy. This contrasts with draft genomes containing high numbers of contigs. Complete, telomere-to-telomere genome assemblies for isolates spanning the phylogenetic breadth of the four *C. auris* clades and the only isolate described to date from Clade V are publicly available (GenBank accession #: GCA_002759435.2, GCA_003013715.2, GCA_002775015, GCA_003014415, and GCA_016809505.1) [14, 15]. The B8441 genome assembly, corresponding to Clade I, has been widely used; however this assembly contains a few gaps between contigs. Selecting a reference genome that is closely related to the isolates under investigations is ideal for analyses of transmission inferences, and previous studies have demonstrated that variant errors increase as the genetic distance between the reference genome and query genome increases [16]. Particularly when analyzing isolates from a single clade, it is optimal to use the best available assembly for that clade for aligning short Illumina reads, which may not align as well to regions that are more divergent between the clades, the sites of genome rearrangements, or genomic regions only present in or missing from some clades [14, 17].

2.2 Repeat Identification and Masking

Fungal genomes often have low complexity DNA regions such as homo-polymeric runs of nucleotides and other repetitive elements such as transposable elements and micro- and minisatellites [18]. Such regions can be difficult to map, due to read alignments to more than one site in the reference genome [19]. To improve the performance of alignment tools, repeat identification and masking is a frequently used step before downstream variant identification [20]; this may be crucial to incorporate in analyses where false-positive variants are more problematic. Here, genome masking is the process of identifying the repeats and annotating them for downstream processes. Examples of annotation include replacing them with "N" or "X" [21]. Another alternative is soft masking where the repeats are in upper case and the DNA in the non-repeat region is in lower case or vice versa [22]. In this workflow, we use MUMmer (v.4.0) (https://mummer4.github.io/) [23] to perform repeat identification. MUMmer alignment of the reference against itself using the *nucmer* and *show-coords* commands and creates a list of inexact repeats with coordinates by setting the --maxmatch and --nosimplify options. Once the coordinates are obtained in the form of a BED file, the *maskfasta* command from Bedtools (v 2.29.2) (https://bedtools.readthedocs.io/en/latest/) [22] is used to mask the repeats using the default hard masking (i.e., replacing with "N") in the reference genome. Using this approach, around 2% of *C. auris* genome is masked.

2.3 Indexing Reference Genome

The reference genome is indexed for faster identification of putative matches [24]. In this workflow, downstream software Burrows–Wheeler Aligner (BWA) (v 0.7.17) (http://bio-bwa.sourceforge.net/) and the Genome Analysis ToolKit (GATK) (v 4.1.4.1) (https://gatk.broadinstitute.org/hc/en-us) are used, and both require an indexed reference for mapping the reads in the form of coordinates [25–27]. For the BWA alignment, the BWA *index* command with the default "IS" algorithm for constructing the suffix array is used. It creates several files (. ann, .bwt, .fai, .pac, and .sa) that are used internally by BWA during alignment.

For the realignment step in GATK, an indexed reference genome file (.fai) and a dictionary file (.dict) are required. The . dict file describes the content of the reference FASTA file, while the .fai file allows for quick access to a reference base at specific genomic coordinate within a FASTA file. For this, the Samtools (v 1.10) (http://www.htslib.org/) *faidx* and Picard GATK (v 2.22.9) (https://broadinstitute.github.io/picard/) *CreateSequenceDictionary* commands are used [28, 29].

2.4 Downsampling Reads

Downsampling refers to the process of randomly discarding a specific fraction of sequencing reads. When the sequencing coverage of samples is beyond the expected threshold (minimum coverage that gives results equivalent to higher coverage depth), downstream processes require more computational power [30]. Hence, to avoid unnecessary computational usage and achieve results with the same sensitivity, downsampling can be applied for samples with higher genome coverage. It should be considered whether genomes with a high guanine-cytosine (GC) bias should be downsampled; in these cases, a higher default coverage may be needed to have enough reads aligning the low GC regions. In this workflow, the Seqtk (v 1.3) [31] (https://github.com/lh3/seqtk) *sample* command is used for downsampling. Either the sample rate (as a fraction, e.g., 0.7) or coverage (e.g., an integer 70 for 70× coverage) can be specified. If sample rate is specified, the number of sampled reads is calculated as *# of Sampled Reads = Sample Rate * # of Reads*. If desired coverage is specified (as a parameter), the number of sampled reads is calculated as *# of Sampled Reads = (Desired Coverage * Reference Length * # of Reads)/ Read Length*. If desired coverage is greater than the actual estimated coverage (*# of Reads * Read Length/Reference Length*), the sample rate is set to 1, preventing downsampling.

2.5 Quality Control of Reads

Data quality can complicate the interpretation of raw sequencing data as biologically relevant information. Hence, it is critical to invest in proper quality control measures at several stages within a workflow; these measures can be at the nucleic acid extraction, library preparation, and post-sequencing stages [32]. Here, we discuss post-sequencing quality control measures. As each sequencing technology comes with its unique error profiles due to inherent chemistry differences [33], we focus on Illumina data here.

First, an initial assessment of the raw reads is performed by calculating basic statistics. One such statistic is the Phred quality score, which is the estimated probability of error for each base call

[34]. Higher Phred scores indicate more confidence in the assignment of the base call by the sequencer. Often, the read quality degrades over the length of the read [35]. Hence, an average Phred score at each position for all the reads and an average Phred quality score of each read provide the distribution of the scores, as well as determine the overall quality of the sequenced reads. Cutoff Phred scores for *C. auris* in this workflow are described in Subheading 5 (*see* **Note 1**). The average Phred score values for the raw as well as the trimmed reads are available in the Report.pdf output (check Quality Report figure) of FaQCs tool used for quality trimming discussed below [36]. Another statistic is the GC content, which is organism-specific and estimates the percentage of nucleotides called as guanine and cytosine. For a specific species, GC content variation may be observed, and this can be due to priming, size selection, PCR artifacts, or contamination of another organism during library preparation or sequencing errors [37–40]. For *C. auris*, GC content is ~45% [15], and the cutoff range for a passing quality metric in this workflow is described in Subheading 5 (*see* **Note 1**). An additional statistic is sequencing coverage. In SNP analysis, sequencing coverage indicates whether the total number of reads is sufficient to obtain a decent read depth (coverage) so that SNPs are called with sufficient sensitivity [41]. Cutoff sequencing coverage for *C. auris* in this workflow is described in Subheading 5 (*see* **Note 1**).

Second, we address low-quality data, trimming, and filtering using FaQCs (v 2.10) [36] with default parameters, including trimming at quality level 5, filtering based on complexity, and presence of ambiguous bases. FaQCs produces textual and graphically illustrated reports that convey information regarding different aspects of the read quality before and after processing.

2.6 Read Alignment

Read alignment, or read mapping, is a crucial step of aligning high-quality, trimmed reads to a specific location on the reference genome [42]. This process helps reveal variants throughout the genome. In this workflow, BWA MEM (v0.7.17) (http://bio-bwa.sourceforge.net/bwa.shtml) is used to align reads to the reference genome [25]. This aligner automatically chooses local over end-to-end alignment, tolerates more error for longer alignments, supports gapped alignment, utilizes paired-end reads to improve mapping quality, and reports reads aligning equally well to multiple sections of a genome with low mapping quality [25].

BWA MEM outputs a sequence alignment map (SAM) file, and Samtools (http://www.htslib.org/) converts this into a binary alignment map (BAM) file. Additionally, Samtools *sort* command is used to sort the alignments over genetic coordinates [28]. Next, Picard GATK(v.2.22.7) (https://broadinstitute.github.io/picard/) identifies and tags duplicate reads [29]. A duplicate read originates from a single DNA fragment and can arise either during sample preparation or during cluster identification by the sequencing instrument. The Picard *MarkDuplicates* command identifies duplicates and flags them with a hexadecimal value (0×0400) or removes them. Along with the processed BAM file, a metrics file indicating the number of duplicates per read is generated.

A cleaning step is then performed by Picard *CleanSam* command where the mapping quality (MAPQ) value is set to zero for unmapped reads and soft clipping is performed, i.e., masking portions of the read that does not align to the reference. Following this, Picard *FixMateInformation* command verifies if the mate pair information is correct and, if required, corrects it for paired read alignments. Next, the downstream SNP caller requires that each read is marked with a read group, sample, and library id; this is to ensure that variants can be mapped to a specific sample. For this, the Picard *AddOrReplaceReadGroups* command tags the Read Group Id for each record. Lastly, after all the processing steps, the BAM file is indexed using the Samtools *index* command to prepare the BAM file for GATK variant calling.

2.7 Variant Calling

In this workflow, variant calling is performed using GATK (v 4.1.4.1) (https://gatk.broadinstitute.org/hc/en-us), which is intended for calling variants in clonal samples (i.e., single individual) [26, 27]. It is aided by an array of command-line tools that can be used independently in a step-by-step manner or as a workflow.

First, processed BAM files from each sample are passed through GATK *HaplotypeCaller* for variant calling. The ploidy parameter is set to haploid as *C. auris* is a haploid fungal species [43]. The caller searches for regions with variations throughout the genome. For each such region, it locally reassembles and realigns each read to the reference region and identifies potential variant sites that are used to determine the haplotypes [27]. The caller performs pairwise alignment using the PairHMM (Pair Hidden Markov Models) algorithm, then calculates the likelihood of each candidate haplotype site under consideration, and assigns the most likely genotype to the sample. For the alignment step, the mode for emitting the reference confidence score (ERC) parameter with the genomic variant call format (GVCF) option is used to obtain individual sample gVCF files as output. All the gVCF files were merged into a single multi-sample gVCF file using the *CombineGVCFs* tool and further passed for genotype calling. Next, the *GenotypeGVCFs* tool performs joint genotyping on a single-input gVCF file (with one or more samples) which contains the genotype likelihood score produced by the "ERC GVCF" while running the *HaplotypeCaller*. The final output is a VCF file with all samples genotyped.

Each VCF file provides numerous statistical metrics that can help filter low-confidence variants. The *VariantFiltration* tool helps filter variants based on these metrics. In this workflow, we apply a conditional filtering expression that keeps a variant only if it passes at least one of the four chosen statistic criteria; thus, variants that fail all four criteria are filtered. The compound filtering expression used is "QD < 2.0 || FS > 60.0 || MQ < 40.0." Here, the QualityByDepth (QD) is calculated as the confidence in a call measured by the QUAL score normalized by the read depth to prevent inflation when there is deep coverage. The FisherStrand (FS) score is a Phred scale probability that there is a strand bias at the site, and the RMSMappingQuality (MQ) is the root mean square mapping quality over all the reads at a given site [44].

Additional filtering is applied through a customized filtering criterion. These include the parameters Genotype Quality (GQ), which denotes the Phred-scaled confidence that the genotype assignment (GT) is correct; allele depth (AD), which denotes the unfiltered number of reads supporting each of the reported alleles; and depth (DP), which denotes the filtered informative reads supporting each of the alleles. The criteria applied to select a variant are a minimum GQ of 50, a minimum of 80% support to the alternate allele in AD, and a minimum DP of 10. All reference calls with GQ = 0 are retained, and none of the reference bases are filtered. After the filtration step, the *SelectVariants* tool extracts only SNPs for the final VCF files. This step also produces a summary file of the number of sites filtered for each category, for each sample; this file is useful to inspect to identify samples with a high number of filtered sites, which may be due to sample or data quality issues.

2.8 Generating a FASTA File

A customized python script (https://github.com/broadinstitute/broad-fungalgroup/tree/master/scripts/SNPs) is used to extract and concatenate all the SNPs per sample and corresponding reference allele at that position, forming a multi-FASTA file. One parameter in the script specifies, for a site, the maximum number of samples that can have an ambiguous nucleotide for the site to be included. In this workflow, we set this parameter to 10% of the total number of samples. For example, if there are a total of 50 samples, the script is set to allow a maximum of five samples to have an ambiguous base at a given site to be included. This multi-FASTA can be used by phylogenetic software of choice to infer the relationship of the analyzed strains.

3 MycoSNP Workflow Implementation

Sharing portable workflows reduces startup efforts and ensures standardization that promotes comparability, accuracy, and reproducibility among various groups performing *C. auris* WGS. In this section, we present MycoSNP (v 0.19), a variant calling workflow designed by *the Scientific Computing and Bioinformatics Support* team at the CDC.

MycoSNP (https://github.com/CDCgov/mycosnp) is implemented using GeneFlow (https://github.com/CDCgov/geneflow2), an open-source Python-based solution for portable scientific workflows. GeneFlow workflows are modular and use human-readable YAML format to describe steps. These workflows can be executed on a standard Linux workstation and high-performance cluster (HPC) environment. MycoSNP is designed to include all the standard WGS SNP analysis tools described in Subheading 2 and custom scripts which are included in each step's source repository. It is condensed into three sub-workflows that can be executed independently (Fig. 1). Steps for each workflow make use of publicly available bioinformatics containers executed with Singularity (https://sylabs.io/singularity).

3.1 BWA Reference Workflow

The BWA Reference workflow prepares the reference sequence for alignment and variant calling (*see* **Notes 2** and **3**) and should be run only once for each reference sequence (https://github.com/CDCgov/mycosnp-bwa-reference).The steps in the workflow (i) include mask repeats in the reference FASTA file, (ii) index the reference FASTA file, and (iii) create a BWA index for sequence alignment (Fig. 2, **step 1**).

3.2 BWA Pre-processing

The BWA pre-process workflow prepares samples (paired-end FASTQ files) for GATK variant calling (*see* **Notes 4–7**) by (i) quality assessment and trimming, (ii) aligning the reads to the BWA index generated by the "BWA Reference" workflow, and (iii) preparing the alignment files (i.e., BAM files) for the GATK

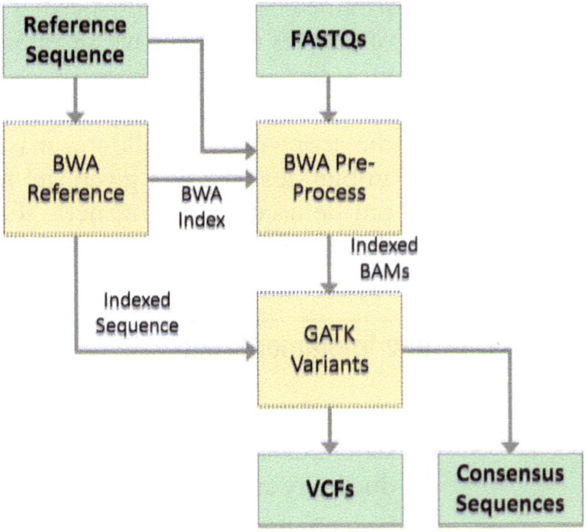

Fig. 1 MycoSNP workflow includes three sub-workflows: BWA Reference, BWA Pre-Process, and GATK Variants

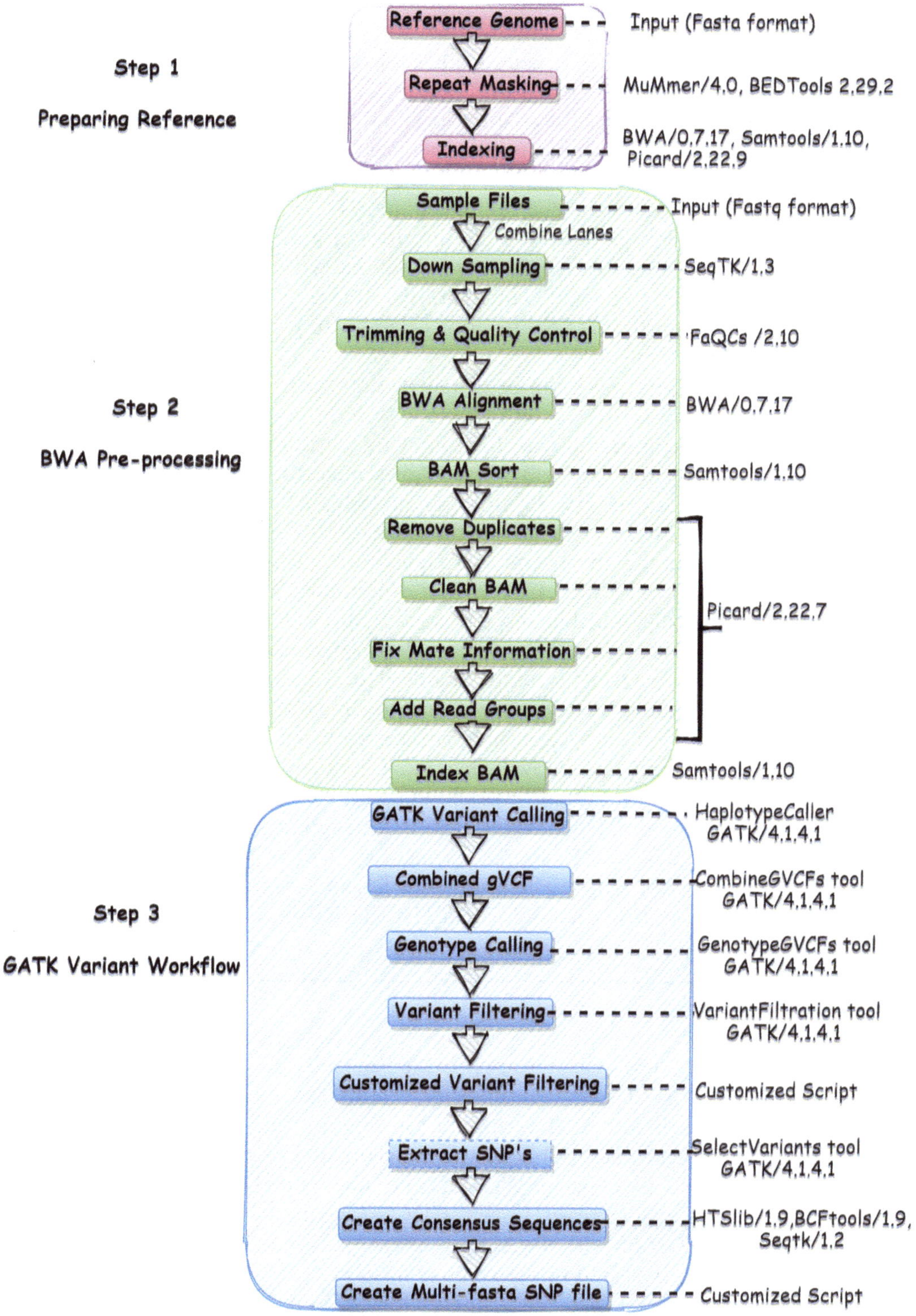

Fig. 2 Whole-genome variant analysis workflow using MycoSNP. Detailed description of each sub-workflow

workflow (https://github.com/CDCgov/mycosnp-bwa-pre-process). This workflow produces two main outputs: a BAM file and a QC report from FaQCs for each sample (Fig. 2, **step 2**).

3.3 GATK Variants

The GATK Variants workflow calls variants and generates a multi-FASTA consensus sequence file (https://github.com/CDCgov/mycosnp-gatk-variants). It considers a batch of multiple samples together when calling variants (i.e., joint genotyping). This workflow produces four outputs: (i) a consensus sequence for each sample, (ii) a combined VCF file with SNPs, (iii) a combined and filtered VCF file with SNPs, and (iv) a consensus multi-FASTA file with only SNP positions (Fig. 2, **step 3**).

3.4 Installation of MycoSNP

MycoSNP can be installed and executed using any computer running the Linux operating system, provided the system has been configured with the following software prerequisites: (i) Git command line tools (https://git-scm.com), (ii) SquashFS (https://tldp.org/HOWTO/html_single/SquashFS-HOWTO/), (iii) Python 3+ (https://www.python.org), (iv) Singularity (https://sylabs.io/singularity), and (V) GeneFlow (https://github.com/CDCgov/geneflow2). Git, Python 3+, Singularity, and SquashFS are generally available and packaged with modern Linux operating systems (e.g., Red Hat, Ubuntu) but may also be installed using those operating systems' standard package managers, if needed. With these tools in place, GeneFlow may be installed using the standard Python 3+ package manager with the following command: "pip3 install geneflow." GeneFlow can then be used to install each of the MycoSNP sub-workflows using the "gf install-workflow . . ." command. Detailed installation instructions are included in the primary MycoSNP source repository (https://github.com/CDCgov/mycosnp), which includes links to instructions for installing and testing each individual MycoSNP sub-workflow.

4 Conclusion

In this chapter, we provided an overview of WGS analysis including SNP calling methods for *C. auris*, an urgent antimicrobial resistant threat. As the capacity for genomic sequencing increases at state and local public health departments, largely in part to SARS-CoV-2 sequencing, WGS of other important pathogens will presumably also increase. In fact, CDC's Mycotic Diseases Branch, in collaboration with the AR Lab Network, is launching FungiNet, a network dedicated to performing WGS of *C. auris* cases. Partners in FungiNet will receive guidance for WGS analysis methods detailed in this chapter. Additional resources will be recommended; these include a *C. auris* benchmark dataset available on GitHub (https://github.com/globalmicrobialidentifier-WG3/datasets) [45], National

Center for Biotechnology Information (NCBI) Pathogen Detection for *C. auris* (https://www.ncbi.nlm.nih.gov/pathogens/), and Microreact, a web application for creating and sharing interactive visualizations of phylogenetic trees and associated metadata [46]. Finally, we hope this information provides the necessary background for *C. auris* WGS analysis that includes SNP calling. We urge the reader to follow the links provided in the text and contact the author for any questions.

5 Notes

1. Currently, CDC adheres to the following quality control metrics for *C. auris* WGS analysis:

QC metrics	Passing value
Average Phred quality score	≥28
Average GC content	42–47.5
Average coverage	≥20×

2. Input samples in the FASTQ (.gz) format and reference genome in the FASTA format are accepted.
3. The reference genome file should be placed in a separate folder.
4. Input FASTQ files from each sample need to be placed in a separate sub-folder within an input folder (with any name) whose path can be provided to the MycoSNP commands.
5. Creating an output folder where all the outputs from each step can be stored is recommended.
6. For each sample, MycoSNP can combine files from different lanes. File naming should convey the lane numbers and should be placed in the same sample sub-folder.
7. SRA dataset can be included, provided the read header format is correct. Use the following command for SRA download for correct read header.

```
fastq-dump --defline-seq '@$sn/$ri' --split-files SRA# --
outdir /path/to/output/dir
```

References

1. U.S. Department of Health and Human Services C (2019) Antibiotic resistance threats in the United States, 2019. CDC. https://www.cdc.gov/drugresistance/pdf/threats-report/2019-ar-threats-report-508.pdf
2. Meis JF, Chowdhary A (2020) Candida auris—"ten years after". J Fungi 6:2
3. Du H, Bing J, Hu T, Ennis CL, Nobile CJ, Huang G (2020) *Candida auris*: epidemiology, biology, antifungal resistance, and virulence. PLoS Pathog 16:e1008921
4. Schelenz S, Hagen F, Rhodes JL, Abdolrasouli A, Chowdhary A, Hall A, Ryan L, Shackleton J, Trimlett R, Meis JF, Armstrong-James D, Fisher MC (2016) First hospital outbreak of the globally emerging *Candida auris* in a European hospital. Antimicrob Resist Infect Control 5:35. https://doi.org/10.1186/s13756-016-0132-5
5. Zhai B, Rolling T, Hohl TM (2021) Exploring *Candida auris* in its habitat. Cell Host Microbe 29(2):150–151. https://doi.org/10.1016/j.chom.2021.01.010
6. Welsh RM, Bentz ML, Shams A, Houston H, Lyons A, Rose LJ, Litvintseva AP (2017) Survival, persistence, and isolation of the emerging multidrug-resistant pathogenic yeast *Candida auris* on a plastic health care surface. J Clin Microbiol 55(10):2996–3005. https://doi.org/10.1128/jcm.00921-17
7. Sexton DJ, Welsh RM, Bentz ML, Forsberg K, Jackson B, Berkow EL, Litvintseva AP (2020) Evaluation of nine surface disinfectants against *Candida auris* using a quantitative disk carrier method: EPA SOP-MB-35. Infect Control Hosp Epidemiol 41(10):1219–1221. https://doi.org/10.1017/ice.2020.278
8. Eyre DW, Sheppard AE, Madder H, Moir I, Moroney R, Quan TP, Griffiths D, George S, Butcher L, Morgan M, Newnham R, Sunderland M, Clarke T, Foster D, Hoffman P, Borman AM, Johnson EM, Moore G, Brown CS, Walker AS, Peto TEA, Crook DW, Jeffery KJM (2018) A *Candida auris* outbreak and its control in an intensive care setting. N Engl J Med 379(14):1322–1331. https://doi.org/10.1056/NEJMoa1714373
9. Chow NA, Gade L, Tsay SV, Forsberg K, Greenko JA, Southwick KL, Barrett PM, Kerins JL, Lockhart SR, Chiller TM, Litvintseva AP (2018) Multiple introductions and subsequent transmission of multidrug-resistant *Candida auris* in the USA: a molecular epidemiological survey. Lancet Infect Dis 18(12):1377–1384. https://doi.org/10.1016/s1473-3099(18)30597-8
10. Chow NA, Muñoz JF, Gade L, Berkow EL, Li X, Welsh RM, Forsberg K, Lockhart SR, Adam R, Alanio A, Alastruey-Izquierdo A, Althawadi S, Araúz AB, Ben-Ami R, Bharat A, Calvo B, Desnos-Ollivier M, Escandón P, Gardam D, Gunturu R, Heath CH, Kurzai O, Martin R, Litvintseva AP, Cuomo CA (2020) Tracing the evolutionary history and global expansion of *Candida auris* using population genomic analyses. mBio 11(2):e03364–03319. https://doi.org/10.1128/mBio.03364-19
11. Lockhart SR, Etienne KA, Vallabhaneni S, Farooqi J, Chowdhary A, Govender NP, Colombo AL, Calvo B, Cuomo CA, Desjardins CA, Berkow EL, Castanheira M, Magobo RE, Jabeen K, Asghar RJ, Meis JF, Jackson B, Chiller T, Litvintseva AP (2017) Simultaneous emergence of multidrug-resistant *Candida auris* on 3 continents confirmed by whole-genome sequencing and epidemiological analyses. Clin Infect Dis 64(2):134–140. https://doi.org/10.1093/cid/ciw691
12. Bakker HC, Switt AI, Cummings CA, Hoelzer K, Degoricija L, Rodriguez-Rivera LD, Wright EM, Fang R, Davis M, Root T, Schoonmaker-Bopp D, Musser KA, Villamil E, Waechter H, Kornstein L, Furtado MR, Wiedmann M (2011) A whole-genome single nucleotide polymorphism-based approach to trace and identify outbreaks linked to a common Salmonella enterica subsp. enterica serovar Montevideo pulsed-field gel electrophoresis type. Appl Environ Microbiol 77(24):8648–8655. https://doi.org/10.1128/aem.06538-11
13. Uelze L, Grützke J, Borowiak M, Hammerl JA, Juraschek K, Deneke C, Tausch SH, Malorny B (2020) Typing methods based on whole genome sequencing data. One Health Outlook 2(1):3. https://doi.org/10.1186/s42522-020-0010-1
14. Muñoz JF, Welsh RM, Shea T, Batra D, Gade L, Howard D, Rowe LA, Meis JF, Litvintseva AP, Cuomo CA (2021) Clade-specific chromosomal rearrangements and loss of subtelomeric adhesins in *Candida auris*. Genetics 218(1):iyab029. https://doi.org/10.1093/genetics/iyab029
15. Muñoz JF, Gade L, Chow NA, Loparev VN, Juieng P, Berkow EL, Farrer RA, Litvintseva AP, Cuomo CA (2018) Genomic insights into multidrug-resistance, mating and virulence in *Candida auris* and related emerging species. Nat Commun 9(1):5346. https://doi.org/10.1038/s41467-018-07779-6
16. Walter KS, Colijn C, Cohen T, Mathema B, Liu Q, Bowers J, Engelthaler DM,

Narechania A, Lemmer D, Croda J, Andrews JR (2020) Genomic variant-identification methods may alter mycobacterium tuberculosis transmission inferences. Microbial. Genomics 6:mgen000418. https://doi.org/10.1099/mgen.0.000418
17. Ambaraghassi G, Dufresne PJ, Dufresne SF, Vallières É, Muñoz JF, Cuomo CA, Berkow EL, Lockhart SR, Luong M-L (2019) Identification of *Candida auris* by use of the updated Vitek 2 yeast identification system, version 8.01: a multilaboratory evaluation study. J Clin Microbiol 57(11):e00884–e00819. https://doi.org/10.1128/jcm.00884-19
18. Wöstemeyer J, Kreibich A (2002) Repetitive DNA elements in fungi (Mycota): impact on genomic architecture and evolution. Curr Genet 41(4):189–198. https://doi.org/10.1007/s00294-002-0306-y
19. Treangen TJ, Salzberg SL (2011) Repetitive DNA and next-generation sequencing: computational challenges and solutions. Nat Rev Genet 13(1):36–46. https://doi.org/10.1038/nrg3117
20. Frith MC, Hamada M, Horton P (2010) Parameters for accurate genome alignment. BMC Bioinform 11(1):80. https://doi.org/10.1186/1471-2105-11-80
21. Tarailo-Graovac M, Chen N (2009) Using RepeatMasker to identify repetitive elements in genomic sequences. Curr Protoc Bioinform. Chapter 4: Unit 4.10. https://doi.org/10.1002/0471250953.bi0410s25
22. Quinlan AR, Hall IM (2010) BEDTools: a flexible suite of utilities for comparing genomic features. Bioinformatics 26(6):841–842. https://doi.org/10.1093/bioinformatics/btq033
23. Delcher AL, Salzberg SL, Phillippy AM (2003) Using MUMmer to identify similar regions in large sequence sets. Curr Protoc Bioinform. Chapter 10: Unit 10.13. https://doi.org/10.1002/0471250953.bi1003s00
24. Jo H, Koh G (2015) Faster single-end alignment generation utilizing multi-thread for BWA. Biomed Mater Eng 26(Suppl 1): S1791–S1796. https://doi.org/10.3233/bme-151480
25. Li H (2013) Aligning sequence reads, clone sequences and assembly contigs with BWA-MEM. arXiv:1303.3997
26. DePristo MA, Banks E, Poplin R, Garimella KV, Maguire JR, Hartl C, Philippakis AA, del Angel G, Rivas MA, Hanna M, McKenna A, Fennell TJ, Kernytsky AM, Sivachenko AY, Cibulskis K, Gabriel SB, Altshuler D, Daly MJ (2011) A framework for variation discovery and genotyping using next-generation DNA sequencing data. Nat Genet 43(5):491–498. https://doi.org/10.1038/ng.806
27. McKenna A, Hanna M, Banks E, Sivachenko A, Cibulskis K, Kernytsky A, Garimella K, Altshuler D, Gabriel S, Daly M, DePristo MA (2010) The Genome Analysis Toolkit: a MapReduce framework for analyzing next-generation DNA sequencing data. Genome Res 20(9):1297–1303. https://doi.org/10.1101/gr.107524.110
28. Li H, Handsaker B, Wysoker A, Fennell T, Ruan J, Homer N, Marth G, Abecasis G, Durbin R (2009) The sequence alignment/map format and SAMtools. Bioinformatics 25(16): 2078–2079. https://doi.org/10.1093/bioinformatics/btp352
29. Institute B Picard Toolkit: Repository. http://broadinstitute.github.io/picard/
30. Sedlar K, Skutkova H, Vitek M, Provaznik I (2016) Set of rules for genomic signal downsampling. Comput Biol Med 69:308–314. https://doi.org/10.1016/j.compbiomed.2015.05.022
31. SEQTK Tookit: Repository. https://github.com/lh3/seqtk
32. Kircher M, Heyn P, Kelso J (2011) Addressing challenges in the production and analysis of illumina sequencing data. BMC Genomics 12(1):382. https://doi.org/10.1186/1471-2164-12-382
33. Pont-Kingdon G, Gedge F, Wooderchak-Donahue W, Schrijver I, Weck KE, Kant JA, Oglesbee D, Bayrak-Toydemir P, Lyon E (2012) Design and analytical validation of clinical DNA sequencing assays. Arch Pathol Lab Med 136(1):41–46. https://doi.org/10.5858/arpa.2010-0623-OA
34. Ewing B, Green P (1998) Base-calling of automated sequencer traces using phred II. Error probabilities. Genome Res 8(3):186–194
35. Simon A (2010) FastQC: a quality control tool for high throughput sequence data. https://www.bioinformatics.babraham.ac.uk/projects/fastqc/
36. Lo C-C, Chain PSG (2014) Rapid evaluation and quality control of next generation sequencing data with FaQCs. BMC Bioinform 15(1): 366. https://doi.org/10.1186/s12859-014-0366-2
37. Hansen KD, Brenner SE, Dudoit S (2010) Biases in Illumina transcriptome sequencing caused by random hexamer priming. Nucleic Acids Res 38(12):e131. https://doi.org/10.1093/nar/gkq224
38. Quail MA, Kozarewa I, Smith F, Scally A, Stephens PJ, Durbin R, Swerdlow H, Turner DJ

(2008) A large genome center's improvements to the Illumina sequencing system. Nat Methods 5(12):1005–1010. https://doi.org/10.1038/nmeth.1270

39. Kozarewa I, Ning Z, Quail MA, Sanders MJ, Berriman M, Turner DJ (2009) Amplification-free Illumina sequencing-library preparation facilitates improved mapping and assembly of (G+C)-biased genomes. Nat Methods 6(4): 291–295. https://doi.org/10.1038/nmeth.1311
40. Nakamura K, Oshima T, Morimoto T, Ikeda S, Yoshikawa H, Shiwa Y, Ishikawa S, Linak MC, Hirai A, Takahashi H, Altaf-Ul-Amin M, Ogasawara N, Kanaya S (2011) Sequence-specific error profile of Illumina sequencers. Nucleic Acids Res 39(13):e90. https://doi.org/10.1093/nar/gkr344
41. Cheng AY, Teo YY, Ong RT (2014) Assessing single nucleotide variant detection and genotype calling on whole-genome sequenced individuals. Bioinformatics 30(12):1707–1713. https://doi.org/10.1093/bioinformatics/btu067
42. Ye H, Meehan J, Tong W, Hong H (2015) Alignment of short reads: a crucial step for application of next-generation sequencing data in precision medicine. Pharmaceutics 7(4):523–541. https://doi.org/10.3390/pharmaceutics7040523
43. Bravo Ruiz G, Ross ZK, Holmes E, Schelenz S, Gow NAR, Lorenz A (2019) Rapid and extensive karyotype diversification in haploid clinical *Candida auris* isolates. Curr Genet 65(5): 1217–1228. https://doi.org/10.1007/s00294-019-00976-w
44. GATK. Hard-filtering germline short variants https://gatk.broadinstitute.org/hc/en-us/articles/360035890471-Hard-filtering-germline-short-variants
45. Welsh RM, Misas E, Forsberg K, Lyman M, Chow NA (2021) *Candida auris* whole-genome sequence benchmark dataset for phylogenomic pipelines. J Fungi (Basel) 7:214. https://doi.org/10.3390/jof7030214
46. Argimón S, Abudahab K, Goater RJE, Fedosejev A, Bhai J, Glasner C, Feil EJ, Holden MTG, Yeats CA, Grundmann H, Spratt BG, Aanensen DM (2016) Microreact: visualizing and sharing data for genomic epidemiology and phylogeography. Microb Genom 2(11): e000093. https://doi.org/10.1099/mgen.0.000093

Chapter 18

Bioinformatic Identification of ABC Transporters in *Candida auris*

Atanu Banerjee, Poonam Vishwakarma, Naveen Kumar Meena, Andrew M. Lynn, and Rajendra Prasad

Abstract

Antifungal resistance mediated by overexpression of ABC transporters is one of the primary roadblocks to effective therapy against *Candida* infections. Thus, identification and characterization of the ABC transporter repertoire in *Candida* species are of high relevance. The method described in the chapter is based on our previously developed bioinformatic pipeline for identification of ABC proteins in *Candida* species. The methodology essentially involves the utilization of a hidden Markov model (HMM) profile of the nucleotide-binding domain (NBD) of ABC proteins to mine these proteins from the proteome of *Candida* species. Further, a widely used tool to predict membrane protein topology is exploited to identify the true transporter candidates out of the ABC proteins. Even though the chapter specifically focuses on a method to identify ABC transporters in *Candida auris*, the same can also be applied to any other *Candida* species.

Key words ABC transporters, Hidden Markov model, NBD-HMM profile, Pfam, HMMER, TOPCONS

1 Introduction

Fungal infections caused by non-*albicans Candida* (NAC) species have witnessed a steep increase in the past decade [1, 2]. Among the NAC species, *Candida auris* has exerted the most profound impact with several outbreaks in critical care settings across the globe [3–5]. Assessment of resistance patterns has revealed a strong propensity of *C. auris* to acquire multidrug resistance (MDR) [6]. Like in *Candida albicans*, drug transporter overexpression shows strong correlation with antifungal resistance phenotypes of *C. auris* [7–10]. The two classes of drug transporters most relevant in the context of fungal pathogens are the ATP-binding cassette

Supplementary Information The online version contains supplementary material available at [https://doi.org/10.1007/978-1-0716-2417-3_18].

Alexander Lorenz (ed.), *Candida auris: Methods and Protocols*, Methods in Molecular Biology, vol. 2517,
https://doi.org/10.1007/978-1-0716-2417-3_18,

(ABC) transporters and the Major facilitator superfamily (MFS) transporters [11]. Out of the two superfamilies, the former is clinically more relevant with a greater number of transporter proteins implicated in antifungal resistance [12]. Among these transporters, *CDR1* in *C. auris* has emerged as a predominant contributor to azole resistance in clinical isolates [7, 9]. Besides, some other ABC transporters have also shown increased expression in clinical and experimentally evolved isolates [6, 8, 13]. In view of the increased relevance of ABC proteins in azole resistance, it is imperative to identify the ABC protein armamentarium present in *C. auris*. To efficiently do this, conventional BLAST-based strategies may not suffice, and more advanced methodologies are needed. One such strategy is to utilize the hidden Markov models. HMMER (current version, 3.3.2) is a software package comprising tools for making probabilistic models of protein sequence domain families—called profile hidden Markov models (profile HMMs). It also includes tools to utilize these profiles to efficiently search sequence databases for homologs and to also create deep multiple sequence alignments for downstream usage. Therefore, to identify ABC protein candidates in a proteome file, one can make use of the HMM profile of the ABC superfamily and use it as a template to search against the entire protein complement of the species. Pfam contains multiple alignments and HMM profiles of complete protein domains [14]. For each protein domain family in Pfam, there exists a seed alignment which is a manually verified multiple alignment of a representative set of sequences [14]. An HMM profile can be easily built from the seed alignment using resources available in the HMMER suite for database searching and alignment purposes. The Pfam record for ABC domain is ABC_tran (PF00005), and it defines the central region that includes the Walker A, B, and Signature motifs and structurally positions them to bind ATP. Thus, the Pfam (PF00005) seed alignment is appropriate to create an ABC system HMM search profile. Once all ABC proteins are mined from the proteome of the organism, a widely used membrane topology prediction tool like TOPCONS [15] can be utilized to extract the transporter candidates out of the ABC proteins. The chapter presents the entire pipeline for the identification and topology prediction of ABC proteins in *C. auris*.

2 Materials

2.1 Data Files

The following two files are required in the analysis:

1. The proteome file of the organism in FASTA format: We will use the proteome file (generated by translating all the ORFs) of the *C. auris* reference strain B8441 present in the *Candida* Genome Database (CGD). The file

("C_auris_B8441_current_orf_trans_all.fasta.gz") is downloadable from http://www.candidagenome.org/download/sequence/C_auris_B8441/current/. The file used was last updated on 9 May 2021.

2. The seed alignment of Pfam ID PF00005 (ABC_tran), downloadable from http://pfam.xfam.org/family/PF00005#tabview=tab3 in FASTA format.

2.2 Computational Resources

1. The HMMER3 suite is downloadable from http://hmmer.org/download.html (*see* **Note 1**).
2. Other system requirements: minimum of 4 GB RAM (preferably 8 GB) and 4–5 GB of free hard disk space.
3. Internet connection and a web browser to download the files, run TOPCONS, and perform BLAST analysis.

3 Methods

3.1 Prediction of Potential ABC Proteins Based on NBD-HMM

To extract the putative ABC proteins within the proteome file, we will utilize a protocol that has been previously reported by us for inventorization of ABC proteins in yeasts [8, 16, 17] (*see* **Note 2**). Herein, a HMM profile created from the seed alignment of ABC_-tran (PF00005) obtained from the Pfam database [14] is used as a query to match against the proteome with the help of the "hmmsearch" function within the HMMER package [18] using the default settings. The HMMER output is further scrutinized to filter out false positives and identify the putative ABC proteins. The steps involved are as follows:

1. The HMM profile is built using the hmmbuild program of HMMER package with default settings while considering all the columns of the alignment. Perform the following steps:
   - Open the terminal.
   - Go to the working directory that contains the Pfam seed alignment file. Run hmmbuild using the command below:

```
hmmbuild abc.hmm PF00005_seed
[abc.hmm: generated output HMM profile]
[PF00005_seed: input alignment file in FASTA format]
```

2. The generated profile is next used to search ABC sequences in the proteome file (target sequence database) using the hmmsearch tool.

- Type the following command in the terminal:

```
hmmsearch abc.hmm C_auris_B8441_current_orf_trans_all.fasta
[abc.hmm: name of the query HMM profile]
[C_auris_B8441_current_orf_trans_all.fasta: name of target
sequence database]
```

- Execution of the above command will show the output in the terminal. To save the terminal output to a file, type the following command:

```
command > hmm_output.txt
[hmm_output.txt: HMM output]
```

3. The "hmm_output.txt" (Supplementary File 1) lists all the sequence top hits along with their E-value and scores (from here onward, referred to as domain score). The hits are sorted by E-value, with the most significant hit appearing first. Since many ABC proteins carry more than one NBD, the E-value and domain score for "full sequence" are further analyzed instead of "best 1 domain" at this stage. The output file also contains the alignment of the candidate sequences with the PF00005

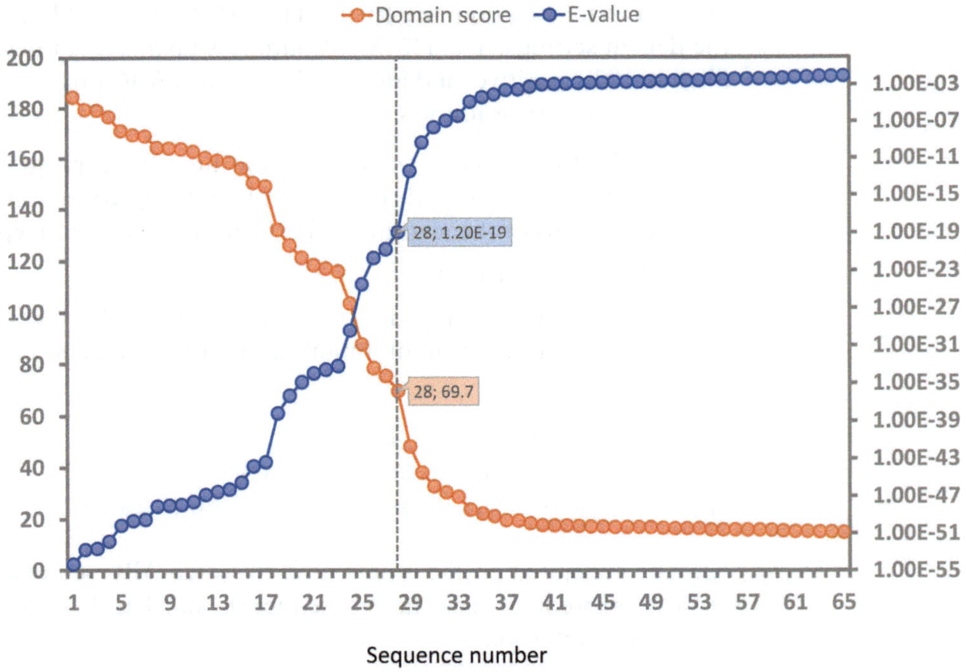

Fig. 1 Plot of E-value and domain score for all the sequences above inclusion threshold in the HMMER output. The vertical line marks the point for sudden change in the domain score and E-value

Table 1
E-value and domain score of all the 28 ABC proteins of *C. auris* identified using HMMER

S. no.	Sequence ID	E-value	Domain score
1	B9J08_005430	4.70E-55	184.5
2	B9J08_004129	1.70E-53	179.5
3	B9J08_004913	2.20E-53	179.2
4	B9J08_004430	1.30E-52	176.6
5	B9J08_002146	7.20E-51	171
6	B9J08_001899	2.20E-50	169.4
7	B9J08_000479	3.10E-50	168.9
8	B9J08_003172	7.90E-49	164.3
9	B9J08_000164	9.60E-49	164.1
10	B9J08_004452	1.20E-48	163.8
11	B9J08_003323	2.40E-48	162.8
12	B9J08_002451	1.40E-47	160.3
13	B9J08_004504	2.90E-47	159.3
14	B9J08_001125	5.20E-47	158.5
15	B9J08_003808	2.80E-46	156.1
16	B9J08_002813	1.50E-44	150.4
17	B9J08_003832	4.10E-44	149
18	B9J08_004446	6.50E-39	132.2
19	B9J08_002232	4.70E-37	126.1
20	B9J08_000705	1.30E-35	121.4
21	B9J08_002706	1.10E-34	118.5
22	B9J08_000657	2.70E-34	117.2
23	B9J08_002389	6.40E-34	116
24	B9J08_003663	4.00E-30	103.6
25	B9J08_005452	3.50E-25	87.6
26	B9J08_001724	2.20E-22	78.5
27	B9J08_000818	1.90E-21	75.5
28	B9J08_000032	1.20E-19	69.7

seed. Even though the output file already denotes proteins that are above the inclusion threshold, a further level of stringency is applied to obtain more accurate candidates. Positive hits above the default threshold are further filtered based on a cutoff defined from the plot of domain score and E-value. It is based on a simple principle that there should be a small difference in the domain scores and E-values between proteins that belong to the same fold, compared to proteins from unrelated folds. This sudden change in the sorted scores is exploited to retrieve proteins belonging to the defined class with minimal false positives. The steps for this analysis are as follows:

- Select all the sequences that are above the inclusion threshold in the HMMER output.
- Prepare a plot (line chart) of E-value and domain score; the domain score can be placed at the primary y-axis, E-value can be put in the secondary y-axis, and the serial number of the sequence can be put in the x-axis. Figure 1 shows the plot generated from this data.
- Carefully inspect the plot to identify a sudden sharp change in the domain score and E-value between two consecutive proteins (*see* **Note 3**). The plot provided in Fig. 1 shows that beyond sequence no. 28 (E-value, 1.20E-19; domain score, 69.7), there is a sudden large drop in the domain score as well as a corresponding significant change in the E-value. Hence all the sequences ranked after sequence no. 28 are probable false positives and hence should be excluded from further analysis. Table 1 lists the sequence IDs of all the 28 proteins and their E-value and domain scores.

3.2 Identification of Putative Transporter Proteins Within the ABC Repertoire of C. auris

Once the final list of predicted ABC proteins is available, the next step is to identify proteins with transport properties. In the case of the ABC superfamily, the ABC proteins with at least a transmembrane domain can be considered as a protein with probable transport capability. To identify the candidate proteins with transmembrane domain(s), many tools are available [15]. However, we will use TOPCONS, which is a widely used program for consensus prediction of membrane protein topology [15]. TOPCONS combines an arbitrary number of topology predictions into one consensus prediction and hence accurately identifies the number and location of the transmembrane helices (TMHs) [15]. Since TOPCONS is available as a webserver, it is a simple-to-use application. The steps to identify the transporter candidates and predict the arrangement of TMHs and/or NBDs in them using TOPCONS and the HMMER output are as follows:

1. Prepare a text file of protein sequences from all the identified ABC proteins in FASTA format and upload at https://topcons.net/pred/. Provide an E-mail ID for notification of the results (*see* **Note 4**).

   The result page will show a tabular summary of the results. The number of TMHs is also available on this page. The results show that the consensus method predicts a number of (20 in

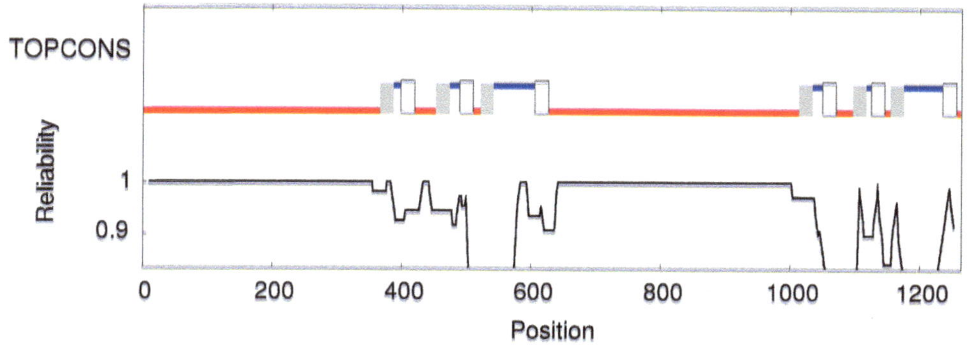

Fig. 2 Representative image of membrane topology output for B9J08_005430 as obtained by running TOPCONS

this case) ABC proteins to contain transmembrane regions. Clicking on each of the "Sequence Name" will show the detailed results including predicted topologies and TMH positions in the proteins, etc. A representative image of TOPCONS-based consensus prediction of the top hit, i.e., B9J08_005430, is shown in Fig. 2.

2. Once the arrangement of the TMHs is decided based on the TOPCONS consensus prediction, the next step is to map the location of the NBDs in the protein sequence. To do this, refer to the "Domain annotation for each sequence (and alignments)" section in the HMMER output file ("hmm_output.txt") which provides details pertaining to the number of domains, their location in the sequence, and statistical parameters. Further, alignment for each domain is also provided. Carefully use the "envelope coordinates" to annotate domain locations on target sequences and compare it with the TOPCONS output for each sequence to predict the arrangement of TMHs and/or NBDs in them. The statistical significance of each NBD should be carefully inspected. This becomes much more important in cases, where more than two domains (not reported for ABC proteins) are detected in the sequence by HMMER. Refer to the conditional E-value, which measures the statistical significance of each domain and the domain score to take a call on their significance. Further, consider the length of each domain (should typically be around 150 amino acids) and the presence of hallmark motifs of ABC proteins in each to confirm their validity (*see* **Note 5**).

   Figure 3 shows the predicted arrangement of TMHs and/or NBDs in all the 28 identified ABC proteins in *C. auris*. It is evident that 20/28 contains transmembrane segments, while the remaining proteins have either one or two NBDs.

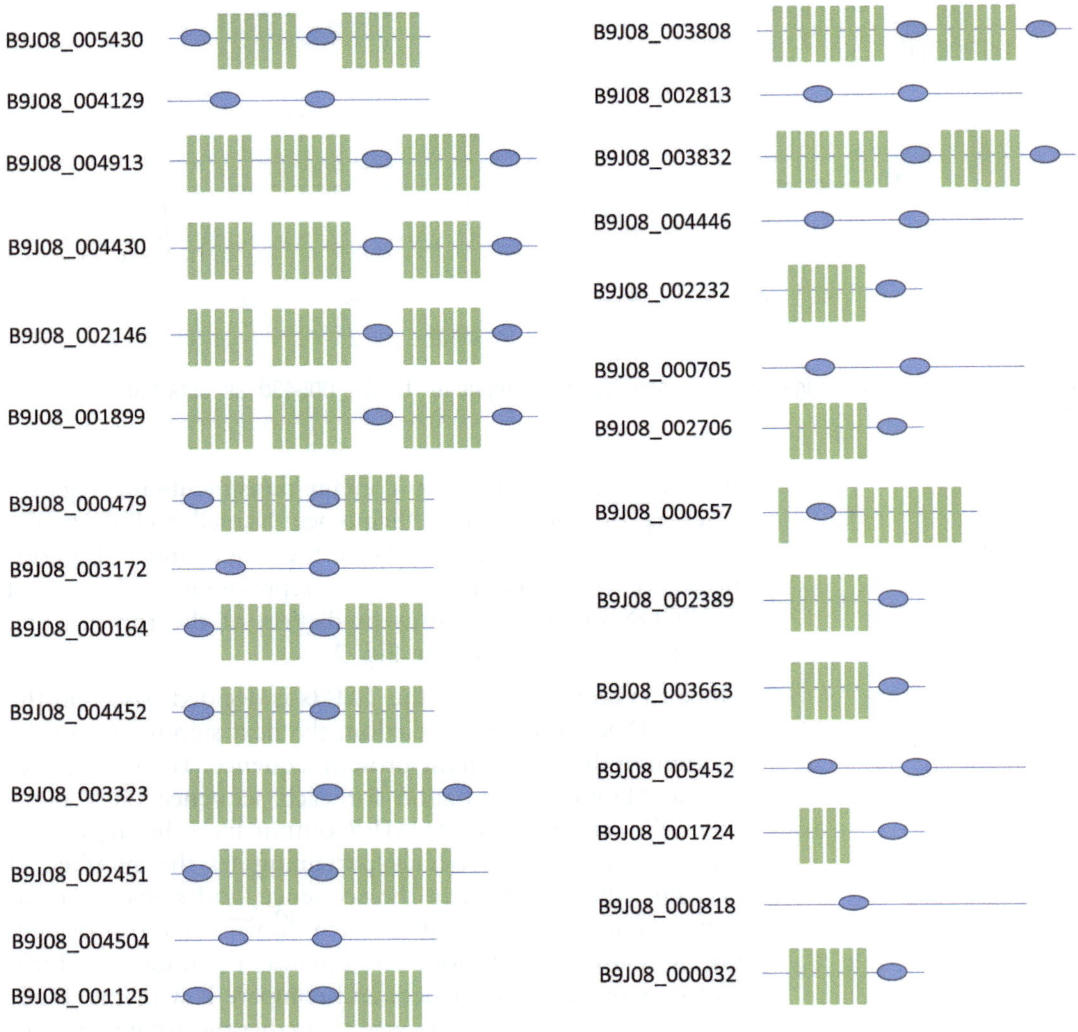

Fig. 3 Cartoon representation of the domain arrangement in ABC proteins of *C. auris.* While the green-colored rectangular bar represents a transmembrane helix (TMH), blue-colored sphere represents the nucleotide-binding domain (NBD). The distance between TMHs and NBDs is not to be scaled with respect to protein sequence length

3. To further confirm the status of ABC proteins with transmembrane domains as transporters, perform BLAST analysis for each of these proteins as query against the proteome of *Saccharomyces cerevisiae* and *Candida albicans.* A considerable proportion of their ABC protein repertoire is experimentally characterized; hence the most plausible transporter proteins can be easily predicted.
4. To perform pBLAST, prepare a text file of protein sequences of the predicted ABC transporter proteins in FASTA format and upload at https://blast.ncbi.nlm.nih.gov/Blast.cgi?PAGE=Proteins.

Table 2
List of predicted ABC transporter proteins of *C. auris* and their orthologs in *Saccharomyces cerevisiae* and *Candida albicans* identified using pBLAST

S. no.	Sequence ID	Predicted orthologs (based on top BLAST hits)	
		S. cerevisiae	*C. albicans*
1	B9J08_005430	YOL075C	Roa1
2	B9J08_004913	Bpt1; Ycf1	CR_08200C_A
3	B9J08_004430	Ybt1	Ycf1
4	B9J08_002146	Ycf1	Ycf1
5	B9J08_001899	Bpt1	Mlt1
6	B9J08_000479	Pdr15	Cdr4
7	B9J08_000164	Pdr5	Cdr1
8	B9J08_004452	Snq2	Snq2
9	B9J08_003323	Yor1	Yor1
10	B9J08_002451[a]	Pdr5	Cdr4
11	B9J08_001125	Snq2	Snq2
12	B9J08_003808	Yor1	Yor1
13	B9J08_003832	Yor1	Yor1
14	B9J08_002232	Atm1	Atm1
15	B9J08_002706	Mdl1	Mdl1
16	B9J08_000657	Adp1	Adp1
17	B9J08_002389	Ste6	Hst6
18	B9J08_003663	Mdl2	Mdl2
19	B9J08_001724	Pxa2	Pxa2
20	B9J08_000032	Pxa1	Pxa1

[a]Protein recently shown to be a functional homolog of Cdr2 by Carolus et al. [6]

5. In the "Choose search set" section, select *Saccharomyces cerevisiae* S288c (taxid:559292) and *Candida albicans* SC5314 (taxid:237561) as the target organisms for the search, and run BLAST (*see* **Note 6**).

   Table 2 lists all the 20 predicted ABC transporters of *C. auris* and their orthologs (identified via pBLAST). The results show that all the 20 proteins show homology to ABC transporters of *S. cerevisiae* and *C. albicans* further supporting their roles as transporters.

4 Conclusions

In this tutorial, we have shown how ABC transporters can be identified in any strain of *C. auris* using HMMER and TOPCONS. Notably, our group has used this same pipeline previously to predict ABC transporters in the *C. auris* CBS10913T strain and obtained the same number of ABC proteins and transporters further emphasizing its utility and reproducibility [8]. This protocol is simple to use and can be exploited to identify ABC proteins, including transporters in any organism, including fungal species. In view of the multiple physiological and clinical implications of ABC transporters from bacteria to human, the protocol presented is extremely useful.

5 Notes

1. HMMER3 is a free to use, open-source software that can be installed locally on UNIX, Linux, and Mac OS/X. The documentation regarding the installation is provided at http://hmmer.org/documentation.html. Windows 10 users can utilize the Windows Subsystem for Linux (WSL) to run HMMER; however, a Linux machine is most preferable for running the HMMER tools. HMMER3 does not have any dependencies other than a C compiler.
2. As an alternative, one can utilize our recently developed webserver, ABC-finder, which is based on the protocol presented in this chapter [19].
3. When facing any difficulty in identifying the sudden change in scores from the plot, the HMMER output file needs to be consulted which lists the E-value and domain scores for best 1 domain as well as full sequence. These parameters can be manually inspected to filter the sequences.
4. While submitting data to the TOPCONS server, one needs to ensure that the E-mail ID is provided. Due to server load, results can take time.
5. In case of any difficulty in positioning the NBDs on the sequence, go to http://pfam.xfam.org/search#tabview=tab1 and submit a FASTA format file containing the sequences to search for matching Pfam families using HMMSCAN. The output contains a pictorial representation of different PF domains identified and their arrangement in the sequence.
6. CGD also lists the orthologs for each gene in other *Candida* and non-*Candida* species. Type the sequence ID in the "Search" bar to retrieve the same. The BLAST results can thus be further cross-checked.

Acknowledgments

This study is supported by funding to AB, AML, and RP from the Department of Biotechnology, Government of India, grant number BT/PR32349/MED/29/1456/2019.

AB acknowledges funding support from Science and Engineering Research Board (SERB) grant number SRG/2019/000514. PV acknowledges the Research Associate fellowship from the Indian Council of Medical Research. NKM thanks the Department of Biotechnology for a Senior Research fellowship award. The authors acknowledge the High-Performance Computational facility (HPCF) at SC&IS, Jawaharlal Nehru University.

References

1. Kumar A, Nair R, Kumar M et al (2020) Assessment of antifungal resistance and associated molecular mechanism in *Candida albicans* isolates from different cohorts of patients in North Indian state of Haryana. Folia Microbiol (Praha) 65:747–754. https://doi.org/10.1007/s12223-020-00785-6
2. Taei M, Chadeganipour M, Mohammadi R (2019) An alarming rise of non-albicans *Candida* species and uncommon yeasts in the clinical samples; a combination of various molecular techniques for identification of etiologic agents. BMC Res Notes 12:779. https://doi.org/10.1186/s13104-019-4811-1
3. Fasciana T, Cortegiani A, Ippolito M et al (2020) *Candida auris*: an overview of how to screen, detect, test and control this emerging pathogen. Antibiotics 9(11):78. https://doi.org/10.3390/antibiotics9110778
4. Lockhart SR, Etienne KA, Vallabhaneni S et al (2016) Simultaneous emergence of multidrug-resistant *Candida auris* on 3 continents confirmed by whole-genome sequencing and epidemiological analyses. Clin Infect Dis 64:134–140. https://doi.org/10.1093/cid/ciw691
5. Forsberg K, Woodworth K, Walters M et al (2019) *Candida auris*: the recent emergence of a multidrug-resistant fungal pathogen. Med Mycol 57:1–12. https://doi.org/10.1093/mmy/myy054
6. Carolus H, Pierson S, Muñoz JF et al (2021) Genome-wide analysis of experimentally evolved *Candida auris* reveals multiple novel mechanisms of multidrug resistance. MBio 12:e03333–e03320. https://doi.org/10.1128/mBio.03333-20
7. Kim SH, Iyer KR, Pardeshi L et al (2019) Genetic analysis of *Candida auris* implicates Hsp90 in morphogenesis and azole tolerance and Cdr1 in azole resistance. MBio 10(1):e02529–e02518. https://doi.org/10.1128/mBio.02529-18
8. Wasi M, Khandelwal NK, Moorhouse AJ et al (2019) ABC transporter genes show upregulated expression in drug-resistant clinical isolates of *Candida auris*: a genome-wide characterization of ATP-binding cassette (ABC) transporter genes. Front Microbiol 10:1445. https://doi.org/10.3389/fmicb.2019.01445
9. Rybak JM, Doorley LA, Nishimoto AT et al (2019) Abrogation of triazole resistance upon deletion of CDR1 in a clinical isolate of *Candida auris*. Antimicrob Agents Chemother 63:e00057–e00019. https://doi.org/10.1128/AAC.00057-19
10. Frías-De-León MG, Hernández-Castro R, Vite-Garín T et al (2020) Antifungal resistance in *Candida auris*: molecular determinants. Antibiotics (Basel, Switzerland) 9:568. https://doi.org/10.3390/antibiotics9090568
11. Prasad R, Banerjee A, Khandelwal NK, Dhamgaye S (2015) The ABCs of *Candida albicans* multidrug transporter Cdr1. Eukaryot Cell 14:1154–1164. https://doi.org/10.1128/EC.00137-15
12. Prasad R, Nair R, Banerjee A (2019) Multidrug transporters of *Candida* species in clinical azole resistance. Fungal Genet Biol 132:103252. https://doi.org/10.1016/j.fgb.2019.103252
13. Muñoz JF, Gade L, Chow NA et al (2018) Genomic insights into multidrug-resistance, mating and virulence in *Candida auris* and related emerging species. Nat Commun 9:5346. https://doi.org/10.1038/s41467-018-07779-6

14. Finn RD, Coggill P, Eberhardt RY et al (2016) The Pfam protein families database: towards a more sustainable future. Nucleic Acids Res 44: D279–D285. https://doi.org/10.1093/nar/gkv1344
15. Tsirigos KD, Peters C, Shu N et al (2015) The TOPCONS web server for consensus prediction of membrane protein topology and signal peptides. Nucleic Acids Res 43:W401–W407. https://doi.org/10.1093/nar/gkv485
16. Wasi M, Khandelwal NK, Vishwakarma P et al (2018) Inventory of ABC proteins and their putative role in salt and drug tolerance in *Debaryomyces hansenii*. Gene 676:227–242. https://doi.org/10.1016/j.gene.2018.07.029
17. Kumari S, Kumar M, Khandelwal NK et al (2018) ABC transportome inventory of human pathogenic yeast *Candida glabrata*: phylogenetic and expression analysis. PLoS One 13:e0202993. https://doi.org/10.1371/journal.pone.0202993
18. Eddy SR (2009) A new generation of homology search tools based on probabilistic inference. Genome Inform 23:205–211
19. Vishwakarma P, Meena NK, Prasad R et al (2021) ABC-finder: a containerized web server for the identification and topology prediction of ABC proteins. Biochim Biophys Acta Biomembr 2021:183640. https://doi.org/10.1016/j.bbamem.2021.183640

Part VI

Exploring Host-Pathogen Interactions of *Candida auris*

Chapter 19

Examining Neutrophil–*Candida auris* Interactions with Human Neutrophils Ex Vivo

Chad J. Johnson and Jeniel E. Nett

Abstract

Neutrophils play a key role in controlling invasive fungal infections. These phagocytes engage and kill fungal pathogens through a variety of effector mechanisms. Here, we describe how to isolate human neutrophils for ex vivo study of neutrophil–*Candida auris* interactions. We detail assays to measure fungal killing, phagocytosis, and reactive oxygen species production.

Key words *Candida auris*, Neutrophil, Phagocytosis, Immunity, Phagocyte, Killing, Reactive oxygen species, Ex vivo

1 Introduction

Neutrophils are crucial for the protection against a variety of invasive fungal infections, including candidiasis [1, 2]. In response to *Candida* in the yeast form, neutrophils typically engage in phagocytosis, engulfing and killing the organisms [3, 4]. Yeast can also trigger neutrophils to generate reactive oxygen species (ROS) through activation of the NADPH oxidase system [3–5]. In prior work, human peripheral blood neutrophils exhibited diminished rates of phagocytosis, ROS production, and killing against *C. auris*, when compared to *C. albicans* [6]. This protocol provides methods to measure these neutrophil responses in the context of encounter with *C. auris*.

The current experiments utilize neutrophils isolated from the human blood. While murine models have served as key in vivo infection models for invasive candidiasis, the inclusion of human samples, when available, is ideal for additional assessment of host–pathogen interactions. As an example, murine and human neutrophils display a species difference in the receptors necessary for the recognition and killing of *Candida* [7, 8]. The current protocol employs a commercial kit that separates neutrophils from other

Alexander Lorenz (ed.), *Candida auris: Methods and Protocols*, Methods in Molecular Biology, vol. 2517,
https://doi.org/10.1007/978-1-0716-2417-3_19,

blood cells by labeling non-neutrophils with antibodies linked to magnetic beads. Because there is little manipulation of the neutrophils, we find less artificial activation of neutrophils for this protocol when compared to those using gradient centrifugation.

2 Materials

2.1 *Candida auris* Propagation

1. Yeast extract–peptone–dextrose (YPD) medium, agar plates: 1% yeast extract, 2% peptone, 2% dextrose, and 2% agar in water (autoclave sterilized on liquid cycle for 20 min).
2. Yeast extract–peptone–dextrose (YPD) medium, broth: 1% yeast extract, 2% peptone, and 2% dextrose in water (autoclave sterilized on liquid cycle for 15 min).
3. Orbital shaker.
4. Commercially purchased physiological Dulbecco's phosphate-buffered saline (DPBS) (-calcium, -magnesium).
5. Centrifuge with swing bucket rotors (50-mL tube compatible rotor).
6. Standard light microscope.
7. Hemocytometer.
8. Vortex mixer.

2.2 Neutrophil Isolation

1. RPMI–FBS medium: commercially purchased RPMI 1640 (without phenol red) supplemented with 2% heat-inactivated fetal bovine serum (FBS) and supplemented with glutamine (0.3 mg/mL).
2. MACSxpress Separator (*see* **Note 1**).
3. Isolation kit for neutrophils: MACSxpress Neutrophil Isolation and MACSxpress Erythrocyte Depletion kits (Miltenyi Biotec Inc., Auburn, CA).
4. 1 M Anticoagulant K^{+}EDTA in DPBS (*see* **Note 2**).
5. Venipuncture supplies: 23-gauge needle, 10–60-mL syringe or vacuum tube blood collection system (depending on volume needed), rubber tourniquet band, 70% isopropyl alcohol to clean the skin surface, adhesive bandage, and sharps waste disposal for safety (*see* **Note 3**).
6. Centrifuge with swing bucket rotors (15- and 50-mL tube compatible rotors).
7. Standard light microscope.
8. Hemocytometer.
9. Viability staining method of choice (*see* **Note 4**).

2.3 Fungal Killing Assay

1. Humidified tissue culture incubator 5% CO_2.
2. Twelve-well flat-bottom microtiter plates.
3. Cell scrapers.
4. Lysis solution: ddH_2O with DNase I at 100 μg/mL (*see* **Note 5**).
5. Individual 5-mL and 1.7-mL tubes.
6. YPD agar plates (*see* Subheading 2.1, **step 1**).
7. Vortex mixer.

2.4 Phagocytosis Assay

1. Humidified tissue culture incubator 5% CO_2.
2. Inverted fluorescence microscope with an appropriate filter set to detect DAPI and FITC.
3. Tissue culture-treated μ-Slides (8 Well) (*see* **Note 6**).
4. 1-mg/mL Calcofluor White stock solution in DPBS.
5. 0.5-mg/mL Calcein AM (Calcein acetoxymethyl) stock in DMSO.

2.5 ROS Assay

1. Fluorescence plate reader.
2. Humidified tissue culture incubator 5% CO_2.
3. Fluorescent oxidative stress dye CMH(2)CFDA (stock at 1.73 mM in DMSO) (*see* **Note** 7).
4. DPBS (*see* Subheading 2.1, **step 4**).
5. 96-well opaque flat-bottom microtiter plates.

3 Methods

3.1 Organisms and Inoculum

1. Add a single colony of *C. auris* from a YPD agar plate to 3-mL YPD liquid medium, and incubate at 30 °C on an orbital shaker at 200 rpm overnight.
2. Dilute 1-mL overnight culture into 20-mL fresh YPD, and propagate for an additional 2 h at 30 °C.
3. Wash yeast cells twice in 10-mL DPBS (centrifuge at 1200 × g for 10 min to pellet). After the final centrifugation step, resuspend pellet in 2-mL DPBS by vortexing (*see* **Note 8**).
4. Enumerate by hemocytometer, and dilute to required concentration in DPBS (*see* Subheadings 3.3, **step 1**; 3.4, **step 1**; and 3.5, **step 2**).

3.2 Human Neutrophil Collection

1. Perform venipuncture to obtain the appropriate blood quantity for experiments (*see* **Notes 3** and **9**).

2. Isolate neutrophils using MACSxpress Neutrophil Isolation and MACSxpress Erythrocyte Depletion kits (Miltenyi Biotec Inc., Auburn, CA) according to the manufacturer's instruction, and suspend neutrophils in 1-mL DPBS/4-mL blood (*see* **Note 10**). One kit bottle utilizes 8-mL blood and yields approximately 2×10^7 neutrophils, although yield may vary by donor (*see* **Note 1**).
3. Enumerate neutrophils using a hemocytometer.

3.3 Killing Assay

1. Dilute rinsed yeast cells in DPBS to 8×10^6 cells/mL, and add to wells (500 μL/well) of a 12-well plate. Dilute neutrophils to 2×10^6 cells/mL in RPMI–FBS, and add 1 mL of cells per well. Include media controls (*see* **Note 11**).
2. Incubate 12-well plates at 37 °C for 4 h in a humidified chamber with 5% CO_2.
3. Very gently remove medium from wells, transferring the contents of each well to individually labeled 5-mL tubes. Gently add 500 μL of lysis solution to each well, and incubate for 20 min at 37 °C.
4. Again, gently remove supernatant from each well (adding to the respective 5-mL tube), replace with 500-μL lysis solution, and incubate for 20 additional min at 37 °C.
5. Carefully use a cell scraper to detach adherent yeast cells from the wells, and move contents to the respective 5-mL tubes. Rinse each well with 500-μL DPBS, and add to the 5-mL tube.
6. Centrifuge collected samples at 1200 × g for 10 min in a swing bucket rotor, and remove supernatant.
7. Suspend the pellet in 700-μL lysis solution, and incubate 20 min at 37 °C (*see* **Note 12**).
8. Centrifuge samples at 1200 × g for 5 min and remove supernatant. Suspend pellet in 1-mL DPBS by vortex-mixing.
9. Serial dilute samples 1:10 in DPBS and plate 10 μL of the dilutions on YPD agar plates.
10. Incubate plate for 24–48 h at 30 °C and enumerate colonies. Comparisons can be made to no neutrophil controls to determine percent killing.

3.4 Phagocytosis Assay

1. In a 1.7-mL labeled microfuge tube, add 900-μL yeast cells (at 1×10^7 cell/mL in DPBS) and 100-μL Calcofluor White (final concentration 100 μg/mL in DPBS). Suspend cells using a vortex mixer, and incubate tubes in the dark for 10 min at room temperature (*see* **Note 13**).
2. Centrifuge yeast at 1200 × g for 2 min. Rinse yeast cells three times with DPBS by removing 950 μL of supernatant, adding 950 μL of DPBS, vortex-mixing, and centrifuging. Following

the final rinse, add 850 μL of DPBS, keeping the overall volume at the original 900 μL. Dilute the stained yeast to 2 × 10^6 cells/mL in DPBS, and keep in the dark until use.

3. Label 1 mL of neutrophils (1 × 10^6 cells/mL in DPBS) by adding 1-μL 0.5-mg/mL Calcein AM stock (final concentration at 0.5 μg/mL). Invert gently several times to mix, and incubate for 10 min at room temperature in the dark.
4. Wash neutrophils two times with DPBS by centrifuging at 200 × g for 2 min to pellet, remove supernatant with a pipette, and suspend neutrophils in 1-mL DPBS by gently pipetting up and down. Suspend the final pellet in 1-mL RPMI–FBS.
5. To the wells of an eight-well μ-Slide, add 100 μL of RPMI–FBS. Follow with the addition of 100-μL labeled *C. auris* (2 × 10^5 cells/well), including controls without *Candida* cells.
6. Add 100-μL labeled neutrophils (1 × 10^5 cells/well), including controls without neutrophils.
7. Incubate the device for 1 h at 37 °C with humidity and 5% CO_2 (*see* **Note 14**).
8. Image the wells of devices at 20–40× magnification, capturing five pre-defined fields of vision per well. We capture brightfield, FITC filter to visualize Calcein AM, and DAPI filter to visualize Calcofluor White at each location (*see* **Note 15**).
9. Using captured images, enumerate the percentage of neutrophils phagocytosing *C. auris* (*see* **Note 16**) (Fig. 1).

3.5 Measurement of ROS

1. Label neutrophils (2 × 10^6 cells/ml) with the fluorescent oxidative stress dye by adding 4.6 μL of CMH(2)CFDA stock at 1.73 mM to 1 mL of neutrophils in DPBS (final CMH(2) CFDA concentration, 8 μM), and incubate at room

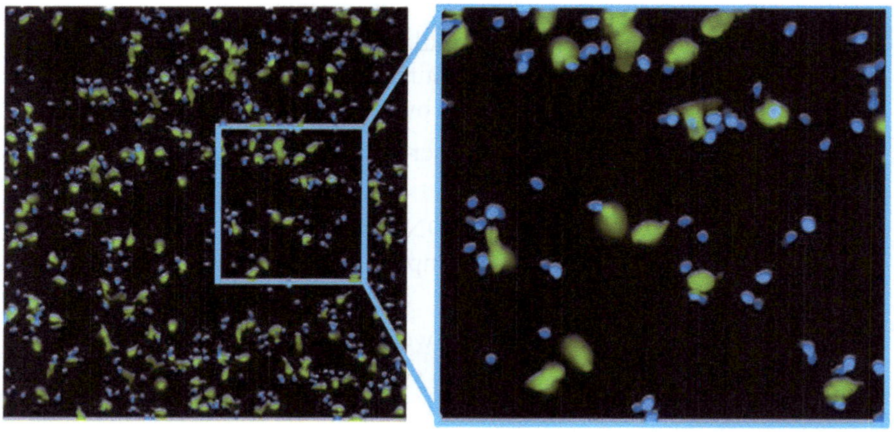

Fig. 1 Human neutrophil interaction with *C. auris*. Isolated neutrophils stained with Calcein AM (green) were co-cultured with Calcofluor White-stained *C. auris* (blue) for 1 h and imaged with fluorescence microscopy

temperature for 10 min in the dark. Rinse labeled neutrophils 3× in 1-mL DPBS (centrifuge 200 × g to pellet, and carefully pipette up and down to suspend the neutrophils), and suspend the final pellet in 1-mL RPMI–FBS.

2. Suspend *C. auris* cells at 4 × 10^6 cells/mL in DPBS.
3. To wells of an opaque flat-bottom 96-well plate, add 50 μL of *C. auris* cells (2 × 10^5 cells/well), including control wells without *Candida* (*see* **Note 17**).
4. Add 100-μL labeled neutrophils (2 × 10^5 cells/well), including control wells without neutrophils (*see* **Note 18**).
5. Immediately record background fluorescence (time = 0) on the plate reader (495/527 nm).
6. Record fluorescence (495/527 nm) every 30 min for 4 h, incubating plate at 37 °C with humidity and 5% CO_2 between reads.
7. Subtract background fluorescence (time = 0) for each condition, and subtract from total fluorescence values prior to data analysis.

4 Notes

1. The MACSxpress system utilizes antibodies labeled with magnetic beads to isolate neutrophils by negative selection, leaving neutrophils unlabeled. We prefer this isolation method because it does not require gradient centrifugation, which can activate neutrophils and alter their phenotypes in assays. However, alternative isolation kits and techniques are available, including EasySep™ Kits or gradient separation-based techniques [9].
2. We use K^+EDTA for studies instead of heparin, as we have found heparin to impact some neutrophil phenotypes.
3. Venipuncture should only be performed by a trained phlebotomist. Venipuncture is a procedure requiring informed consent and approval by an Institutional Review Board.
4. We routinely observe >99% viable cells using MACSxpress system.
5. DNase degrades DNA released from neutrophils following lysis and prevents clumping of fungal cells which can impact the plating efficiency.
6. We find the eight-well μ-Slide works well with our microscope, but the protocol could be adjusted for other devices with coverslip bottoms.
7. Other fluorescent probes and non-fluorescent methods exist to measure reactive oxygen species [10, 11].

8. Time the washing of the culture so that yeasts are not held in DPBS for more than 1 h.
9. If using a syringe for drawing blood, place 2 μL of K^+EDTA stock per mL of blood to be drawn into the syringe prior to venipuncture.
10. When feasible, use 1-mL pipette tips, and pipette gently to prevent neutrophil activation.
11. This is an effector target ratio of 1:2, but other ratios and concentrations can be explored.
12. This step lyses any neutrophils that were removed from wells with the first supernatant removal.
13. This concentration is used during labeling to create a visible pellet during rinse steps, with the concentration to be adjusted later.
14. If the microscope has the capacity to maintain incubation conditions, this step could be skipped, and time-lapse microscopy could be performed starting at time = 0. We use 60-s intervals and show videos at five frames per second.
15. If possible, use an automated stage with preset cube changes.
16. To determine the percent of viable neutrophils, the number of neutrophils retaining the Calcein AM stain in the field of view can be compared to the Calcein AM-labeled neutrophils in the neutrophil-only control. To label dead neutrophils, incubate samples with the membrane-impermeable dye propidium iodide (3 μM) for 15 min.
17. We include three replicate wells for each condition per plate.
18. This is a 1:1 effector-to-target ratio, but other ratios can be explored as long as the neutrophil numbers per well are maintained.

Acknowledgments

This work was supported by the National Institutes of Health (R01AI145939 and R21AI159583), the Burroughs Wellcome Fund (1012299), and the Doris Duke Charitable Foundation (2017074).

References

1. Lehrer RI, Cline MJ (1969) Interaction of *Candida albicans* with human leukocytes and serum. J Bacteriol 98:996–1004
2. Mansour MK, Levitz SM (2002) Interactions of fungi with phagocytes. Curr Opin Microbiol 5:359–365
3. Branzk N, Lubojemska A, Hardison SE et al (2014) Neutrophils sense microbe size and selectively release neutrophil extracellular traps in response to large pathogens. Nat Immunol 15:1017–1025

4. Johnson CJ, Kernien JF, Hoyer AR et al (2017) Mechanisms involved in the triggering of neutrophil extracellular traps (NETs) by *Candida glabrata* during planktonic and biofilm growth. Sci Rep 7:13065
5. Gazendam RP, Van Hamme JL, Tool AT et al (2014) Two independent killing mechanisms of *Candida albicans* by human neutrophils: evidence from innate immunity defects. Blood 124:590–597
6. Johnson CJ, Davis JM, Huttenlocher A et al. (2018) Emerging fungal pathogen *Candida auris* evades neutrophil attack. mBio 9:(4): e01403–18
7. Li X, Utomo A, Cullere X et al (2011) The beta-glucan receptor Dectin-1 activates the integrin Mac-1 in neutrophils via Vav protein signaling to promote *Candida albicans* clearance. Cell Host Microbe 10:603–615
8. Ferwerda B, Ferwerda G, Plantinga TS et al (2009) Human dectin-1 deficiency and mucocutaneous fungal infections. N Engl J Med 361:1760–1767
9. Kuhns DB, Priel DAL, Chu J et al (2015) Isolation and functional analysis of human neutrophils. Curr Protoc Immunol 111:7–23
10. Gomes A, Fernandes E, Lima JL (2005) Fluorescence probes used for detection of reactive oxygen species. J Biochem Biophys Methods 65:45–80
11. Duanghathaipornsuk S, Farrell EJ, Alba-Rubio AC et al (2021) Detection technologies for reactive oxygen species: fluorescence and electrochemical methods and their applications. Biosensors (Basel) 11:5136–5144

Chapter 20

Modeling *Candida auris* Colonization on Porcine Skin Ex Vivo

Chad J. Johnson, J. Z. Alex Cheong, Dhanansayan Shanmuganayagam, Lindsay R. Kalan, and Jeniel E. Nett

Abstract

Candida auris spreads person to person in hospitals and other healthcare facilities. The heightened capacity for *C. auris* to colonize skin contributes to the difficulty in eradicating this drug-resistant and deadly pathogen in nosocomial settings. Models for the study of *C. auris* skin colonization are critical for understanding this virulence trait. In light of the similarities between the skin properties of humans and pigs, pigs represent an ideal model for the investigation of skin–*C. auris* interactions. Here, we describe how to utilize porcine skin for ex vivo studies of *C. auris* colonization.

Key words *Candida auris*, Colonization, Biofilm, Skin, Sweat, Porcine, Ex vivo

1 Introduction

C. auris efficiently colonizes the skin and readily transmits person to person in hospitals and other healthcare settings [1–3]. Patients can become persistently colonized, even when decolonization is attempted [3–7]. This capacity for persistence on the skin remains an obstacle for controlling nosocomial outbreaks. Furthermore, colonization places patients at risk for invasive candidiasis, a disease with mortality reported for up to 60% of patients [1]. Patients with invasive candidiasis frequently require indwelling devices, such as vascular catheters and gastrostomy feeding tubes [1, 4]. *C. auris* likely gains access to the blood through these and other devices. Understanding skin colonization and identifying measures to disrupt this process are important for both the control of outbreaks and prevention of invasive disease.

As a model to study *C. auris* colonization, we utilize the pig skin because of its many similarities to the human skin [8]. As examples, the skins of humans and pigs have similar thicknesses, cellular distributions, and repair processes [9–12]. Miniature pig

Alexander Lorenz (ed.), *Candida auris: Methods and Protocols*, Methods in Molecular Biology, vol. 2517, https://doi.org/10.1007/978-1-0716-2417-3_20,

models more closely mimic human skin composition than conventional swine, which have been propagated for increased muscle mass [9, 13]. We adapted the current protocol for the use of Wisconsin Miniature Swine™, but it can easily be adjusted for utilization of other breeds or animals [8, 13]. We describe how to process the skin from animals that have been sacrificed for other reasons, should they be available. However, biopsies from live animals could be used in this ex vivo model.

2 Materials

2.1 Harvesting and Cleansing of the Pig Skin

1. Biosafety cabinet.
2. Full-thickness excised porcine skin (*see* **Note 1**).
3. Ice and cooler (if needed for transport of the excised skin).
4. Electric razor, razor, or scissors to trim hair (*see* **Note 2**).
5. 70% ethanol.
6. Paper towels.
7. Dulbecco's phosphate-buffered saline (DPBS) (-calcium, -magnesium).

2.2 Obtaining Ex Vivo Samples by Biopsy

1. Humidified tissue culture incubator 5% CO_2.
2. Sterile forceps.
3. 70% ethanol.
4. 12-mm punch biopsy.
5. Single-edge razor blade.
6. 12-well flat-bottom plate.
7. Dulbecco's Modified Eagle Medium (DMEM) supplemented with heat-inactivated fetal bovine serum (FBS) (10%), penicillin (1000 U/ml), and streptomycin (1 mg/ml).
8. DMEM–FBS: DMEM supplemented with 10% FBS only.
9. 1% agarose in DPBS.

2.3 Candida auris Propagation

1. Solid yeast extract–peptone–dextrose (YPD) medium: 1% Bacto™ yeast extract, 2% Bacto™ peptone, 2% Bacto™ agar, and 2% dextrose.
2. Liquid yeast extract–peptone–dextrose (YPD) medium: 1% Bacto™ yeast extract, 2% Bacto™ peptone, and 2% dextrose.
3. Orbital shaker.
4. Dulbecco's phosphate-buffered saline (DPBS) (-calcium, -magnesium)
5. Vortex mixer.

6. Standard light microscope.
7. Hemocytometer.
8. Synthetic sweat medium (*see* **Note 3**).

2.4 Assessment of *C. auris* Colonization on Porcine Skin Samples

1. Humidified tissue culture incubator 5% CO_2.
2. Paraffin wax.
3. Sterile forceps in 70% ethanol.
4. Clean paper towels.
5. DPBS.
6. YPD agar plates (for determination of viable burden) (*see* Subheading 2.3, **step 1**).
7. Vortex mixer.
8. DPBS containing 4% formaldehyde.

3 Methods

3.1 Harvesting and Cleansing of the Pig Skin

Day 1

1. Place full-thickness skin in biosafety cabinet on coated bench covers.
2. Scrub the skin with paper towels. Alternate between DPBS and 70% ethanol washes. Repeat until area and paper towels appear clean.
3. Coarsely trim the hair on the pig skin (*see* **Note 4**).
4. Scrub the pig skin with paper towels. Alternate between DPBS and 70% ethanol until all hair particles are removed, ending with a DPBS wash.
5. Finely trim the hair on the pig skin (*see* **Note 4**).
6. Scrub the pig skin with paper towels. Alternate between DPBS and 70% ethanol until all hair particles are removed, ending with a DPBS wash.

3.2 Obtaining Ex Vivo Samples by Biopsy

Day 1

1. Fill each well of a 12-well plate with 3 mL DMEM supplemented with FBS and antibiotics.
2. Use a 12-mm biopsy punch to cut into the pig skin to the depth of the punch. Use the forceps to pull the biopsy away from the pig skin, and separate from the skin using a razor (Fig. 1a).
3. Lay the skin sample perpendicular on a clean surface. Use a razor to cut the fatty subcutaneous layer from the dermis.
4. Place the skin sample in a well of the 12-well plate.

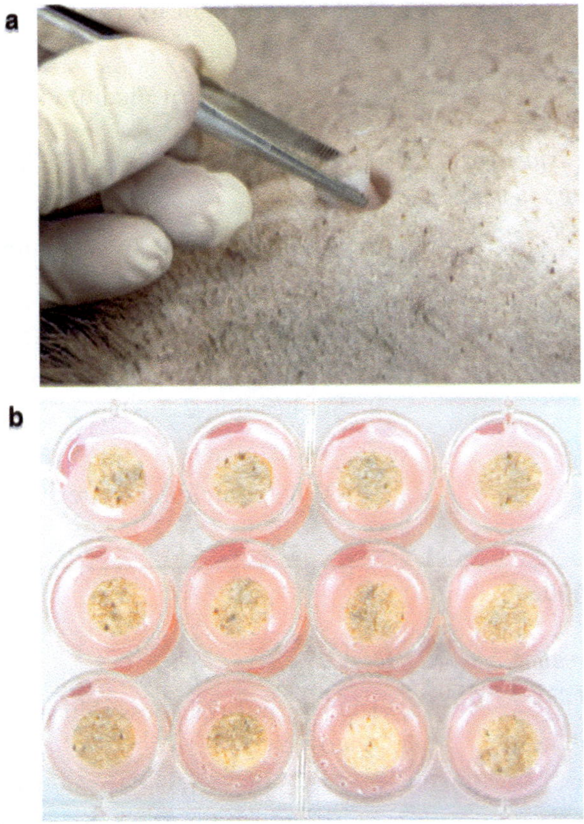

Fig. 1 (**a**) Following washing and shaving, skin samples are obtained with a 12-mm biopsy punch followed by razor excision at the depth of the punch. (**b**) Skin samples are placed in semi-solid tissue culture medium in a 12-well plate, and paraffin is applied to create a barrier at the epidermal surface

5. Repeat for additional biopsies.
6. Incubate samples at 37 °C with 5% CO_2 for 18 h.
7. Warm DMEM–FBS in a 40 °C water bath.
8. Melt 1% agarose in DPBS in a microwave until liquid.
9. Measure the required volume of DMEM–FBS into a warm sterile Erlenmeyer flask (each well requires 3 mL of a 60:40 mixture of 1% agarose in DPBS/DMEM–FBS. This equates to 1.8 mL of agarose and 1.2 mL of medium per well).
10. Add the required volume of molten 1% agarose, and mix by pipetting up and down.
11. Working rapidly, move 3 mL of molten media mixture into each well of a 12-well plate. Close the lid of plate and let cool until solidified (Fig. 1b).
12. Incubate the plate with semi-solid media mixture in a humidified 37 °C incubator with 5% CO_2 for overnight.

3.3 Organisms and Inoculum

Day 1

1. Add a single colony of *C. auris* to 3 mL YPD in a 15-mL conical tube, and incubate on a slant at 30 °C on an orbital shaker at 200 rpm overnight.

Day 2

2. Dilute overnight culture 1:20 in YPD, and propagate for an additional 2 h at 30 °C on an orbital shaker at 180 rpm.
3. Wash yeast cells twice in 10-mL DPBS (centrifuge at 1200 × g for 10 min to pellet). After the final centrifugation step, resuspend pellet in 2 mL DPBS by vortexing (*see* **Note 5**).
4. Enumerate by a hemocytometer, and dilute to 1×10^7 cell/mL in artificial sweat or media of choice (*see* **Notes 3** and **6**).

3.4 Assessment of *C. auris* Colonization on Porcine Skin Samples

Days 2–3

1. Melt paraffin (9.6 mL/12-well plate) (*see* **Note 7**).
2. Transfer skin samples from overnight incubation 12-well plate to 20 mL DPBS in a 50-mL conical tube, and rinse by inverting tube (12 samples/tube).
3. Dry DPBS from the skin surface using clean paper towels, and place skin samples, epidermal side up, in the wells of the 12-well plate containing semi-solid DMEM.
4. Pipette 800 μL paraffin wax around each skin sample at the interface of the epidermis and semi-solid media. Leave a space at the well edge to allow for gas exchange (*see* **Note 8**).
5. Pipette 10 μl of *C. auris* (10^7 cells/mL) to the center of the skin.
6. Incubate plates for 24 h at 37 °C with humidity and 5% CO_2 (*see* **Note 9**).
7. Samples can be processed for viable burden and/or histopathology, as described below, or for scanning electron microscopy (*see* **Note 10**).

3.5 Viable Burden

1. To determine viable burden of colonization, transfer samples to individual tubes containing 10-mL DPBS, and vortex on high for 1 min.
2. Serial dilute (1:10) samples and plate on YPD plates.
3. Incubate plates for 24 h at 30 °C and count colonies.

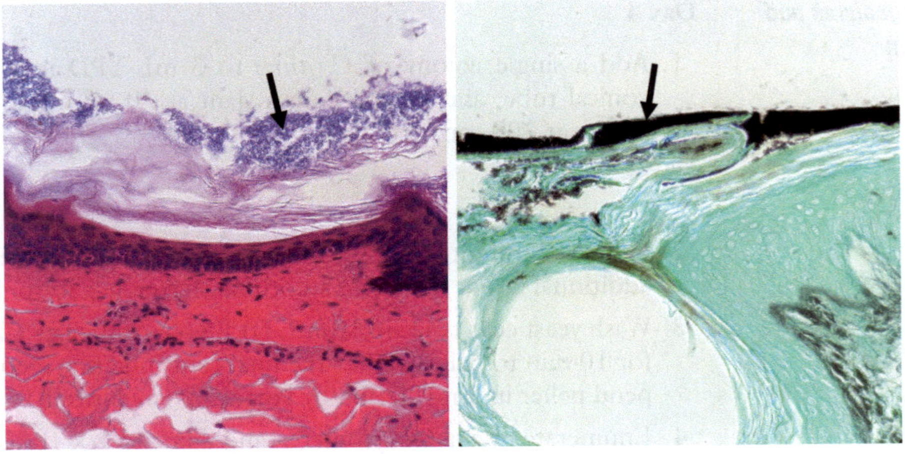

Fig. 2 Following *C. auris* colonization for 24 h, skin samples were fixed, embedded, sectioned for histopathology, stained with hematoxylin and eosin (left) or Grocott's methenamine silver stain (right), and imaged. *C. auris* colonizes the epidermis without invasion of deeper tissues. *C. auris* can be visualized as colonies are adherent to the epidermal surface and are marked with arrows. Organisms are stained purple by hematoxylin and eosin (left) and black by Grocott's methenamine silver stain (right)

3.6 Histopathology

1. Place individual samples gently in 10 mL of DPBS containing 4% formaldehyde.
2. Incubate samples at 4 °C for 24 h.
3. Samples can be fixed, embedded, sectioned, and stained by hematoxylin and eosin (H&E) or Grocott's methenamine silver (GMS) stain to visualize colonization (*see* **Note 11**) (Fig. 2).

4 Notes

1. This protocol is designed to use the skin from animals that have been sacrificed under approved protocols. If animals are not sacrificed, full-thickness skin samples may be obtained directly from animals by biopsy after cleaning the skin surface. Work with animals requires approval by an Institutional Animal Care and Use Committee.
2. We use an electric razor followed by manual razor shaving.
3. Synthetic sweat media is based on the composition of human sweat [8, 14]. Media composition on a %w/v: sodium chloride 2.923, lactic acid 0.1441, urea 1.201, glycine 0.125, L-leucine 0.026, L-cysteine 0.003, L-serine 0.303, L-alanine 0.081, L-arginine 0.001, L-histidine 0.043, L-valine 0.032, L-isoleucine 0.020, L-lysine 0.021, L-phenylalanine 0.021, L-tyrosine 0.034, L-glutamine 0.003, L-aspartic acid 0.055, D-glucose 0.180, creatinine 0.004, sodium pyruvate 0.055,

potassium H carbonate 0.120, NaH2PO4 0.002, calcium sulfate 0.057, squalene 0.002, cholesterol 0.0008, fatty acids (25% lauric acid, 25% palmitic acid, 50% myristic acid) 0.0164. Alternative media, such as RPMI, may be substituted.

4. For initial coarse trimming, we find that an electric razor against the grain of the hair works well. If using scissors, cutting perpendicular to the grain works best. For the final fine trimming, using a manual razor against the grain of the hair works well. Alternatively, the hair may be left longer if the exposed hair is desired.
5. Time washing of culture so that yeasts are not held in DPBS for more than an hour.
6. Pellet diluted samples at 1200 × g for 5 min. Remove supernatant, and resuspend in artificial sweat to remove remaining DPBS.
7. Place the paraffin tube in a water-filled beaker on a hot plate set to 170 °C setting.
8. Prevent 1-mL pipette tips from clogging by pipetting up and down in molten paraffin several times before the initial transfer.
9. If incubating longer than 24 h, skin biopsies should be transferred to new semi-solid media and resealed with paraffin.
10. To assess the structural architecture of adherent *C. auris*, samples can also be processed for scanning electron microscopy, as described [8].
11. Skin samples can be processed for histopathology with fixation, paraffin embedding, sectioning, and staining. To examine the skin tissue, consider hematoxylin and eosin (H&E) staining [15]. Grocott's methenamine silver (GMS) staining can be utilized to assess fungal colonization [16].

Acknowledgments

This work was supported by the National Institutes of Health (R01AI145939 and R21AI159583) and the Burroughs Wellcome Fund (1012299).

References

1. Lockhart SR, Etienne KA, Vallabhaneni S et al (2017) Simultaneous emergence of multidrug-resistant *Candida auris* on 3 continents confirmed by whole-genome sequencing and epidemiological analyses. Clin Infect Dis 64: 134–140
2. Rudramurthy SM, Chakrabarti A, Paul RA et al (2017) *Candida auris* candidaemia in Indian ICUs: analysis of risk factors. J Antimicrob Chemother 72:1794–1801
3. Schelenz S, Hagen F, Rhodes JL et al (2016) First hospital outbreak of the globally emerging *Candida auris* in a European hospital. Antimicrob Resist Infect Control 5:35
4. Adams E, Quinn M, Tsay S et al (2018) *Candida auris* in healthcare facilities, New York,

USA, 2013–2017. Emerg Infect Dis 24: 1816–1824
5. Mulet Bayona JV, Tormo Palop N, Salvador Garcia C et al (2020) Characteristics and management of candidaemia episodes in an established *Candida auris* outbreak. Antibiotics (Basel) 9:558
6. Biswal M, Rudramurthy SM, Jain N et al (2017) Controlling a possible outbreak of *Candida auris* infection: lessons learnt from multiple interventions. J Hosp Infect 97: 363–370
7. Eyre DW, Sheppard AE, Madder H et al (2018) A *Candida auris* outbreak and its control in an intensive care setting. N Engl J Med 379:1322–1331
8. Horton MV, Johnson CJ, Kernien JF et al (2020) *Candida auris* forms high-burden biofilms in skin niche conditions and on porcine skin. mSphere 5:e00910–19
9. Liu Y, Chen JY, Shang HT et al (2010) Light microscopic, electron microscopic, and immunohistochemical comparison of Bama minipig (*Sus scrofa domestica*) and human skin. Comp Med 60:142–148
10. Eaglstein WH, Mertz PM (1978) New methods for assessing epidermal wound healing: the effects of triamcinolone acetonide and polyethelene film occlusion. J Invest Dermatol 71: 382–384
11. Sullivan TP, Eaglstein WH, Davis SC et al (2001) The pig as a model for human wound healing. Wound Repair Regen 9:66–76
12. Summerfield A, Meurens F, Ricklin ME (2015) The immunology of the porcine skin and its value as a model for human skin. Mol Immunol 66:14–21
13. Schomberg DT, Tellez A, Meudt JJ et al (2016) Miniature swine for preclinical modeling of complexities of human disease for translational scientific discovery and accelerated development of therapies and medical devices. Toxicol Pathol 44:299–314
14. Callewaert C, Buysschaert B, Vossen E et al (2014) Artificial sweat composition to grow and sustain a mixed human axillary microbiome. J Microbiol Methods 103:6–8
15. Feldman AT, Wolfe D (2014) Tissue processing and hematoxylin and eosin staining. Methods Mol Biol 1180:31–43
16. Kain R (2017) Histopathology. Methods Mol Biol 1508:185–193

Chapter 21

A Human Ex Vivo Skin Model to Study *Candida auris* Biofilms

Dora E. Corzo-León, Catherine Mark, Donna M. MacCallum, and Carol A. Munro

Abstract

Candida auris can persist for long periods on hospital surfaces and on the skin. *C. auris* has the ability to form drug-resistant biofilms, which can substantially impact on patient outcome. In comparison to *Candida albicans*, *C. auris* has a lower capacity to form biofilms in in vitro models and a higher capacity when tested on animal skin models. Intraspecies variation is shown to exist, with some clinical isolates having greater biofilm capabilities than others. There is a need for models that closely mimic the real niches where infection occurs on human patients. This protocol describes, in detail, a human skin model to study *C. auris* biofilm formation using catheterized and non-catheterized skin.

Key words Human skin model, Catheterized skin, *Candida auris*, Biofilm, Scanning electron microscopy sonication, Colony-forming units

1 Introduction

Invasive infections due to azole- and multidrug-resistant *Candida auris* have been reported in more than 30 countries worldwide in the last decade [1]. *C. auris* is associated with persistence in hospital environments and on plastic surfaces, even after surface treatment with disinfectant agents [2, 3]. Also, *C. auris* is a persistent skin colonizer, and sustained presence of yeasts results in continuous hospital transmission and infection outbreaks [4]. Intravenous catheters (IVC) are commonly used healthcare medical devices used to deliver therapeutic agents that cannot be administered by a different route. If fungi colonize the skin surrounding the IVC, the catheter becomes a direct port of entry for pathogens into the bloodstream, and biofilms formed on the catheter surface act as a reservoir for drug-resistant fungal cells [5].

C. auris biofilm formation contributes to its persistence, over long periods, on surfaces and the skin [6]. The ability of *C. auris* to form biofilms has been reported to be lower compared to

Alexander Lorenz (ed.), *Candida auris: Methods and Protocols*, Methods in Molecular Biology, vol. 2517,
https://doi.org/10.1007/978-1-0716-2417-3_21,

C. albicans when grown on in vitro models [7]. In contrast, the biofilm formation capacity of *C. auris* is higher than *C. albicans* on animal skin models (with or without indwelling catheters) [7, 8]. This highlights the need for models closely mimicking the real niches where infection occurs allowing better characterization of host–pathogen interactions. This protocol describes a human model of catheterized and non-catheterized skin to study biofilm formation by *C. auris*.

2 Materials

2.1 Fungal Strains and Growth Media

1. *Candida auris* strain.
2. YPD broth and agar: 1% (w/v) yeast extract, 2% (w/v) mycological peptone, 2% (w/v) glucose, and 2% (w/v) technical agar for agar plates.
3. Dulbecco's phosphate-buffered saline (PBS).

2.2 Equipment

1. Static incubators (set at 35 °C).
2. Shaking incubators (set at 37 °C).
3. Neubauer hemocytometer.
4. L-shaped cell spreaders (39 × 140 mm).
5. Sterile centrifuge tubes 15 mL and 50 mL.
6. Sterile microcentrifuge tubes 2 mL.
7. Sterile polystyrene Petri dishes.

2.3 Ex Vivo Human Skin Models to Study Fungal Infection

1. Human skin tissue samples without stretch marks and without adipose layer obtained from either abdominal or breast surgeries.
2. Ice box, dry ice, and/or cooling packs for transportation.
3. Sterile forceps, surgical scissors, and scalpels.
4. Sterile polystyrene Petri dishes.
5. Sterile surgical blades, stainless steel size 15.
6. Disposable sterile needles (19 G, 2″, 1.1 × 50 mm).
7. Sterile six-well plate for cell culture.
8. Supplemented DMEM: Dulbecco's Modified Eagle Medium supplemented with 1% (v/v) penicillin (10,000 units/ml)/streptomycin (10 mg/mL) solution and 10% heat-inactivated fetal bovine serum (HI-FBS).
9. Sterile pipette tips (1000 μL, 200 μL, 10 μL).
10. CO_2 laboratory incubator (5% CO_2 and 37 °C).
11. Nitrile gloves.

12. Laboratory disinfectant.
13. 70% ethanol.
14. 24-G Neoflon intravenous catheters.
15. Ultrasonic water bath.

2.4 Recovery and Long-Term Storage of Samples

1. RNA preservation medium (*see* **Note 1**).
2. Sterile microcentrifuge tubes, 2 mL and 1.5 mL (clear and dark).
3. Sterile polystyrene Petri dishes.
4. Stereo microscope.
5. OCT embedding compound.
6. Disposable base molds (15 × 15 × 5 mm).
7. Dry ice.
8. Isopentane (laboratory-grade reagent).
9. Glutaraldehyde buffer: 2.5% glutaraldehyde and 0.1 M sodium cacodylate trihydrate (pH adjusted to 7.2–7.4 with NaOH) in ddH_2O.

2.5 Scanning Electron Microscopy (SEM)

1. Cacodylate buffer: 0.1 M sodium cacodylate trihydrate in ddH_2O, pH adjusted to 7.2–7.4 with NaOH.
2. 1% osmium tetroxide in ddH_2O.
3. 100% ethanol.
4. 100% hexamethyldisilazane (HMDS).
5. Scanning electron microscope.

3 Methods

3.1 Inoculum Preparation

1. One colony from a *C. auris* 24 h YPD agar plate incubated at 35°C is inoculated into 10 mL of YPD broth and incubated at 37 °C overnight in a shaking incubator at 200 rpm (*see* **Note 2**). The *Candida auris* strain used to validate this protocol was VPCI479/P/13 (Figs. 1, 2, 3, and 4) [9].
2. After incubation, the broth culture is centrifuged for 4 min at room temperature at 3700 × g.
3. The supernatant is discarded, and the cell pellet washed twice with PBS and resuspended in 1 mL of PBS. The final fungal cell density is determined by counting cells with a hemocytometer after dilution.
4. The cell suspension is adjusted to 1×10^8 yeast cells per mL, so that skin explants are infected with 1×10^6 yeasts in 10 μL (*see* **Note 3**), which is verified by viable cell counts (colony-forming units) on YPD agar.

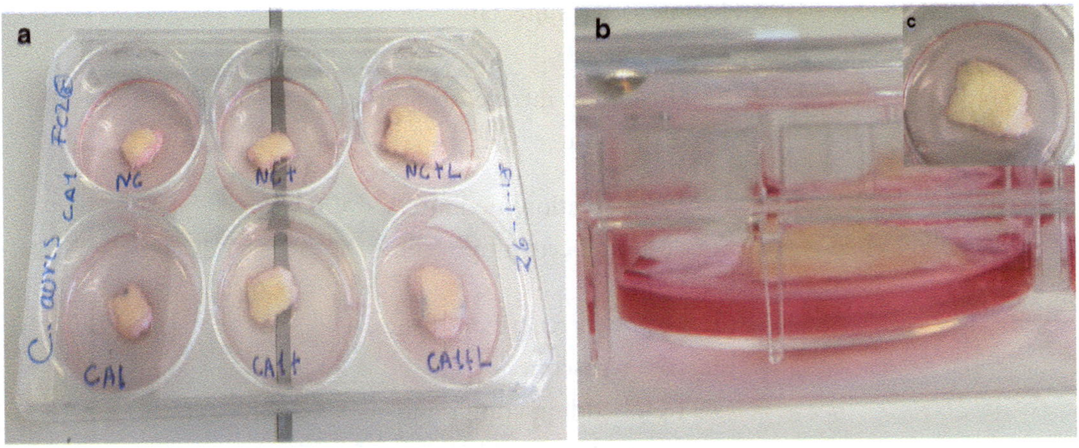

Fig. 1 Skin explants in supplemented DMEM with an air–liquid interphase. (**a**) 1 cm^2 skin explants placed in individual wells of a six-well plate. (**b**, **c**) Skin explants kept with an air–liquid interphase, where the upper epidermis is not in contact with DMEM

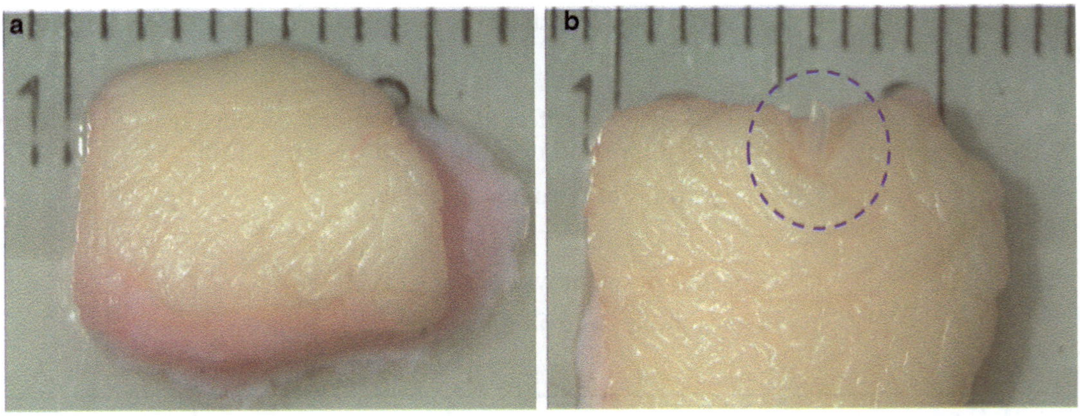

Fig. 2 Skin model of catheterized and non-catheterized skin. (**a**) A wounded skin and (**b**) a catheter inserted in the skin, catheter within the dashed purple circle. The ruler is used to indicate the scale, with the space between each bar corresponding to 1 mm

3.2 Preparation and Infection of Skin Explants

1. Before starting with the preparation of skin explants, working surfaces are cleaned with laboratory disinfectant (follow manufacturer's instructions for applying an appropriate dilution).
2. As reported previously by our group [10], the human skin model is set up as described in **steps 3–13**.
3. Explants are transported and maintained at 4 °C until processed, which occurs within 36 h of surgery.
4. The skin is washed with supplemented DMEM and kept moist in a Petri dish in the same medium.
5. The explant is cut into 1 cm^2 pieces.

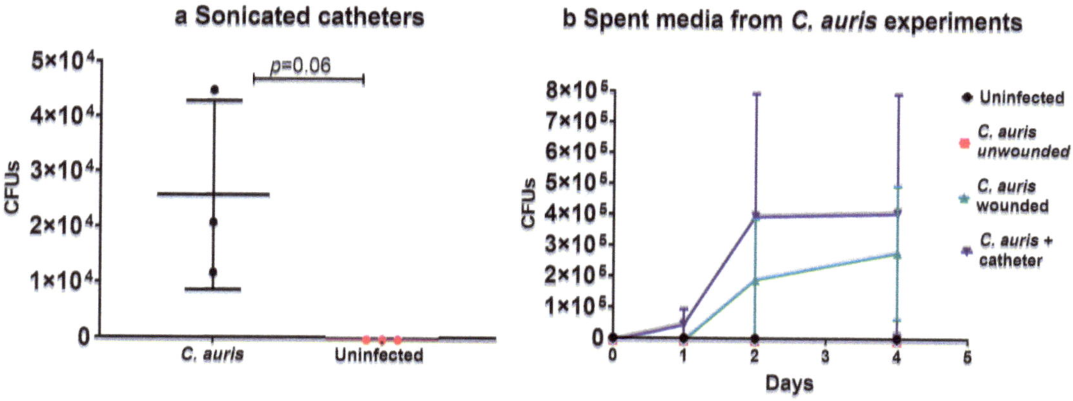

Fig. 3 Quantification of fungi on catheters and in spent media around the skin infected with *C. auris* after 4 days. (**a**) CFUs recovered from sonicated catheters ($n = 3$ biological replicates analyzed in duplicate). Data are mean ± standard deviation, analyzed by student t-test. (**b**) CFUs in spent media at different time points ($n = 2$). Data are mean and range of two experiments analyzed in duplicate (no statistical analysis performed)

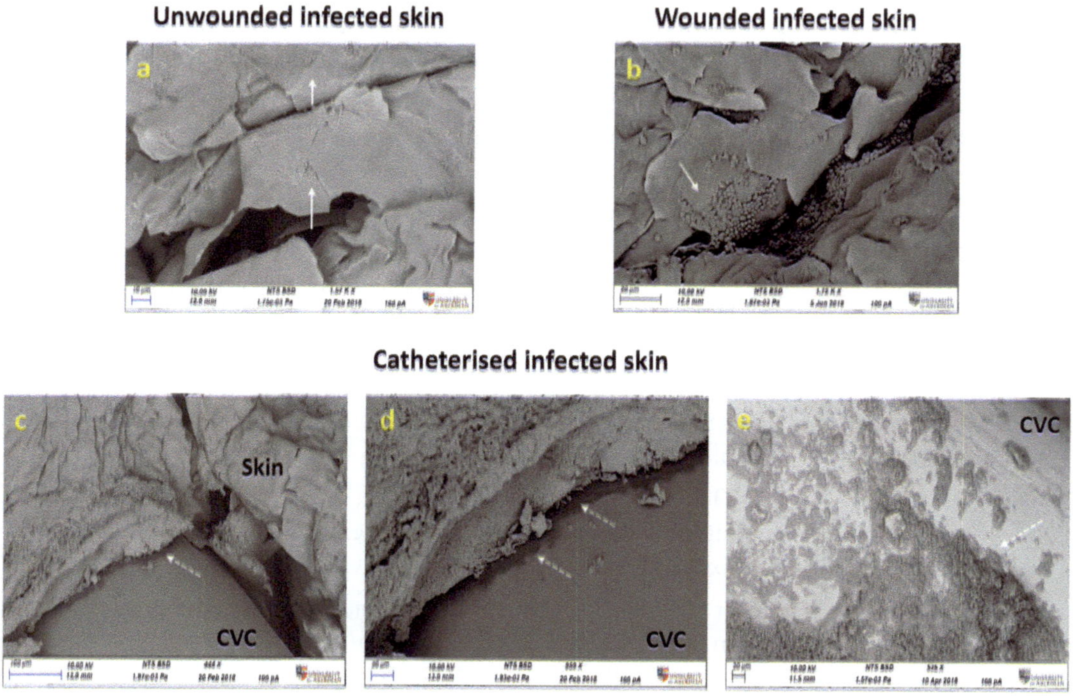

Fig. 4 *C. auris* biofilm formed on infected skin explants and on catheters inserted into the skin on day 4 post-infection. Images were taken by SEM. (**a**) An unwounded infected skin (scale bars 10 μm) and (**b**) an infected wounded skin (scale bar 20 μm). White arrows indicate *C. auris* yeasts. Images are representative of two biological replicates. (**c**) The catheter (CVC) inserted into the skin at low magnification (scale bar 100 μm) and (**d, e**) biofilm formation observed around and on the catheter (scale bars 20 μm)

6. Three different conditions are used to study *C. auris* with the skin explant model: (i) unwounded, (ii) wounded, and (iii) catheterized.
7. For the wounded sample, the surface of each piece of the skin is gentlly wounded using a needle, without penetrating the entire skin thickness, by pricking six to ten times with a needle to approximately 2–3 mm depth.
8. Each piece of skin is placed in an individual well of a six-well plate. An air–liquid interphase is maintained by adding 1 mL of supplemented DMEM (Fig. 1).
9. The explant (wounded or unwounded) is inoculated by applying 10 μL of fungal suspension directly on to the epidermis (*see* **Note 4**). Uninfected skin controls are included in all experiments.
10. For the catheterized explant, a 24-G Neoflon catheter is inserted through the epidermis, reaching midway through the dermis. The catheter is inserted at one edge of the skin (Fig. 2). After the catheter is inserted, the fungal inoculum is applied to the opposite edge. Uninfected controls are included in experiments.
11. A well containing medium only is added as a negative control and serves as a control for contamination.
12. Explants are incubated at 37 °C and 5% CO_2 in a CO_2 incubator for 4 days.
13. The medium is changed every 24 h, and the spent medium is stored in 2 mL tubes for subsequent analysis. Recovered spent medium is stored at −80°C.

3.3 Recovery and Long-Term Storage of Samples

1. After 4 days of incubation, skin samples are recovered and placed in a Petri dish.
2. Depending on the experiment, catheters are also recovered, either still inserted in the skin explant or separated from skin explants.
3. If recovered separately from skin explants, catheters are washed in PBS to remove planktonic fungi and then placed in 1 mL sterile PBS in a microcentrifuge tube.
4. Catheters are sonicated using an ultrasonic water bath at an amplitude of 100 μm for 60 s (equivalent to 200–240 W).
5. Serial dilutions (tenfold) of the resulting cell suspension are prepared, and 100 μL of each is plated onto YPD agar to determine viable cell counts by counting colony-forming units (CFUs). Plates are incubated at 35 °C for 24 h to determine CFUs (Fig. 3).

6. CFUs can also be determined from the recovered medium used for skin culture. After the medium is recovered, serial tenfold dilutions are prepared, and 100 μL samples plated on YPD agar as described in **step 5**.
7. Prior to processing, the macroscopic appearance of explants is evaluated by the eye, but images are also captured on a stereo microscope.
8. Recovered skin samples are processed depending on further analyses and planned analyses (described in **steps 9–11**).
9. For histology, tissue samples are placed into molds, embedded in OCT compound, and flash-frozen with dry ice and isopentane. These samples are stored at −20 °C for immediate analysis (up to 30 days) or at −80 °C for long-term storage (up to 1–2 years). Samples may be paraffinized for long-term storage (years to decades) (*see* **Note 5**).
10. For scanning electron microscopy (SEM), tissue samples are fixed in glutaraldehyde buffer overnight at 4 °C. Fixing is started as soon as skin samples are removed from the growth medium.
11. For RNA–protein extraction, tissue samples are cut into smaller pieces and placed in a microcentrifuge tube with RNA preservation medium for subsequent RNA extraction. RNA preservation medium to sample ratio volume used is 5:1 (*see* **Note 6**). These samples are stored at −80 °C for long-term storage (indefinitely) or at −20 °C for more immediate use (up to a month).

3.4 Confirmation of Biofilm Formation on Catheters and Yeast Growth on Skin Explants by SEM

1. Tissues for SEM are immediately fixed in glutaraldehyde buffer and incubated overnight at 4 °C.
2. Samples are washed twice with cacodylate buffer for 5 min.
3. Tissues are treated with 1% osmium tetroxide for 1 h and then washed three times with cacodylate buffer for 5 min each. Samples are dehydrated in an ethanol gradient (70%, 80%, 90%), with samples incubated for 10 min in each concentration. Samples are incubated three times in 100% ethanol for 10 min. Finally, samples are incubated for 3 min in HMDS and placed in a desiccator for 25 min to remove any residual water.
4. Samples are observed and images captured using a scanning electron microscope. Images are generated by detection of secondary electrons (SE1) and backscatter electron detection (NTS BSD) (*see* **Note 7**) and captured at 10 kV resolution and at different magnifications (Fig. 4).

4 Notes

1. RNA preservation medium is commercially available as RNAlater® Tissue storage reagent.
2. 35 °C and 37 °C temperatures are used to culture *C. auris* following recommendations of reference laboratories [11, 12] including the UK National Mycology Reference Laboratory (MRL), Public Health England, Bristol, United Kingdom [13].
3. It is important to use an inoculum volume of 10 μL to ensure that it remains on the skin surface and does not contaminate the surrounding growth medium.
4. The use of a small volume avoids contact with the dermis and leakage into the surrounding medium.
5. OCT-embedded samples can be processed for a wide range of histological and fluorescent microscopy assays.
6. For RNA extraction, it is recommended that the skin is processed to the smallest size possible, ideally to the consistency of a paste or powder, as this improves RNA yield. We suggest that cut samples are ground in a mortar and then flash-frozen with liquid nitrogen to ground the tissue to powder. Alternatively, after liquid nitrogen freezing, bead beating can be used to pulverize the tissue samples.
7. SEM images generated by NTS BSD have better quality than SE1 if the catheters make images too bright on SE1.

Acknowledgments

This project was funded by a Wellcome Trust Strategic Award for Medical Mycology and Fungal Immunology (097377) and Medical Research Council Centre for Medical Mycology (MR/N006364/1), previously at the University of Aberdeen, now at the University of Exeter.

References

1. Chen J, Tian S, Han X et al (2020) Is the superbug fungus really so scary? A systematic review and meta-analysis of global epidemiology and mortality of *Candida auris*. BMC Infect Dis 20(1):827. https://doi.org/10.1186/s12879-020-05543-0
2. Ledwoch K, Maillard JY (2018) *Candida auris* dry surface biofilm (DSB) for disinfectant efficacy testing. Materials (Basel) 12(1):18. https://doi.org/10.3390/ma12010018
3. Kean R, Sherry L, Townsend E, Ramage G et al (2018) Surface disinfection challenges for *Candida auris*: an *in-vitro* study. J Hosp Infect 98(4):433–436. https://doi.org/10.1016/j.jhin.2017.11.015
4. Eyre DW, Sheppard AE, Jeffery KJM et al (2018) A *Candida auris* outbreak and its control in an intensive care setting. N Engl J Med 379(14):1322–1331. https://doi.org/10.1056/NEJMoa1714373

5. Akbari F, Kjellerup BV (2015) Elimination of bloodstream infections associated with *Candida albicans* biofilm in intravascular catheters. Pathogens 4(3):457–469. https://doi.org/10.3390/pathogens4030457
6. Short B, Brown J, Kean R et al (2019) *Candida auris* exhibits resilient biofilm characteristics *in vitro*: implications for environmental persistence. J Hosp Infect 103(1):92–96. https://doi.org/10.1016/j.jhin.2019.06.006
7. Horton MV, Johnson CJ, Nett JE et al (2020) *Candida auris* forms high-burden biofilms in skin niche conditions and on porcine skin. mSphere 5(1):e00910–19. https://doi.org/10.1128/mSphere.00910-19
8. Vila T, Montelongo-Jauregui D, Jabra-Rizk MA et al (2020) Comparative evaluations of the pathogenesis of *Candida auris* phenotypes and *Candida albicans* using clinically relevant murine models of infections. mSphere 5(4): e00760–20. https://doi.org/10.1128/mSphere.00760-20
9. Sharma C, Kumar N, Chowdhary A et al (2016) Whole genome sequencing of emerging multidrug resistant *Candida auris* isolates in India demonstrates low genetic variation. New Microbes New Infect 13:77–82. https://doi.org/10.1016/j.nmni.2016.07.003
10. Corzo-León DE, Munro CA, MacCallum DM (2019) An *ex vivo* human skin model to study superficial fungal infections. Front Microbiol 10:1172. https://doi.org/10.3389/fmicb.2019.01172
11. Rhodes J, Abdolrasouli A, Farrer RA, Cuomo CA, Aanensen DM, Armstrong-James D, Fisher MC, Schelenz S (2018) Genomic epidemiology of the UK outbreak of the emerging human fungal pathogen *Candida auris*. Emerg Microbes Infect 7(1):43. https://doi.org/10.1038/s41426-018-0045-x
12. Ben-Ami R, Berman J, Novikov A, Bash E, Shachor-Meyouhas Y, Zakin S, Maor Y, Tarabia J, Schechner V, Adler A, Finn T (2017) Multidrug-Resistant *Candida haemulonii* and *C. auris*, Tel Aviv, Israel. Emerg Infect Dis 23(1):195–203. https://doi.org/10.3201/eid2302.161486
13. Borman AM, Szekely A, Johnson EM (2017). Isolates of the emerging pathogen *Candida auris* present in the UK have several geographic origins. Med Mycol 55(5):563–567. https://doi.org/10.1093/mmy/myw147

Chapter 22

In Vivo Skin Colonization and Decolonization Models for *Candida auris*

Janet Herrada, Ahmed Gamal, Lisa Long, and Mahmoud A. Ghannoum

Abstract

Candida auris infections present a critical problem to the healthcare system in many parts of the world. This yeast clinically manifests as a disseminated candidiasis which can be life-threating for susceptible individuals, as well as cutaneous and wound infections. Moreover, *C. auris* can colonize the skin and act as a nidus of infection. Importantly, this emerging yeast unlike other *Candida* spp. has demonstrated multidrug resistance; thus its eradication can be challenging. Animal models are important to gain insight into the pathogenesis of this infection, as well as play a significant role in drug development. In this chapter, we describe two *C. auris* animal models: a cutaneous infection guinea pig model and a skin decolonization mouse model.

Key words *Candida auris*, Mouse models, Animal model, Guinea pig model of cutaneous infection, Disseminated candidiasis

1 Introduction

Candida auris is an emerging fungus that presents a serious global threat mainly because it is multidrug resistant which makes it difficult to treat, which has led the Centers for Disease Control and Prevention (CDC) to designate it as an urgent-need resistant organism. Some *C. auris* strains are resistant to all three commercially available classes of antifungals (azoles, polyenes, and echinocandins) [1]. Importantly, *C. auris* has the ability to colonize the skin and persist on environmental surfaces (e.g., bedding, tables, counters, etc.) in healthcare settings [2]. In some patients *C. auris* can enter the bloodstream and disseminate throughout the body, causing serious invasive infections and high attributable mortality. Patients who have been in the intensive care unit for long periods or have a central venous catheter placed and have previously received antibiotics or antifungal medications, appear to be at higher risk of infection with this yeast [3]. Additionally, this

Alexander Lorenz (ed.), *Candida auris: Methods and Protocols*, Methods in Molecular Biology, vol. 2517,
https://doi.org/10.1007/978-1-0716-2417-3_22,

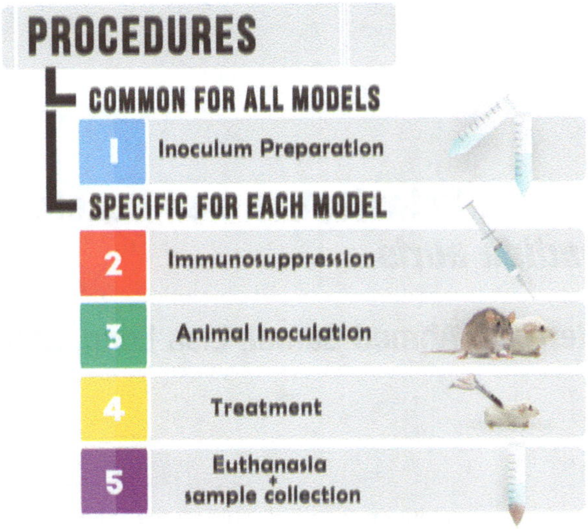

Fig. 1 Summary of steps utilized in both the guinea pig and mouse models of *C. auris* skin colonization and decolonization

pathogen is difficult to identify using conventional methodologies and thereby under-reported. Misidentification may lead to inappropriate management resulting in outbreaks of invasive infections [4]. There is an urgent need to develop novel antifungal agents that target this resistant threat. Having optimized animal models for this emerging yeast is crucial for (1) discovery of drugs that are efficacious, (2) evaluating adverse effects (toxicity, drug interactions), and (3) gaining insight into the pathogenesis/virulence of this organism.

In this chapter, we describe *C. auris* animal models: cutaneous infection guinea pig model and a skin colonization mouse model (Fig. 1).

2 Materials

2.1 C. auris Inoculum Preparation

1. *C. auris* strain of choice.
2. Disposable hemocytometers.
3. Sterile 50-mL polypropylene conical tubes.
4. Serological pipette.
5. Adjustable volume micropipettes.
6. Microscope (white light).
7. Vortex mixer.
8. Centrifuge.
9. Manual hand tally counter.

10. Orbital platform shaker.
11. General-purpose incubator.
12. Sterile phosphate-buffered saline (PBS) pH 7.2 (1×): 210-mg/L potassium phosphate monobasic, 9.0-g/L sodium chloride, and 726-mg/L sodium phosphate dibasic (*see* **Note 1**).
13. 0.85% saline (NaCl) solution in water.
14. Sabouraud dextrose broth (SDB): 20.0-g/L dextrose, 5.0-g/L pancreatic digest of casein, and 5.0-g/L peptic digest of animal tissue (*see* **Note 2**).
15. Potato dextrose agar (PDA) with antibiotics: 4.0-g/L potato starch, 20.0-g/L dextrose, 0.125-g/L chloramphenicol, 0.03-g/L gentamicin (sulfate salt), and 15.0-g/L agar (*see* **Note 3**).
16. Stir bar.
17. Aluminum foil.
18. 2-mL microcentrifuge tubes.
19. Sterile polystyrene Petri dishes.
20. Granulated agar (solidifying agent).
21. Distilled water.

2.2 Guinea Pig Immunosuppression and Inoculation

1. Male albino guinea pigs, 450–500 g.
2. Prednisolone.
3. Scalp vein set 23-G × ¾" for subcutaneous (SC) administration.
4. 3-mL syringes.
5. Adjustable volume pipettes.
6. Vortex mixer.
7. Sterile pipette tips.
8. Ketamine/xylazine combination (80-mg/mL ketamine +5-mg/mL xylazine): 8.0-mL ketamine (100 mg/mL), 1.0-mL xylazine (100 mg/mL), and 1.0-mL sterile water.
9. 0.03-mg/mL buprenorphine: 1-mL buprenorphine (0.3 mg/mL) and 9-mL sterile water.
10. 5-mg/mL atipamezole solution.
11. Clippers.
12. 2.5-cm^2 stencil square.
13. Permanent marker.
14. 5″ curved hemostats.
15. 150-grit sand paper.

16. Water-circulating pad or hot snaps to provide supplemental heat during anesthesia.
17. Petrolatum ophthalmic ointment.
18. 18-G needles.

2.3 Guinea Pig Treatments

1. Drug to be administered.
2. Adjustable volume pipette.
3. Sterile pipette tip.
4. 3-mL syringes attached to a 20-G × 1.2″ disposable animal feeding for oral treatment.
5. 1-mL syringe and needle (23–25 G; 1/2–5/8″) for intraperitoneal (IP) administration.

2.4 Guinea Pig Euthanasia and Sample Collection

1. Ketamine/xylazine combination (*see* Subheading 2.2, **step 8**).
2. 390-mg/mL pentobarbital solution.
3. 2.5-cm^2 stencil square.
4. Sterile forceps.
5. 10-mm biopsy punches.
6. 4-mm biopsy punches.
7. 1-mL syringes with 27-G × ½" needle for ketamine/xylazine administration.
8. 3-mL syringes with 23-G × 1″ needle for intracardiac puncture.
9. 5-mm Stainless steel beads.
10. 2-mL microcentrifuge tubes.
11. 10% neutral buffered formalin.
12. PBS (*see* Subheading 2.1, **step 12**).
13. Tissue homogenizer.
14. PDA plates (*see* Subheading 2.1, **step 15**).

2.5 Colonization of the Mouse Skin

1. Female BALB/c mice with a body weight of approximately 20 g.
2. *C. auris* strain of choice.
3. Small clippers.
4. Hair removal lotion containing aloe and lanolin.
5. 1.5-cm^2 stencil square.
6. Autoclave tape.
7. Cotton-tip applicator.
8. Sterile water.
9. 2″ × 2″ squares of gauze.

10. Permanent marker.
11. 10-mL syringe.
12. Water-circulating blanket.

2.6 Mouse Decolonization Treatment

1. Treatment drugs.
2. Adjustable volume pipette.
3. Sterile pipette tips.

2.7 Mouse Euthanasia and Sample Collection

1. CO_2 tank and euthanasia chamber.
2. Sterile forceps.
3. Sterile scissors.
4. 10-mm biopsy punches.
5. 4-mm biopsy punches.
6. 2-mL microcentrifuge tubes.
7. 10% neutral-buffered formalin.
8. PBS (*see* Subheading 2.1, **step 12**).
9. 5-mm stainless steel beads.
10. Tissue homogenizer.
11. PDA plates (*see* Subheading 2.1, **step 15**).

3 Methods

3.1 Candida auris Cutaneous Infection Guinea Pig Model

Guinea pigs have been extensively used as a model for experimental cutaneous infections due to the similarities to the human skin in terms of cellular and molecular composition as well as hair follicle anatomy [5]. Here we describe a *C. auris* cutaneous infection guinea pig model. Animals are treated with prednisolone to induce local immunosuppression of the affected skin area [6]. Drug treatment is administered via topical applications.

3.1.1 Inoculum Preparation

The procedure for inoculum preparation is the same for challenging animals in the two models described in this chapter. However, the final concentration of each inoculum used is specific to each model (*see* **Note 4**).

1. Subculture *C. auris* strain of choice on PDA plates (we used clinical isolate MRL #35368 to evaluate this protocol), streaking for isolation so that single colonies can be obtained.
2. Invert the subcultured plates and incubate at 37 ° C for 48 h.
3. Subculture the *C. auris* isolate again by transferring four single colonies into each of four 50-mL conical tubes each containing 30 mL of SDB, replace the lid but leave it loose to allow for air flow. Leave a fifth 50-mL conical tube with 30-mL SDB

un-inoculated as a sterility control. Incubate overnight at 37 ° C on a shaker set to 180 rpm.

4. Next, remove all tubes from the incubator, inspect the sterility control tube which should not have any growth present. If growth is present in the sterility control, then the inoculum is contaminated, and a fresh batch of broth will have to be made and the procedure repeated. If the sterility control tube contains no growth, then it can be used to counterbalance the centrifuge if needed or discarded. Centrifuge all the tubes at 1450 × g for 5 min to obtain a pellet.
5. Remove the supernatant from all the tubes, resuspend the pellets in 10 mL of PBS, and resuspend the pellet using a vortex mixer. Then centrifuge the tubes again (as in **step 4**). Repeat this step for a total of three times to wash the cells.
6. After the third wash, discard the supernatants of all the tubes except one. Resuspend culture by vortex-mixing this tube, and pour the contents into the tube from which the supernatant has been removed to combine the pellets. Repeat this with each pellet in the remaining tubes. Vortex-mix after every addition of a pellet to end up with only one tube which will be your zero dilution (Fig. 2).

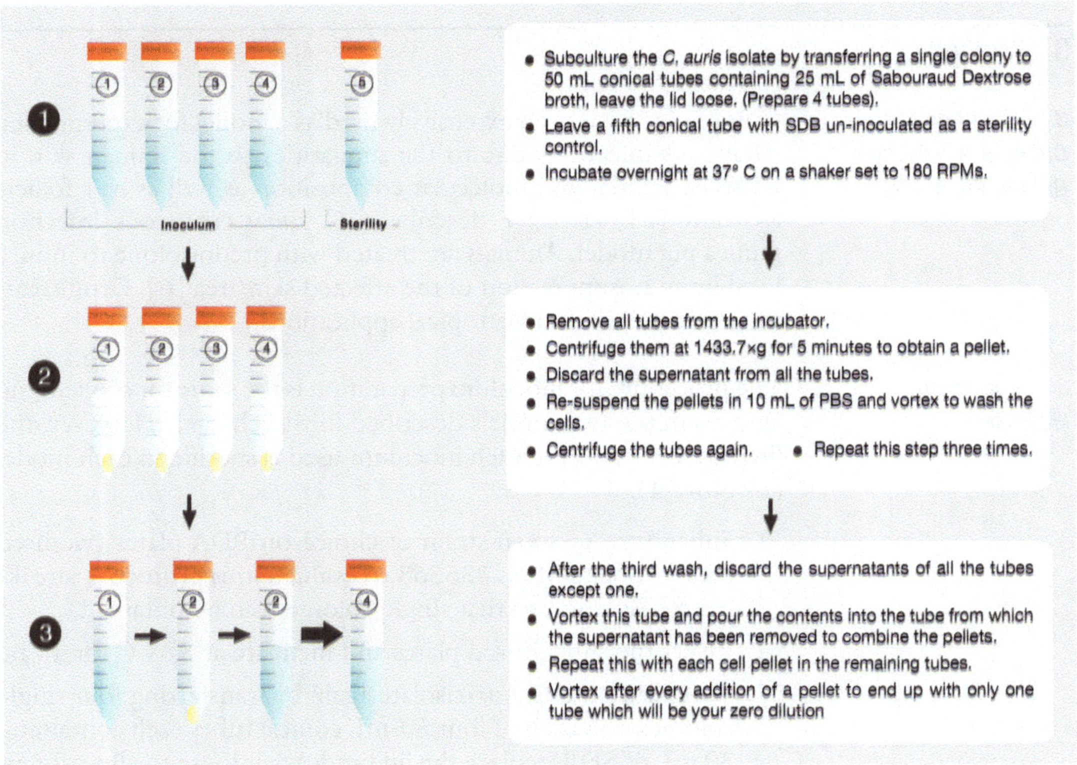

Fig. 2 Step-by-step procedure for culturing, washing, and preparing the *C. auris* inoculum

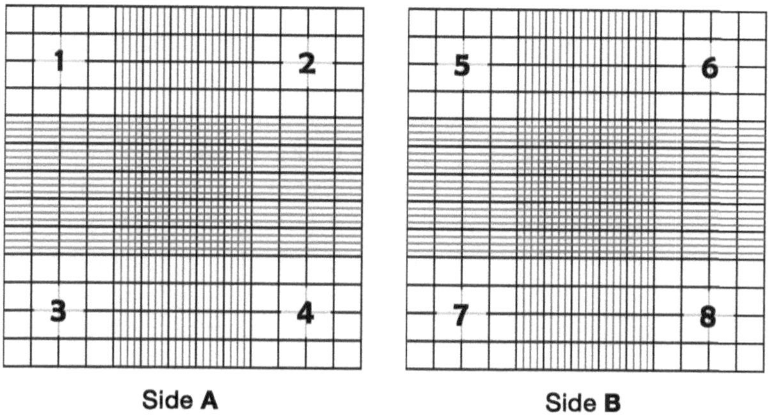

Fig. 3 Schematic of a hemocytometer showing two chambers (Side **a** and Side **b**) with corner squares (1–8) used for counting *C. auris* cells

7. Centrifuge the zero dilution at 1450 × g for 5 min, discard the supernatant, and add an appropriate volume of PBS to suspend your pellet. To calculate this volume, use the following formula:

$$\#\text{animals} \times \text{the volume of inoculum per animal } (0.1 \text{ mL}) \times 1.5 \text{ (which will add 50\%extra volume for contingencies)}$$

For example, if you are challenging 40 mice, then the volume of inoculum is 40 × (0.1 mL/mouse) = 4 mL × 1.5 = 6.0-mL total volume needed. Resuspend using a vortex mixer.

8. Prepare four tenfold serial dilutions in 2-mL microcentrifuge tubes, and label them −1, −2, −3, and −4.
9. Vortex-mix the −4 tube and carefully pipette out 9 μL, place this amount into both chambers of a disposable hemocytometer. Do not overfill.
10. Examine the hemocytometer using a microscope. Count all the yeast cells in the eight corners of each side of the hemocytometer (Fig. 3, squares 1–8). Cells that are overlapping the outside boundary lines are not counted.
11. Average the counts of the squares. Based on the volume of the chambers of the hemocytometer, this represents the average count × 10^4 cells/mL (*see* **Note 4**).

3.1.2 Immunosuppression and Inoculation

1. Allow guinea pigs to acclimate for a minimum of 5 days prior to use.
2. Local immunosuppression (in the area to be inoculated) is induced by subcutaneous (SC) injection of 1 mL of prednisolone (30 mg/kg) per guinea pig administered three times: one

day prior inoculation and on days 1 and 3 post inoculation. The use of an 18-gauge needle facilitates drawing up prednisone (*see* **Note 5**).

3. On the day of inoculation, prepare 10^9/mL (10^8 in 100 μl per animal) C. *auris* inoculum to challenge guinea pigs. Inoculations are performed in a biosafety cabinet.
4. Anesthetize the guinea pigs with a cocktail of ketamine/xylazine, 0.05 mL/100 g intraperitoneally. Administer 0.05-mg/kg buprenorphine subcutaneously, an analgesic opioid partial agonist, as preemptive analgesia to minimize any discomfort from the procedure (*see* **Note 6**).
5. After approximately 5 min, check for anesthetic depth by verifying the absence of response to a painful stimulus by checking the pedal withdrawal reflex or pinching of the ear.
6. Once the animal is anesthetized, apply ophthalmic ointment to both eyes to prevent dryness and damage to the cornea.
7. In order to prevent hypothermia which can prolong recovery and potentially be fatal, place the anesthetized animal on a warm surface inside the biosafety cabinet, and position the guinea pig in the right lateral recumbence.
8. Use an electrical shaver to clip the hair of an area on the back of the guinea pigs (left side).
9. Use curved hemostats to pluck the remaining hair of the shaved area.
10. Mark the plucked area of the skin using a stencil of 2.5 cm^2 (Fig. 4).

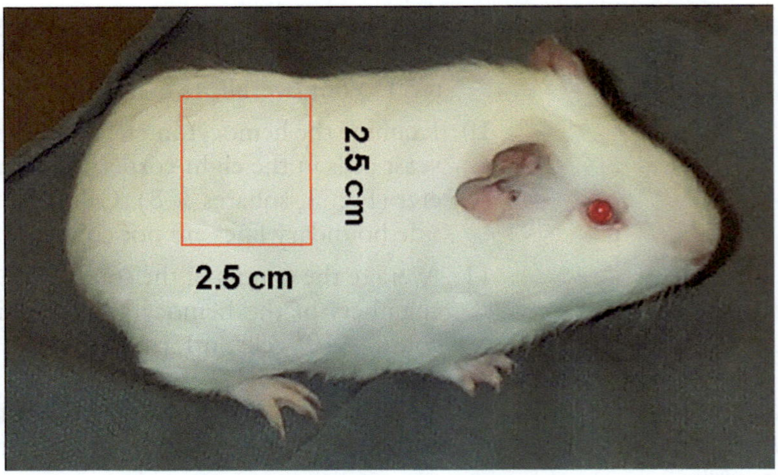

Fig. 4 Approximate location of the area of the skin to be abraded and inoculated. Within this square the hair will be shaved, plucked, and skin-abraded with sterile sand paper

11. Use sterile sand paper to abrade the skin within the stenciled area. The abrasion should be mild, and the skin should be irritated without bleeding.
12. Apply onto the abraded area of the skin 100-μl suspension containing 10^8 *C. auris* cells (inoculum prepared as indicated in Subheading 3.1.1), and using a sterile pipette tip, spread the inoculum thoroughly to cover the entire marked area with *C. auris*.
13. Transfer the guinea pigs to home cage setup with heat supply (water-circulating blanket or hot snaps). The heat source should be covered with a drape, do not place the guinea pigs directly on the heat source. Allow the animals to recover from anesthesia (*see* **Note 7**).
14. Check animals approximately every 5 min, and monitor the following parameters: mucous membrane color, respiratory rate and breathing pattern, heart rate, and posture (*see* **Note 8**). Once the animals are awake and able to maintain a sternal position, remove the heat supply from the cages. Keep monitoring the animals until they are fully ambulatory (Table 1).
15. Atipamezole, an α2-adrenergic receptor antagonist, may be administered at a dose of 0.5–1 mg/kg subcutaneously, as needed, to reduce the sedative effect of xylazine, and the unpleasant side effects of bradycardia, hypotension, and poor tissue perfusion, and to promote rapid recovery.

Table 1
Anesthesia recovery chart

Date: ________ **PI:** __________ **IACUC #:** _______ **Species:** Guinea Pig

Procedure: Skin abrasion and inoculation

Anesthesia: Ketamine/Xylazine (IP)

	POSTURE			COLOR MUCUS MEMBRANE			RESPIRATION			HEART RATE/PULSE			
Time	**NR**	**Lifts Head**	**Sternal**	**Pink**	**Pale**	**Blue**	**Fast**	**Slow**	**Reg**	**Fast**	**Slow**	**Reg**	**Initials**

NR = nonresponsive

Reg = within the appropriate range

NR nonresponsive, *Reg* within the appropriate range

3.1.3 Treatments

1. Start animal treatment 24 h post challenge, and continue for 7 days. Treatment dose and route will be dependent on the pharmacokinetic/pharmacodynamics of the compound to be evaluated.
2. For topical treatment apply 100 μL of test compound, and spread throughout the entire marked square on the skin using a sterile pipette tip. Treatment may also be given intraperitoneally in a volume of 100 μl or oral in a volume of 200 μL.

3.1.4 Clinical Evaluation

1. On day 8 post challenge, guinea pigs are anesthetized in the same way described above (*see* Subheading 3.1.2, **steps 2–5**).
2. Photographs of the test area can be taken.
3. Perform a clinical assessment of local changes of the infected skin area, and score each area as follows:

   0 = No lesions.

   1 = Few slightly erythematous places on the skin.

   2 = Well-defined redness, swelling with bristling hairs.

   3 = Large areas or marked redness incrustation, scaling, bald patches, ulcerated in places.

   4 = Partial damage to the integument and loss of hair.

   5 = Extensive damage to the integument and complete loss of hair at the site of infection.
4. The efficacy of each treatment is determined based on the average clinical score. Differences in mean clinical scores of the infected skin area are compared to the untreated/vehicle controls and other treatment groups using a one-way ANOVA with a post hoc Bonferroni. A *P*-value of <0.05 is considered statistically significant.

3.1.5 Euthanasia and Mycological Evaluation

1. On day 8 post challenge after the clinical evaluations, the anesthetized guinea pigs are sacrificed with 1-mL/guinea pig intracardiac (IC) injection of pentobarbital. Confirm death by palpating for the absence of heart beats, toe pinch, and eye reflex.
2. After death of animals has been confirmed, use a 10-mm biopsy punch to remove a skin sample from the infected square, and place the sample in a 2-mL microcentrifuge tube.
3. Weigh the collected skin samples.
4. Add one steel bead to each tube containing the collected sample, and homogenize in 1-mL sterile 0.85% saline in a tissue homogenizer (*see* **Note 9**).
5. Prepare tenfold dilutions from each homogenized sample.
6. Plate 100 μL of each dilution in the middle of a PDA plate. Use an L-shaped spreader to disperse the sample on the entire surface of the plate.

7. Invert and incubate the plates at 37 °C for 48 h to allow the growth of fungal colonies.
8. Count the colony-forming units (CFUs), and express the tissue fungal burden as log CFUs/gram of tissue. The mean CFU burden on the skin is compared between treatment and control groups using a one-way ANOVA with a post hoc Bonferroni. A *P*-value of <0.05 is considered statistically significant.

3.1.6 Histological Evaluation

1. Take a 4-mm biopsy punch from one animal per group for histopathology examination.
2. Place the sample in 10% neutral buffered formalin.
3. Biopsies are processed for periodic acid–Schiff staining and microscopically observed for fungal elements (Fig. 6).

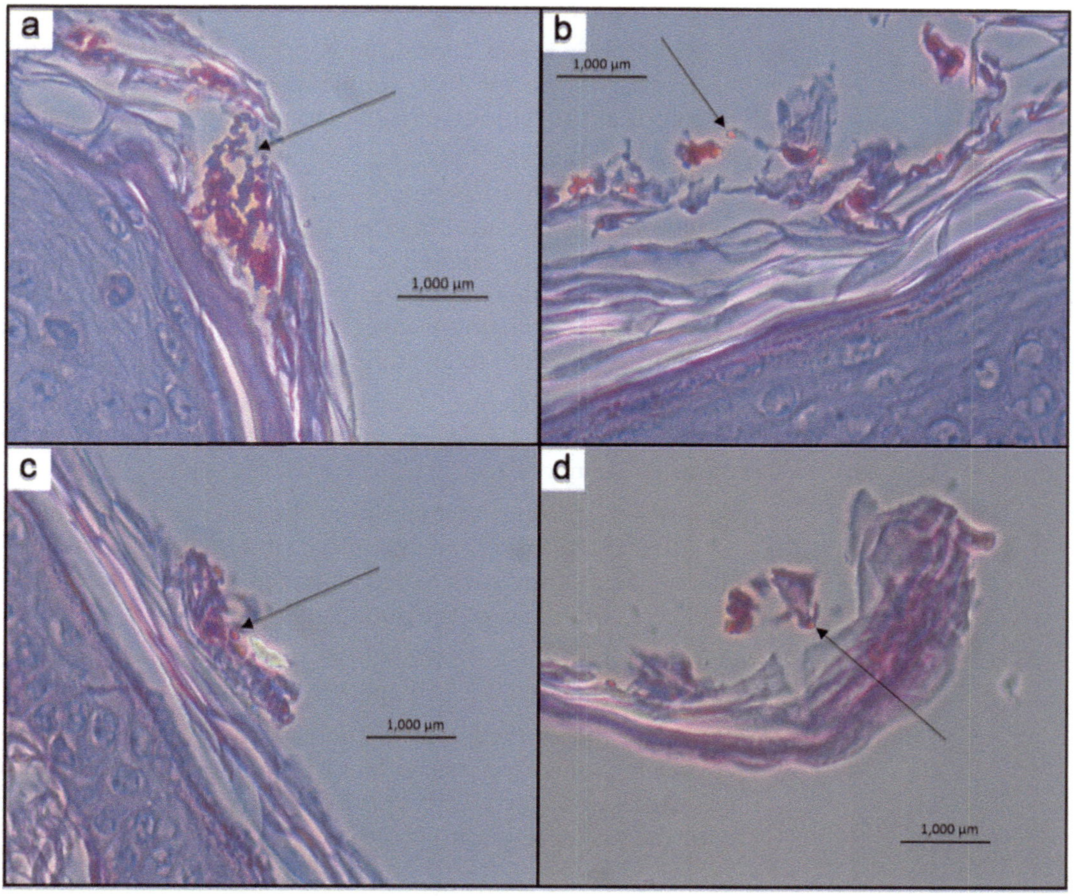

Fig. 6 Examples of periodic acid–Schiff (PAS) stain of the guinea pig skin showing the presence of *C. auris* (indicated by arrows) on (**a**) day 1, (**b**) day 3, (**c**) day 5, and (**d**) day 7 post inoculation

3.2 Candida auris Skin Decolonization Mouse Model

A novel skin decolonization mouse model was recently established by our group with the purpose of determining if topical antifungal agents would be able to decolonize the skin; this is a very important step that would help in the development of drugs that are not only efficacious in the treatment of patients' skin colonized with *C. auris* but also drugs that are intended for skin decolonization of *C. auris* [7]. This is critical as it has been shown that *C. auris* colonize the skin and can act as a nidus of infection and contamination of hospital environment [2].

3.2.1 Colonization of the Skin

Colonization of the skin is achieved by inoculation of the pinna of the ears and skin of the back of the mouse. To ensure colonization, it is necessary to inoculate the mice four times (Table 2).

1. Two days prior to the first skin inoculation, shave all the hair from the back of all mice.
2. Next depilate the shaved area with hair removal lotion (dispense lotion in a 10-mL syringe).
3. Leave the hair removal lotion in contact with the skin of the back of the mouse for 25 s (double this time may be necessary to complete the removal of hair). Do not rub the hair removal lotion into the skin.
4. Gently wipe off the hair removal lotion from the back of the mouse with a wet gauze, then squeeze a few drops of water on the back and dry by tapping it (do not rub). Make sure not to leave traces of hair removal lotion on the back of the mouse. Place a water-circulating blanket under the mouse cage during the entire procedure to avoid hypothermia.
5. On the day of challenge (day 0), use a 1.5-cm^2 stencil to mark the area of the back (this step does not need to be applied to the pinna of the mice ears).
6. In order to disrupt the integrity of the skin and to encourage *C. auris* colonization, apply a piece of autoclave tape to the marked area of the back (again, this step does not need to be applied to the pinna of the mice ears).
7. Allow the tape to sit on the skin of the back of the mouse for about 5 s, then remove it. Repeat this step four times.
8. Inoculate the marked area of the back of the mouse with a cotton-tip applicator soaked in *C. auris* inoculum (10^8 yeast cells/mL) until the entire square is visibly wet. Rub the skin with the soaked applicator five times.

Table 2
Colonization and decolonization treatment schedule

Day	0	1	2	3	4	5	6	7	8	9	10	11	12	13	14
	Inoc		Inoc		Inoc		Inoc	Tx	Tx	Tx	Tx	Tx	Tx	Tx	Euthanize

9. Inoculate the pinna of the ears in the same fashion as in **step 8** (five rubs with the soaked swab per ear).
10. The skin of the ears and back are similarly inoculated on days 2, 4, and 6 for a total of four inoculations.

3.2.2 Decolonization Treatment and Clinical Evaluation

1. Beginning on day 7, after colonization of the skin is established, the test compounds are administered once a day for 7 days. Dosing will depend on the pharmacokinetic/pharmacodynamics of the compound to be tested.
2. Apply 5 μL of the test compound on each ear, and spread it with a sterile pipette tip.
3. Apply 40 μL of the test compound to the area of the back of the mice, and spread it with a sterile pipette tip.
4. A clinical assessment of the treated skin area is performed throughout the experiment, and the skin of the ears and back of mice should show a normal appearance with no signs of infection or inflammation (Fig. 5).

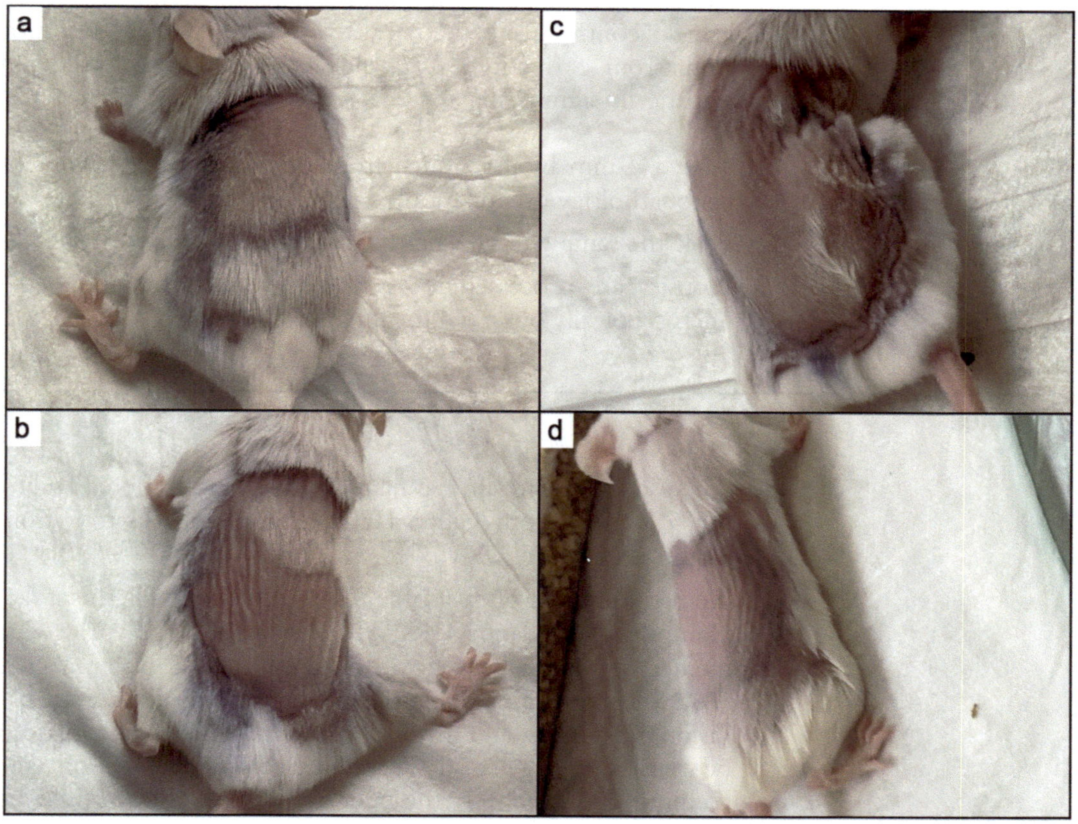

Fig. 5 Clinical appearance of the skin of treated mice inoculated with *C. auris* and treated with (**a**) terbinafine hydrochloride (1%) formulated with Advanced Penetration Technology (APT), (**b**) clotrimazole (1%) formulated with APT, (**c**) vehicle, and (**d**) untreated

3.2.3 Euthanasia and Mycological Evaluation

1. Euthanize mice by CO_2 asphyxiation and cervical dislocation 1 day after the last day of treatment (day 14).
2. Aseptically collect a 10-mm skin biopsy punch from the colonized skin from the back and the pinna of the right and left ears.
3. Place the samples in pre-weighed 2-mL microcentrifuge tubes containing a steel bead.
4. Add 1 mL of sterile 0.85% saline to each sample, and homogenize in a tissue homogenizer (*see* **Note 9**).
5. Prepare tenfold dilutions from tissue homogenate. Place 100 μL of each dilution in the middle of a PDA plate, and use an L-shaped spreader to disperse the sample on the entire surface of the plate.
6. Invert and incubate the inoculated plates at 37 °C for 48 h to allow the colonies to grow.
7. Count the CFUs. Express the tissue fungal burden as log CFU/gram of tissue.
8. Differences in average log CFU on skin samples from the ears and back are compared between treated groups and untreated/vehicle control using a one-way ANOVA with a post hoc Bonferroni correction. A *P*-value of <0.05 was considered statistically significant.

3.2.4 Histological Evaluation

1. Take a 4-mm biopsy punch from one animal per group for histopathology examination.
2. Place the sample in 10% neutral buffered formalin.
3. Biopsies are processed for periodic-acid–Schiff staining and microscopically observed for fungal elements.

4 Notes

1. We use commercially available phosphate-buffered saline pH 7.2 (1×) from Gibco (Life Technologies Corporation, Grand Island, NY, USA).
2. We use commercially available SDB from Oxoid (Basingstoke, Hampshire, England, UK). Fill a graduated polypropylene pitcher or a glass beaker with 1 L of distilled water. Weigh 30-g SDB powder and transfer to the pitcher, place a stir bar inside the pitcher, and mix well using a stir plate. The mix can be transferred to a glass flask, and heat to boiling to dissolve the medium completely. If this method is used, care should be taken so to not overboil and spill the mix. Cover the top of the pitcher with aluminum foil, and autoclave at 121 °C for 40 min. Allow to cool and store at 4–8 °C.

3. We use commercially available PDA from Difco (Becton, Dickinson and Company, Sparks, MD, USA). Fill a graduated polypropylene pitcher or a glass beaker with 1 L of distilled water. Weigh 39 g of PDA powder, 1 g of granulated agar, 0.125 g of chloramphenicol, and 0.03 g of gentamicin and transfer to the pitcher, place a stir bar inside the pitcher, and mix well using a stir plate. Cover the top of the pitcher with aluminum foil, and autoclave at 121 °C for 40 min. Pour the media into sterile Petri plates. The media needs to be stirred before plates are poured, if not some of the plates may not harden correctly.
4. We perform all animal inoculations in a biosafety cabinet. Follow your institutional and national guidelines while handling *C. auris*.

   For the inoculum preparation, follow this step-by-step example:

   - If the total count in the squares marked 1–8 = 348.
   - Average count: $348/8 = 48 \times 10^4$ cells/mL.
   - Number of animals = 40.
   - Volume of inoculum per animal = 0.1 mL (100 μl).
   - Concentration of yeast cells per animal = 3×10^7 *C. auris* cells.

   Step 1. *Volume of inoculum needed:*

   (# animals × the volume of inoculum per animal) × (1.5) which will add 50% extra volume for contingencies:

$$40 \times (0.1\ \text{mL/mouse}) = 4\ \text{mL} \times 1.5$$
$$= 6.0\ \text{mL total volume needed.}$$

   Step 2. *Number of yeast cells/mL:*

   (Concentration of inoculum/100 μL) × 10:

$$\left(3 \times 10^7\ \text{cells}/100\ \mu\text{L}\right) \times 10 = 3 \times 10^8\ \text{cells/mL}$$

   Step 3. *Total number of yeast cells needed:*

   (Concentration of cells/mL) × (volume of inoculum needed):

$$\left(3 \times 10^8\ \text{cells/mL}\right) \times (6.0\ \text{mL}) = 18 \times 10^8\ \text{cells total}$$

   Step 4. *Count of yeast cells from dilution – 4 in the hemocytometer:*

$$\text{Average count} : 348/8 = 48.$$

   Thus, the count in −4 dilution is 48×10^4 yeast cells/mL.
   The count in −3 dilution is 48×10^5 yeast cells/mL.
   The count in −2 dilution is 48×10^6 yeast cells/mL.
   The count in −1 dilution is 48×10^7 yeast cells/mL.

The count in 0 dilution is 48×10^8 yeast cells/mL (working dilution).

Step 5. *Volume needed from working dilution*:

(Total number of *yeast cells* needed)/(*yeast cells* count in working dilution):

$$(18 \times 10^8 \text{ cells})/(48 \times 10^8 \text{ cells/mL}) = 0.375 \text{ mL } (375 \text{ µL})$$

Step 6. *Prepare inoculum*:

Amount of PBS needed:

(Total volume of inoculum) – (volume from working dilution):

$$6 \text{ mL} - 0.375 \text{ mL} = 5.625 \text{ mL PBS}$$

To complete the preparation of the inoculum, pipette 375 µL from the working dilution into a 15-mL conical tube, and add 5.625 mL of PBS.

5. The subcutaneous injection in the left side of the back of guinea pig is easier to perform with two people since in this area the skin is tighter and the animals may resist. The use of a 23-G × ¾" scalp vein set with extension is recommended to facilitate administration. Also, the substance to be injected may be warmed up to be more comfortable for the animals. We use an 18-gauge needle only to draw up the prednisone into the syringe.
6. The duration of the ketamine/xylazine anesthesia is approximately 30 min. Although this is a short procedure, if extended anesthesia is needed after 30 min, another half of the anesthetic dose may be administered.
7. Thermal support is critical in the anesthesia recovery of small animals which have a greater body surface area-to-body mass ratio. Warming devices such as water-circulating blankets should be used with a blanket or towel between the animal and the circulating water device to protect the animals against burns.
8. For the anesthetic monitoring of guinea pigs, follow your guidelines for the use of animals in research. We monitor anesthesia and record parameters approximately every 5 min until the animals are responsive and regain sternal posture (Table 1).
9. Tubes containing the tissue sample, saline, and bead are placed in TissueLyser adapter set and shaken at a very high speed, originating the tissue samples to be disrupted and homogenized by the beating and grinding effect of the beads. We employ a TissueLyser (QIAGEN) set for 5 min at 30 Hz. Other types of homogenizers need to be operated according to the manufacturer's instructions.

References

1. Miesel L, Lin KY, Ong V (2019) Rezafungin treatment in mouse models of invasive candidiasis and aspergillosis: insights on the PK/PD pharmacometrics of rezafungin efficacy. Pharmacol Res Perspect 7(6):e00546. https://doi.org/10.1002/prp2.546
2. Kean R, Sherry L, Townsend E, McKloud E, Short B, Akinbobola A, Mackay WG, Williams C, Jones BL, Ramage G (2018) Surface disinfection challenges for Candida auris: an in-vitro study. J Hosp Infect 98(4):433–436. https://doi.org/10.1016/j.jhin.2017.11.015
3. Niwano Y, Seo A, Kanai K, Hamaguchi H, Uchida K, Yamaguchi H (1994) Therapeutic efficacy of lanoconazole, a new imidazole antimycotic agent, for experimental cutaneous candidiasis in guinea pigs. Antimicrob Agents Chemother 38(9):2204–2206. https://doi.org/10.1128/aac.38.9.2204
4. Maebashi K, Itoyama T, Uchida K, Suegara N, Yamaguchi H (1994) A novel model of cutaneous candidiasis produced in prednisolone-treated guinea-pigs. J Med Vet Mycol 32(5):349–359. https://doi.org/10.1080/02681219480000471
5. Baumbach CM, Schrodl W, Nenoff P, Uhrlass S, Mulling CKW, Michler JK (2020) Modeling dermatophytosis: Guinea pig skin explants represent a highly suitable model to study Trichophyton benhamiae infections. J Dermatol 47(1):8–16. https://doi.org/10.1111/1346-8138.15150
6. Ghannoum M, Isham N, Angulo D, Borroto-Esoda K, Barat S, Long L (2020) Efficacy of Ibrexafungerp (SCY-078) against Candida auris in an in vivo guinea pig cutaneous infection model. Antimicrob Agents Chemother 64(10):e00854–e00820. https://doi.org/10.1128/AAC.00854-20
7. Ghannoum M, Herrada J, McCormick TS, Long L (2021) A novel transdermal application for clearing skin colonization by Candida auris. Antimicrob Agents Chemother 65(5):e02303–e02320. https://doi.org/10.1128/AAC.02303-20

Chapter 23

The Use of *Galleria mellonella* Larvae to Study the Pathogenicity and Clonal Lineage-Specific Behaviors of the Emerging Fungal Pathogen *Candida auris*

Andrew M. Borman

Abstract

Candida species are the most common fungal causes of disseminated infections in humans. Although such infections are associated with high morbidity and mortality, it is widely accepted that virulence, antifungal susceptibility, and disease outcome vary according to individual *Candida* species. In this respect, the emerging pathogen *Candida auris* has received much attention due to its propensity to cause widespread nosocomial outbreaks, to exhibit high virulence in several infection models, and to develop resistance to multiple classes of antifungal drugs. Although mammalian models of infection have long been viewed as the gold standard for studies on fungal virulence, comparative pathogenicity, and evaluation of antifungal drug efficacy, the larvae of the greater wax moth *Galleria mellonella* have shown considerable promise as an alternative invertebrate model of infection. *Galleria* larvae are inexpensive, are easily maintained in the laboratory, tolerate incubation at human physiological temperatures, possess cellular and humoral immune systems that share many features with mammals, and allow investigation of pathogenicity/virulence using multiple different reading endpoints. Here, I describe in detail the methods that can be used to study the virulence/pathogenicity of *Candida auris* in *G. mellonella*.

Key words Pathogenesis, Virulence, *Galleria mellonella*, *Candida auris*, Clades, Clonal lineages

1 Introduction

The incidence of invasive candidosis, which includes candidemia and deep tissue infections, continues to rise and is associated with significant mortality rates [1, 2]. Although *Candida albicans* remains the most common cause of invasive candidosis, the prevalence of non-*albicans* species has increased dramatically over recent years [1–3]. First described in 2009 in association with otitis [4], *Candida auris* has emerged as an important nosocomial health risk, with large outbreaks in numerous hospitals worldwide [5–10]. Almost all isolates of *C. auris* exhibit fluconazole resistance, with individual isolates rapidly developing resistance to additional

Alexander Lorenz (ed.), *Candida auris: Methods and Protocols*, Methods in Molecular Biology, vol. 2517, https://doi.org/10.1007/978-1-0716-2417-3_23,

antifungal drugs/classes upon treatment ([3, 5–7]; reviewed in [10]). Importantly, whole genome sequencing (WGS) of isolates from disparate geographic regions identified at least four distinct clades of *Candida auris* [7], corresponding to populations found in South Africa, South Asia (India, Pakistan, Kuwait), East Asia (Korea, Japan), and South America (Colombia, Venezuela), with a potential fifth clade from Iran [11]. Moreover, WGS provided strong evidence for clonality within each of the clades with individual isolates within a clade varying by less than 100 SNPs, but vast inter-clade genetic distances (several hundred thousand SNP differences), evidence that the four major clonal lineages/clades had simultaneously emerged on distinct continents.

Nosocomial outbreaks caused by *C. auris* have occurred in numerous countries worldwide and are both difficult and expensive to control [12–15]. In the United Kingdom, we have witnessed the separate introduction of multiple isolates from three of the four clonal lineages [16] and have reported significant differences in phenotypic behavior and antifungal susceptibility depending on the lineage [17, 18]. Notably, isolates of the South African lineage exhibit cell division/separation defects that result in an "aggregative" growth form [18], elevated resistance to antifungal drugs as compared to isolates of the S. Asian lineage [17], and altered abilities to grow in the presence of actidione.

For many years, mammalian infection models (particularly murine models) have been viewed as the gold standard for studies on fungal virulence, pathogenesis, host–pathogen interactions, and evaluation of antifungal drug efficacy. However, invertebrate models of fungal infection, including the larvae of both *Drosophila* and the greater wax moth *Galleria mellonella* are an increasingly popular choice as they circumvent many of the logistical, economic, and ethical issues that confront animal experimentation ([19–24]; reviewed in [25]). Infection of *Galleria mellonella* is easy due to their size, multiple inoculations can be performed, inoculum concentrations can be accurately controlled, and infected larvae can be maintained at human physiological temperatures [18–27]. In addition, wild-type larvae are susceptible to infection by human-pathogenic fungi, and the larval immune system shares many similarities with mammalian counterparts at both the cellular and humoral levels [25, 28]. Indeed we recently showed that differences in pathogenicity between different clades of *C. auris* that were initially detected in *G. mellonella* [18] were faithfully reproduced in an immunocompromised murine model of infection [29]. This chapter details the use of the *Galleria mellonella* larval infection model to study pathogenicity of *Candida auris* isolates and highlights the utility of this popular invertebrate model for the study of pathogen–host interactions.

2 Materials

1. Sabouraud's (SAC) medium (glucose peptone agar): 10-g mycological peptone, 40-g glucose, and 15-g agar in 1-L water, sterilized by autoclaving at 121 °C for 15 min, supplemented with 0.05-mg/mL chloramphenicol once autoclaved and cooled. Pre-poured media are available commercially. Store at 4 °C.
2. The final (sixth) instar larvae that weigh approximately 300 mg each and that are free of grey markings. Larvae should be maintained at room temperature in the dark for at least 24 h after receipt/transport and should be inoculated within 48 h of receipt (*see* **Note 1**).
3. A 700-series 10-μL Hamilton syringe fitted with a 26-gauge blunt small removable needle.
4. Sterile 1.5-mL type D screw-top microtubes.
5. Sterile phosphate-buffered saline pH 7.4 without magnesium or calcium (PBS): 8 g of NaCl, 0.2 g of KCl, 1.44 g of Na_2HPO_4, and 0.24 g of KH_2PO_4, dissolved in 1-L sterile water and pH adjusted to 7.4 with HCl.
6. Standard sterile 90-mm Petri dishes.
7. Universal polystyrene 30-mL containers.
8. Polypropylene storage containers (7500-mL capacity) with snap-on lids.
9. Alcohol-based disinfectant wipes (70% propan-2-ol).
10. Neubauer hemocytometer counting chamber and cover slips.
11. High-level, medical-grade laboratory disinfectant; follow the manufacturer's instructions and local health and safety guidelines for use and disposal.
12. Sterile plastic 10-μL inoculating loops.
13. Disposable sterile scalpels.
14. Disposable plastic sterile forceps.
15. Calcofluor white fluorescent enhancer.
16. Potassium hydroxide (10% w/v): 100 g of KOH per liter of sterile water.
17. Fluorescent microscope fitted with a UV light: V-2A filter (excitation 380–420 nm, dichromatic mirror 430 nm, and barrier filter 450 nm).
18. Standard 25 mm × 75 mm plain glass microscope slides.
19. Standard 22 mm × 22 mm plain glass coverslips.
20. Standard benchtop laboratory vortex mixer.
21. PBS supplemented with 0.37% (v/v) mercaptoethanol: 185 μL of β-mercaptoethanol in 50 mL of sterile PBS.

3 Methods

3.1 Harvesting of Fungal Cultures and Preparation of Yeast Inocula

1. Prepare pure cultures of each *Candida auris* isolate by inoculating onto SAC medium, and incubate plates inverted at 30 °C for a minimum of 24–48 h (*see* **Note 2**).
2. Harvest cultures by scraping the surfaces of yeast colonies with sterile loops and suspending harvested yeast cells in 1-mL volumes of sterile PBS in screw-top microtubes. Preparations should be washed twice in sterile PBS by centrifugation at 1000 × *g* for 30 s, removal of PBS supernatant, and resuspension in 1 mL of fresh PBS.
3. After the final wash, mix suspensions well on a vortex mixer at a full speed for 30 s to achieve a single-cell suspension. Allow suspensions to settle for 10 min before removal of the supernatant to fresh sterile screw-top microtubes (*see* **Note 3**).
4. Quantify the inoculum density using a hemocytometer, and adjust to 1×10^6 cells/μL by dilution in fresh sterile PBS. Make further tenfold dilutions of this suspension to achieve the final fungal inocula required (*see* **Note 4**).
5. Store the adjusted inocula at room temperature and use within 2 h. Immediately before inoculation, mix inocula with a 5-s pulse on a vortex mixer to resuspend any cells that have settled to the bottom of the tubes.

3.2 Inoculation of Galleria Larvae

1. Individual larvae are inoculated in the left rear proleg with 1×10^6 yeast cells in PBS (final inoculum volume, 10 μL) using a 10-μL Hamilton syringe fitted with a 26-gauge blunt needle. At least ten larvae should be inoculated per isolate per experiment, and for comparison of different clades/species, experiments should ideally employ a minimum of five to ten independent isolates of each *Candida* test species or clade. Inoculate control groups of larvae with 10 μL of sterile PBS in exactly the same manner.
2. Gently wipe the ventral surface of each larva with an alcohol-based disinfectant wipe to remove surface microorganisms prior to inoculation.
3. Wash the Hamilton syringe in between individual larvae by rinsing in sterile PBS and in between batches of five larvae by rinsing once with sterile PBS, once with laboratory disinfectant, and twice more with sterile PBS.
4. Place the inoculated larvae in their groups of 10 in labeled sterile Petri dishes inside sealed polypropylene storage contain-

ers containing a layer of paper roll moistened with sterile distilled water. Incubate inoculated larvae at 30, 35, 37, or 42 °C as required.

5. Score larvae for viability or signs of infection at 4–8 h intervals as described below.

3.3 Analysis of *C. auris* Pathogenicity in Inoculated Larvae

3.3.1 Using Larval Survival/Death as Primary Endpoint

1. At regular time intervals, remove all inoculated larvae from incubators, and score each individual larva per inoculation group for viability. This is typically performed by gently prodding each larva with a sterile pipette tip or inoculating loop and measuring response to physical stimuli. Larvae which respond are scored as viable and returned to their Petri dishes, and those that do not are scored as dead and discarded or used for histological analyses (*see* **Note 5**). Systems have been proposed to detect more subtle differences in larval health; see larval health index scoring system (*see* Subheading 3.3.2).
2. Plot any resulting differences in survival using Kaplan–Meier plots [30], and assess the statistical significance of any differences using Mann–Whitney (two-sample Wilcoxon) or log-rank tests.

3.3.2 Additional Health Endpoints: Activity, Melanization, Cocoon Formation, and the Larval Health Index Scoring System

1. As an alternative to live/dead scoring as primary endpoint, a more sophisticated approach can be employed, based on the larval health index scoring system proposed by Loh et al. [31] and summarized in Table 1.
2. At each time point, inspect larvae as per Subheading 3.3.1, but include quantitative scores for degree of movement with/without stimuli, melanization, and cocoon formation (*see* **Note 6**).

3.3.3 Fungal Burden in Infected Larvae

1. At appropriate times post-infection (typically at 12-h intervals after inoculation), sacrifice representative larvae from each inoculation group by decapitation with a sterile disposable scalpel, and squeeze internal body contents (fat body, solid internal structures, and hemolymph) into a sterile screw-top microtube (*see* **Note 7**).
2. Homogenize in a standard volume of sterile PBS by pounding/mixing with the handle of a sterile inoculating loop and vortex mixing.
3. Make serial tenfold dilutions of larval extract in sterile PBS, and plate 50-μL volumes of each dilution onto fresh Sabouraud agar.
4. Incubate inoculated plates at 48 h at 37 °C, enumerate fungal colonies, and calculate fungal burden in cfu/larva (*see* **Note 8**).
5. If larval contents are separated before homogenization (hemolymph and hemocytes versus internal organs/fat body), fungal burden assessments can be made on individual larval

Table 1
The larval health index scoring system proposed by Loh et al. [31]

Variable	Scoring guidelines	Score
Survival	Alive	2
	Dead	0
Larval activity	Moves without stimuli	3
	Moves when stimulated	2
	Minimal movement with stimuli	1
	No detectable movement	0
Melanization	No melanization	4
	<3 melanin spots on beige larvae	3
	>3 melanin spots on beige larvae	2
	Black spots on brown larvae	1
	Larvae completely black	0
Cocoon formation	Full cocoon	1
	Partial cocoon	0.5
	No cocoon	0

compartments (and even in hemolymph versus hemocytes if the latter are harvested from hemolymph by centrifugation at 500 × *g* for 5 min).

*3.3.4 Measuring Hemocyte Density (see **Note 9**)*

1. Determine hemocyte density as described previously [28]. Bleed representative larvae from each inoculation group by pricking with a sterile hypodermic needle and gently squeezing two drops of hemolymph (approximately 40 μL/larva) into a micro-centrifuge tube that has been pre-chilled to prevent melanization.
2. Dilute the collected hemolymph in PBS supplemented with 0.37% (v/v) β-mercaptoethanol, and assess cell density using a hemocytometer.
3. Calculate hemocyte numbers, and express as the number of hemocytes per mL of hemolymph.

3.4 Physical Examination of Fungal Cell Morphology in Infected Larvae

1. At appropriate times post-infection (typically at 12-h intervals after inoculation), sacrifice representative larvae from each inoculation group by decapitation with a sterile disposable scalpel, and squeeze internal body contents (fat body, solid internal structures, and hemolymph) into a sterile screw-top

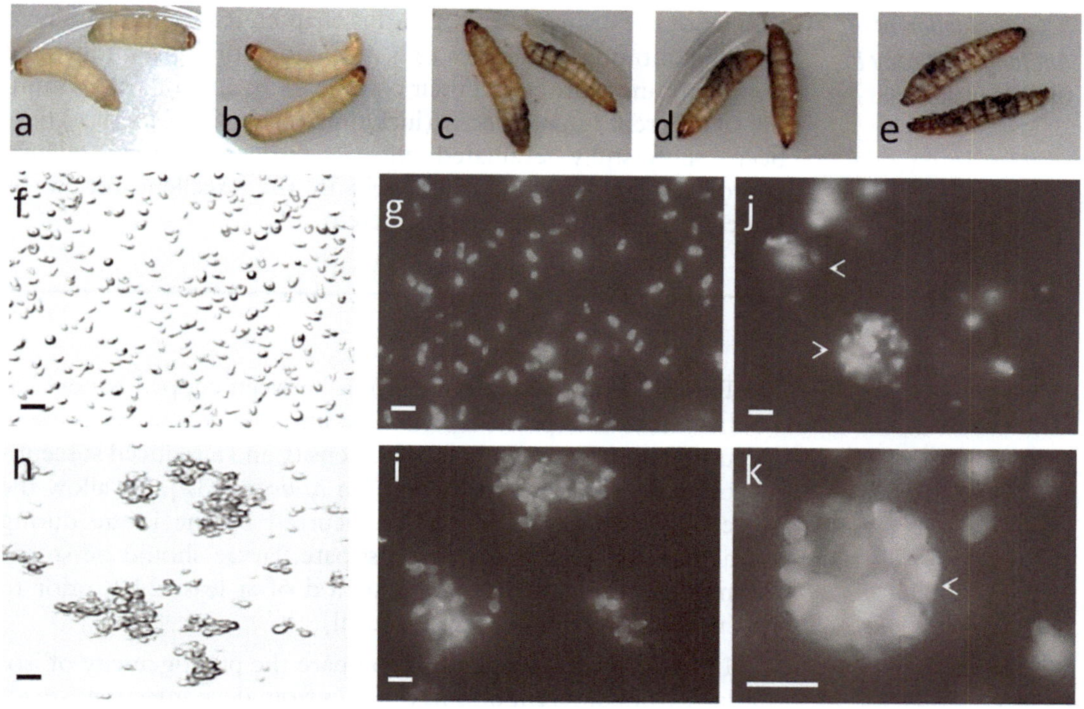

Fig. 1 *Galleria mellonella* larvae infected with *Candida auris* strains. Larvae were injected with PBS (**a**) or infected with 1×10^6 cfu/larva of *Candida auris* strain NCPF 8971 (S. Asian lineage, non-aggregative), panels (**b**) through (**e**), incubated at 37 °C and photographed after 6 h (**b**), 12 h (**c**), 24 h (**d**), and 48 h (**a** and **e**). Microscopic appearance of *C. auris* after growth on Sabouraud agar (light microscopy; **f** and **h**) and in *G. mellonella* 24 h post-infection (Calcofluor and UV microscopy; **g** and **i**). Panels (**f**) and (**g**) = NCPF 8971 (S. Asian lineage, non-aggregative) and panels (**h**) and (**i**) = NCPF 8977 (S. African lineage, aggregative). Panels (**j**) and (**k**) depict *G. mellonella* hemocytes (arrow heads) engorged with a differing number of *C. auris* cells 24 h post-infection with NCPF 8971. Scale bar = 10 μm

microtube (*see* **Note 7**). Hemolymph and internal organs may be harvested separately if required (*see* Subheadings 3.3.3 and 3.3.4).

2. Add 1 drop of 10% KOH and 1 drop of Calcofluor to collected sample, mix well (hemolymph) or squash under a microscope cover slip (larval fat body), and examine under a fluorescent microscope [18, 27]. In this way, the fungal cell morphology/phenotype (single cell versus aggregative form) can be easily accessed in vivo (*see* Fig. 1) (*see* **Note 10**).
3. If known volumes of hemolymph per larva are examined, a semi-quantitative analysis of fungal burden can be achieved by counting the number of fluorescent yeast cells per X number of microscopy fields.

3.5 Evaluation of Antifungal Efficacy in Galleria mellonella

While not the primary subject of this chapter, it should be pointed out that antifungal efficacies of a wide range of conventional and novel antifungal drugs and their combinations against a large number of different fungal species (including many *Candida* spp.) have been successfully evaluated in *G. mellonella* larvae. These approaches are covered extensively in the excellent review by Jemel et al. [23] and references therein.

4 Notes

1. Physical stress (shaking for 2 min between cupped hands) has been shown to prime the *Galleria* immune system, resulting in transient increases in hemocyte density and a reduced susceptibility to infections with *Candida albicans* [32]. To allow the effects of any minor stresses incurred by the larvae during acquisition or transport to dissipate, larvae should be stored motionless in the dark for a period of at least 24 h prior to inoculation (a desk drawer is ideal).
2. In experiments designed to compare the pathogenicity of isolates from different origins (isolates from deep infections versus superficial colonization) and different clonal lineages/clades or to compare *Candida auris* with other pathogenic yeast species, typically five to ten isolates per "comparator" are the minimum required.
3. Isolates of the South African lineage of *C. auris* grow as large aggregates of cells on conventional solid mycological media due to failure to release daughter cells and complete abscission after cell division [16–18] (Fig. 1). Since these aggregates arise due to a defect in cytokinesis, they cannot be physically dispersed by mechanical disruption or treatment with detergents. Isolates from the South and East Asian lineages on occasion produce smaller aggregates. These aggregates are difficult to inject into larvae as they block the syringes employed and additionally make it impossible to accurately quantify inoculum concentrations. To ensure that suspensions for inoculation into larvae are comparable, it is essential to leave the initial suspension of yeast cells to settle for at least 10 min to allow sedimentation of aggregates. If doubt remains as to whether all aggregates have been removed by sedimentation, a 5-μL sample of supernatant can be examined by light microscopy prior to preparation of the final yeast cell suspension for inoculation.
4. In most experiments conducted with *Candida* species and *Galleria mellonella* at 37 °C, the ideal starting inoculum is 10^6 yeast cells per larva, i.e., 10 μL of a suspension of 10^5 cells/μL. This will usually result in maximal larval death between 48 and 96 h, depending on the isolate/clonal lineage. If lower incubation temperatures are employed, larval death is

generally reduced/delayed, and higher concentrations of yeast inocula may be required (e.g., 5×10^6 or 10^7 cells/larva). Conversely, if endpoints other than cell death are important and extended larval infection times are required, lower concentrations should be employed.

5. Traditional scoring of viability is based on larval response to stimuli and is dichotomous: movement = alive, no movement = dead. This well-defined, but slightly crude, endpoint allows the calculation of a standardized LD_{50} value or the "virulence index" based on the time to death of 50% of the experimental cohort normalized to that of the wild-type or control group and expressed as a log value [21, 22]. To be of value, the intervals between monitoring of larval death should be relatively short (~6 h), which can be laborious and time-consuming in experiments employing a large number of larvae/test conditions.
6. The larval health index scoring system proposed by Loh et al. [31] includes quantitative scores for the degree of movement or melanization of inoculated larvae and also the ability of the larvae to complete cocoon formation. Although it allows a more subtle scoring system for measuring the outcomes of experimental *Galleria* infections than the simple alive versus dead endpoint traditionally employed, there are several potential confounders to its use. Notably, the degree of melanization has been reported to be pathogen-specific [22], and in our experience when using fresh larvae, even PBS-control inoculated larvae do not progress to cocoon formation before the end of a typical fungal pathogenicity experiment. Additionally, this expanded scoring algorithm is even more time-intensive than the traditional alive versus dead approach.
7. Measurement of fungal load in infected larvae can be extended to investigate individual larval compartments (cadaver, hemolymph, fat body, hemocoel), by separately draining hemolymph after decapitation (followed by centrifugation to separate hemocytes from hemocoel) and squeezing to remove fat body/internal organs (followed by homogenization). Initial calculations of the number of larvae per inoculum group should be adjusted based on the numbers of additional larvae that will be sacrificed at each time point for fungal burden studies.
8. Although fungal burden studies traditionally rely upon plating of serial dilutions and colony counting, the use of quantitative PCR is an attractive alternative, and *Candida auris*-specific real-time PCR assays have been described [33, 34]. However, results should be interpreted with an element of caution since PCR-based approaches will not distinguish between growing (alive) *C. auris* cells and DNA released from yeast cells that have been killed by the larval immune system [33].

9. Previous studies have demonstrated that infection of *Galleria* with sub-lethal fungal inocula resulted in an upregulation in hemocyte density that could partially protect larvae from subsequent lethal challenges ([35, 36]; discussed in [28]). Typically, a transient increase in hemocyte numbers is seen post-infection of naïve larvae with lethal doses of fungal pathogens (possible due to recruitment of hemocytes from sites of sequestration in the larval body) followed by a rapid decline in hemocyte density as fungal infection progresses [28].
10. An alternative to examining fungal cell proliferation in larvae that have been sacrificed and harvested of hemolymph/internal structures involves histological examination of the entire larvae after formalin fixation and conventional histological processing. Protocols for this approach have been published [37].

Acknowledgments

We thank the staff of the UK National Mycology Reference Laboratory for their interest in these approaches.

References

1. Nucci M, Marr KA (2005) Emerging fungal diseases. Clin Infect Dis 41:521–526
2. Pappas PG, Lionakis MS, Arendrup MC, Ostrosky-Zeichner L, Kullberg BJ (2018) Invasive candidiasis. Nat Rev Dis Primers 4: 18026. https://doi.org/10.1038/nrdp.2018.26
3. Quindós G, Marcos-Arias C, San-Millán R, Mateo E, Eraso E (2018) The continuous changes in the aetiology and epidemiology of invasive candidiasis: from familiar *Candida albicans* to multiresistant *Candida auris*. Int Microbiol 21:107–119. https://doi.org/10.1007/s10123-018-0014-1
4. Satoh K, Makimura K, Hasumi Y, Nishiyama Y, Uchida K, Yamaguchi H (2009) *Candida auris* sp. nov., a novel ascomycetous yeast isolated from the external ear canal of an inpatient in a Japanese hospital. Microbiol Immunol 53: 41–44. https://doi.org/10.1111/j.1348-0421.2008.00083.x
5. Chowdhary A, Anil Kumar V, Sharma C, Prakash A, Agarwal K, Babu R, Dinesh KR, Karim S, Singh SK, Hagen F, Meis JF (2014) Multidrug-resistant endemic clonal strain of *Candida auris* in India. Eur J Clin Microbiol Infect Dis 33:919–926. https://doi.org/10.1007/s10096-013-2027-1
6. Magobo RE, Corcoran C, Seetharam S, Govender NP (2014) *Candida auris*-associated candidemia, South Africa. Emerg Infect Dis 20:1250–1251. https://doi.org/10.3201/eid2007.131765
7. Lockhart SR, Etienne KA, Vallabhaneni S, Farooqi J, Chowdhary A, Govender NP, Colombo AL, Calvo B, Cuomo CA, Desjardins CA, Berkow EL, Castanheira M, Magobo RE, Jabeen K, Asghar RJ, Meis JF, Jackson B, Chiller T, Litvintseva AP (2017) Simultaneous emergence of multidrug-resistant *Candida auris* on 3 continents confirmed by whole-genome sequencing and epidemiological analyses. Clin Infect Dis 64:134–140. https://doi.org/10.1093/cid/ciw691
8. Borman AM, Johnson EM (2020) *Candida auris* in the UK: introduction, dissemination, and control. PLoS Pathog 16:e1008563. https://doi.org/10.1371/journal.ppat.1008563
9. Cortegiani A, Misseri G, Fasciana T, Giammanco A, Giarratano A, Chowdhary A (2018) Epidemiology, clinical characteristics, resistance, and treatment of infections by *Candida auris*. J Intensive Care 6:69. https://doi.org/10.1186/s40560-018-0342-4

10. Jeffery-Smith A, Taori SK, Schelenz S, Jeffery K, Johnson EM, Borman A, Candida auris Incident Management Team, Manuel R, Brown CS (2017) *Candida auris*: a review of the literature. Clin Microbiol Rev 31: e00029-17. https://doi.org/10.1128/CMR.00029-17
11. Chow NA, de Groot T, Badali H, Abastabar M, Chiller TM, Meis JF (2019) Potential fifth clade of *Candida auris*, Iran, 2018. Emerg Infect Dis 25:1780–1781. https://doi.org/10.3201/eid2509.190686
12. Taori SK, Khonyongwa K, Hayden I, Athukorala GDA, Letters A, Fife A, Desai N, Borman AM (2019) *Candida auris* outbreak: mortality, interventions and cost of sustaining control. J Infect 79:601–611. https://doi.org/10.1016/j.jinf.2019.09.007
13. Eyre DW, Sheppard AE, Madder H, Moir I, Moroney R, Quan TP, Griffiths D, George S, Butcher L, Morgan M, Newnham R, Sunderland M, Clarke T, Foster D, Hoffman P, Borman AM, Johnson EM, Moore G, Brown CS, Walker AS, Peto TEA, Crook DW, Jeffery KJM (2018) A *Candida auris* outbreak and its control in an intensive care setting. N Engl J Med 379:1322–1331. https://doi.org/10.1056/NEJMoa1714373
14. Schelenz S, Hagen F, Rhodes JL, Abdolrasouli A, Chowdhary A, Hall A, Ryan L, Shackleton J, Trimlett R, Meis JF, Armstrong-James D, Fisher MC (2016) First hospital outbreak of the globally emerging *Candida auris* in a European hospital. Antimicrob Resist Infect Control 5:35. https://doi.org/10.1186/s13756-016-0132-5
15. Mulet Bayona JV, Tormo Palop N, Salvador García C, Herrero Rodríguez P, Abril López de Medrano V, Ferrer Gómez C, Gimeno Cardona C (2020) Characteristics and management of candidaemia episodes in an established *Candida auris* outbreak. Antibiotics (Basel) 9:558. https://doi.org/10.3390/antibiotics9090558
16. Borman AM, Szekely A, Johnson EM (2017) Isolates of the emerging pathogen *Candida auris* present in the UK have several geographic origins. Med Mycol 55:563–567. https://doi.org/10.1093/mmy/myw147
17. Szekely A, Borman AM, Johnson EM (2019) *Candida auris* isolates of the Southern Asian and South African lineages exhibit different phenotypic and antifungal susceptibility profiles *in vitro*. J Clin Microbiol 57:e02055-18. https://doi.org/10.1128/JCM.02055-18
18. Borman AM, Szekely A, Johnson EM (2016) Comparative pathogenicity of United Kingdom isolates of the emerging pathogen *Candida auris* and other key pathogenic *Candida* species. mSphere 1:e00189-16. https://doi.org/10.1128/mSphere.00189-16
19. Lionakis MS (2011) *Drosophila* and *Galleria* insect model hosts: new tools for the study of fungal virulence, pharmacology and immunology. Virulence 2:521–527. https://doi.org/10.4161/viru.2.6.18520
20. Kavanagh K, Sheehan G (2018) The use of *Galleria mellonella* larvae to identify novel antimicrobial agents against fungal species of medical interest. J Fungi (Basel) 4:113. https://doi.org/10.3390/jof4030113
21. Ames L, Duxbury S, Pawlowska B et al (2017) *Galleria mellonella* as a host model to study *Candida glabrata* virulence and antifungal efficacy. Virulence 8:1909–1917
22. Champion OL, Titball RW, Bates S (2018) Standardization of *G. mellonella* larvae to provide reliable and reproducible results in the study of fungal pathogens. J Fungi (Basel) 4: 108. https://doi.org/10.3390/jof4030108
23. Jemel S, Guillot J, Kallel K, Botterel F, Dannaoui E (2020) *Galleria mellonella* for the evaluation of antifungal efficacy against medically important fungi, a narrative review. Microorganisms 8:390. https://doi.org/10.3390/microorganisms8030390
24. Durieux MF, Melloul É, Jemel S et al (2021) *Galleria mellonella* as a screening tool to study virulence factors of *Aspergillus fumigatus*. Virulence 12:818–834. https://doi.org/10.1080/21505594.2021.1893945
25. Borman AM (2018) Of mice and men and larvae: *Galleria mellonella* to model the early host-pathogen interactions after fungal infection. Virulence 9:9–12. https://doi.org/10.1080/21505594.2017.1382799
26. Mowlds P, Kavanagh K (2008) Effect of pre-incubation temperature on susceptibility of *Galleria mellonella* larvae to infection by *Candida albicans*. Mycopathologia 165:5–12. https://doi.org/10.1007/s11046-007-9069-9
27. Borman AM, Szekely A, Linton CJ, Palmer MD, Brown P, Johnson EM (2013) Epidemiology, antifungal susceptibility, and pathogenicity of *Candida africana* isolates from the United Kingdom. J Clin Microbiol 51: 967–972. https://doi.org/10.1128/JCM.02816-12
28. Sheehan G, Kavanagh K (2018) Analysis of the early cellular and humoral responses of *Galleria mellonella* larvae to infection by *Candida albicans*. Virulence 9:163–172. https://doi.org/10.1080/21505594.2017.1370174
29. Forgács L, Borman AM, Prépost E et al (2020) Comparison of *in vivo* pathogenicity of four *Candida auris* clades in a neutropenic

bloodstream infection murine model. Emerg Microbes Infect 29:1160–1169. https://doi.org/10.1080/22221751.2020.1771218

30. Kaplan EL, Meier P (1958) Nonparametric estimation from incomplete observations. J Am Stat Assoc 53(282):457–481

31. Loh JM, Adenwalla N, Wiles S, Proft T (2013) *Galleria mellonella* larvae as an infection model for group A *Streptococcus.* Virulence 4: 419–428. https://doi.org/10.4161/viru.24930

32. Mowlds P, Barron A, Kavanagh K (2008) Physical stress primes the immune response of *Galleria mellonella* larvae to infection by *Candida albicans.* Microbes Infect 10:628–634. https://doi.org/10.1016/j.micinf.2008.02.011

33. Leach L, Zhu Y, Chaturvedi S (2018) Development and validation of a real-time PCR assay for rapid detection of *Candida auris* from surveillance samples. J Clin Microbiol 56: e01223-17. https://doi.org/10.1128/JCM.01223-17

34. Arastehfar A, Fang W, Daneshnia F, Al-Hatmi AM, Liao W, Pan W, Khan Z, Ahmad S, Rosam K, Lackner M, Lass-Flörl C, Hagen F, Boekhout T (2019) Novel multiplex real-time quantitative PCR detecting system approach for direct detection of *Candida auris* and its relatives in spiked serum samples. Future Microbiol 14:33–45. https://doi.org/10.2217/fmb-2018-0227

35. Bergin D, Murphy L, Keenan J, Clynes M, Kavanagh K (2006) Pre-exposure to yeast protects larvae of *Galleria mellonella* from a subsequent lethal infection by *Candida albicans* and is mediated by the increased expression of antimicrobial peptides. Microbes Infect 8:2105–2112

36. Fallon JP, Troy N, Kavanagh K (2011) Pre-exposure of *Galleria mellonella* larvae to different doses of *Aspergillus fumigatus* conidia causes differential activation of cellular and humoral immune responses. Virulence 2: 413–421. https://doi.org/10.4161/viru.2.5.17811

37. Kloezen W, van Helvert-van Poppel M, Fahal AH, van de Sande WW (2015) A *Madurella mycetomatis* grain model in *Galleria mellonella* larvae. PLoS Negl Trop Dis 9:e0003926. https://doi.org/10.1371/journal.pntd.0003926

Chapter 24

Drosophila melanogaster as a Rapid and Reliable In Vivo Infection Model to Study the Emerging Yeast Pathogen *Candida auris*

Sebastian Wurster, Nathaniel D. Albert, and Dimitrios P. Kontoyiannis

Abstract

While mammalian models remain the gold standard to study invasive mycoses, mini-host invertebrate models have provided complementary platforms for explorative investigations of fungal pathogenesis, host–pathogen interplay, and antifungal therapy. Specifically, our group has established *Toll*-deficient *Drosophila melanogaster* flies as a facile and cost-effective model organism to study candidiasis, and we have recently expanded these studies to the emerging and frequently multidrug-resistant yeast pathogen *Candida auris*. Our proof-of-concept data suggest that fruit flies could hold a great promise for large-scale applications in antifungal drug discovery and the screening of *C. auris* (mutant) libraries with disparate pathogenic capacity. This chapter discusses the advantages and limitations of *D. melanogaster* to study *C. auris* candidiasis and provides a step-by-step guide for establishing and troubleshooting *C. auris* infection and antifungal treatment of *Toll*-deficient flies along with essential downstream readouts.

Key words *Drosophila melanogaster*, Fruit flies, Mini-host model, Virulence, Pathogenicity, Antifungal treatment

1 Introduction

Mammalian hosts, especially mouse models, remain the gold standard for in vivo characterization of fungal pathogenesis, biomarker development, and preclinical evaluation of therapeutic innovations [1, 2]. However, despite the acceleration of research on the molecular mechanisms of pathogenic fungi [3], large-scale in vivo studies (e.g., characterization of fungal mutant libraries) in mammalian models remain laborious and slow. Bridging the gap between high-throughput in vitro platforms [4–6] and mammalian hosts, invertebrate animal models such as the fruit fly (*Drosophila melanogaster*) have greatly contributed to our understanding of fungal pathogenesis and host defense [7–10]. Similarly, fruit fly infection models have provided a facile and rapid screening

Alexander Lorenz (ed.), *Candida auris: Methods and Protocols*, Methods in Molecular Biology, vol. 2517,
https://doi.org/10.1007/978-1-0716-2417-3_24,

platform for antifungal compounds and advanced the dissection of antifungal immunopharmacology and resistance [7–10]. A recent study by our group has expanded the use of fruit fly infection models to the emerging, enigmatic, and frequently multidrug-resistant yeast pathogen *Candida auris* [11]. In this chapter, we are discussing major advantages and limitations of *D. melanogaster* as a mini-host model for *C. auris* research and provide step-by-step guidance for establishing and troubleshooting *C. auris* infection and antifungal treatment in *Toll (Tl)*-deficient fruit flies.

The many advantages of *D. melanogaster* as an infection model include its small size, relatively low cost, and short generation time of 8–12 days [9]. As flies are invertebrates, ethical concerns and regulatory burden are lower compared to mammalian models or vertebrate mini-hosts such as zebrafish. Trained investigators can easily infect and assess hundreds of flies in a single experiment, thus allowing for multi-group comparisons and side-by-side screening of several fungal strains, mutants with disparate pathogenicity, or investigational treatments [10].

Another major advantage of the fly model is the ability to simulate different routes of infection [10]. Depending on the pathogen of interest and its common route of transmission, injection-based, ingestion-based, or skin surface infection assays have been developed [10]. Among these infection protocols, direct inoculation of yeast cells into the hemolymph can be considered the pathophysiologically most relevant mode of infection for *C. auris*, given the pathogen's significance as a cause of bloodstream infections [12, 13]. Although microinjection protocols using a capillary needle and dispensing apparatus for infection of flies have been developed [14] and may be adapted for *C. auris*, we favor pricking of flies with a needle/insect pin dipped into yeast cell suspensions [11]. In our experience, the pricking method is more cost-effective, less prone to cross-contamination, and faster than microinjection, thus resulting in shortened anesthesia of flies. Although the pricking method provides limited inoculum control compared with microinjection, experienced investigators can achieve outstanding reproducibility of survival outcomes of *C. auris*-infected flies [11] (Fig. 1).

Despite direct injection of the pathogen into the hemolymph, fruit flies have potent immune defense mechanisms to eliminate most opportunistic fungi without developing lethal infection [8–11, 14]. As in mammalians, *Tl* receptors serve as a pivotal pattern recognition pathway, triggering a proinflammatory response to invading fungi and eliciting a humoral immune response through release of antifungal peptides [8–10, 14]. Because of the low susceptibility of wild-type flies to most fungal pathogens, including *C. auris*, *Tl* mutant flies are frequently employed to establish robust and reproducible infection [10, 11, 15, 16] and are recommended for the protocol described in this chapter.

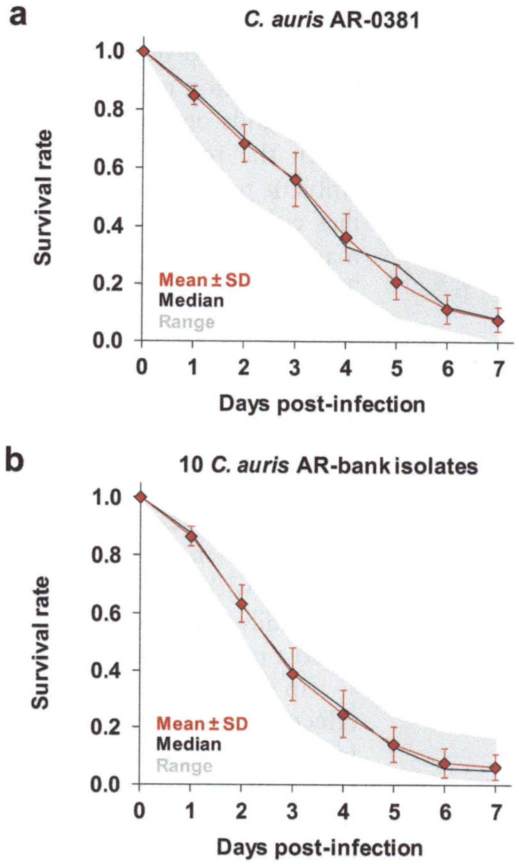

Fig. 1 Reproducibility of survival outcomes in *C. auris*-infected flies. (**a**) Technical reproducibility of survival outcomes in untreated *Tl*-deficient *D. melanogaster* flies infected with *C. auris* strain AR-0381. Mean, median, and range of survival rates were calculated based on nine independent replicates with 22–30 flies per experiment, performed by the same investigator over a 16-month period. (**b**) Strain-dependent variability of survival outcomes in untreated *Tl*-deficient *D. melanogaster* flies infected with ten different *C. auris* isolates (AR-0381-AR-0390). Mean, median, and range of survival rates were calculated based on *aggregate* survival data from three independent experiments with a total of 80 flies per cohort. These analyses utilize previously published raw data (Ref. [11]; Figures 2, 3, and S2)

As another advantage of the fruit fly model, simple administration of orally bioavailable antifungal agents can be achieved by drug supplementation of the fly food [10, 11, 15] (*see* Subheading 3.6). Although drug exposures cannot be precisely adjusted and quantitative bioassays [15] are imprecise and cumbersome, our published experience indicated good concordance of in vitro susceptibility profiles and in vivo efficacy of azole antifungals against representative *C. auris* isolates in flies [11].

However, some disadvantages and limitations of the fruit fly model need to be considered. Unlike wax moth (Galleria mellonella) infection models, manipulation and infection of flies require some hands-on training (*see* **Note 1**) and microscopic equipment. Given the recommended housing conditions with a 25–29 °C ambient temperature and 12-hourly alternating light-dark cycles, flies either require a dedicated room or specialized incubator for maintenance [10]. Furthermore, the temperature window of 25–29 °C is considerably lower than the human body temperature and might therefore pose limitations to the recapitulation of temperature-dependent virulence factors [11, 12]. As all invertebrate hosts, flies lack advanced adaptive immunity and thus provide a highly simplified representation of the complex antifungal immune defense of mammalians [14, 17]. Therefore, studies of immunotherapeutic interventions targeting adaptive immunity are generally not feasible in flies. Nonetheless, owing to the infinite genetic amenability of *Drosophila melanogaster*, a myriad of mutants with various natures and degrees of immune dysfunction are readily available as a primary screening system for determinants of host immunity to yeast pathogens [14, 18, 19].

In summary, despite some limitations, flies are an easy and cost-effective surrogate model for comparative virulence studies and in vivo efficacy testing of orally bioavailable (investigational) antifungal drugs. Given the increasing interest in virulence determinants of *C. auris* [12, 13, 20, 21] and the development of molecular tools to create large-scale *Candida* mutant collections [22], fruit flies might have a particular value as an in vivo screening and prioritization platform to elucidate pathogenicity-associated genetic factors and potential therapeutic targets (Fig. 2). In the future, several potential refinements could lead to further exciting advances in the use of flies in studies of *C. auris* candidiasis. On the one hand, although not our routine method for pathogen inoculation into the hemolymph, volume-controlled microinjection [23] could allow for reproducible high-inoculum infection of flies with *C. auris* that could potentially facilitate establishment of lethal infection in immunocompetent flies. Additionally, the prior successful use of fruit flies to study various bacterial agents of bloodstream infection, e.g., *Staphylococcus aureus* and *Pseudomonas aeruginosa*, may pave the way for the development of dual or sequential infection protocols [14, 24, 25]. Furthermore, recapitulation of metabolic or pharmacological risk factors for *C. auris* candidiasis, including obesity and diabetes mellitus, corticosteroids, or antibiotic pre-treatment [26, 27], is feasible in the fly model [8, 28, 29]. Importantly, efforts to establish more advanced mini-host models of *C. auris* infection need to be accompanied by improvements in mammalian models (Refs. [30, 31], and

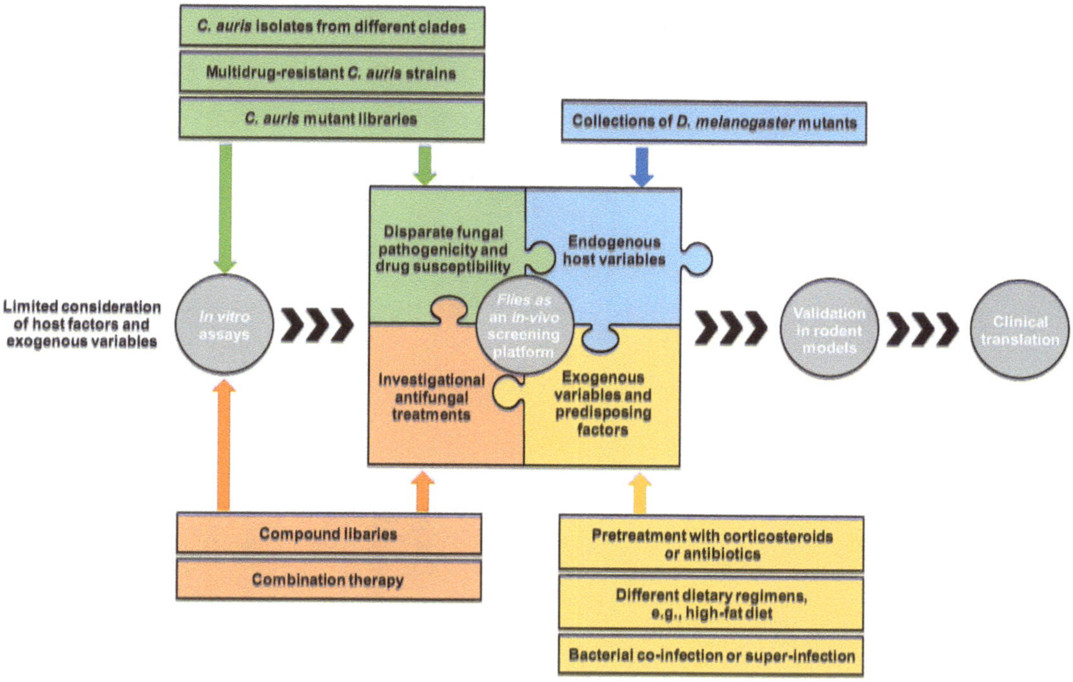

Fig. 2 *D. melanogaster* as an in vivo screening and prioritization platform for fungal virulence determinants and novel therapeutics

https://pubmed.ncbi.nlm.nih.gov/35224979/) for in-depth dissection of *C. auris* pathogenesis, evaluation of diagnostic biomarkers, and validation of investigational treatments (Fig. 2).

2 Materials

2.1 Housing and Manipulation of Flies

1. Adult female $Tl^{r632}/Tl^{I\text{-}RXA}$ Toll-deficient flies (*see* **Notes 2–4**).
2. Stereoscopic microscope.
3. Controllable carbon dioxide flow pad connected to a CO_2 line (Fig. 3).
4. Light source to illuminate the fly pad.
5. Fly incubator with high humidity capacity (60–75%), adjustable temperature, and a 12-h light/12-h dark cycle.
6. Modified Bloomington cornmeal fly food, either from a commercial source (*see* **Note 5**) or prepared according to the following recipe: 1.59% w/v inactivated dry yeast, 0.92% w/v soy flour, 6.71% w/v yellow cornmeal, 7.06% v/v light corn syrup, 0.44% v/v propionic acid (13.4-M stock concentration), and 0.64% w/v agar (gel strength 920 g/cm^2); boil ingredients in water for 10 min at >90 °C while stirring the mixture, and poor 10 mL per vial in 28.5 × 95-mm polystyrene vials.

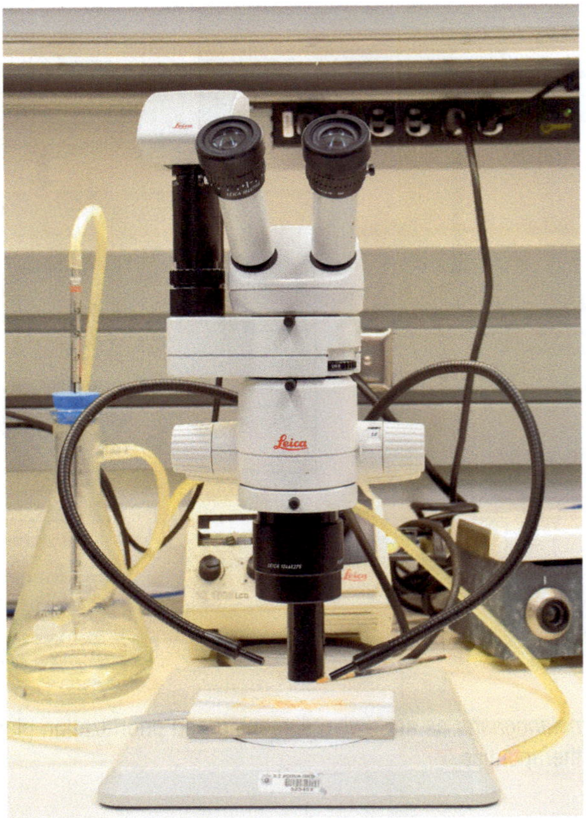

Fig. 3 Fly manipulator station with an attached CO_2-flow fly pad used to anesthetize flies

7. Antifungal agents of choice (*see* **Note 6**).
8. Sterile DMSO for preparation of antifungal drug stocks.
9. Inactive dry yeast flakes.
10. Spatula.
11. Stainless steel insect pin, size 000, tip diameter 0.03 mm, rod diameter 0.25 mm.
12. Pin vise to hold the insect pin.
13. Small insect manipulation brush or size 0 paintbrush.
14. Bunsen burner.
15. Fly morgue filled with 70% ethanol.

2.2 Culture of C. auris Isolates and Inoculum Preparation

1. *Candida auris* isolates (*see* **Note 7**).
2. Yeast extract–peptone–dextrose (YPD) agar plates, either from a commercial source or prepared according to the following recipe: 2% w/v dextrose, 2% w/v peptone, 1% w/v yeast extract, and 2% w/v agar in ddH_2O; autoclave.
3. Liquid YPD medium, either using a commercial preparation or prepared according to the following recipe: 2% w/v dextrose,

2% w/v peptone, 1% w/v yeast extract, and 0.032% w/v tryptophan in ddH_2O; autoclave.

4. Sterile phosphate-buffered saline (PBS), either using a commercial formulation or prepared according to the following recipe: 0.8% w/v NaCl, 0.02% w/v KCl, 0.144% w/v Na_2HPO_4, and 0.024% w/v KH_2PO_4 in ddH_2O, adjust the pH to 7.4 by addition of HCl; autoclave or filter-sterilize.
5. Glycerol for isolate banking.
6. Sterile disposable 100 × 15-mm Petri dishes.
7. Autoclavable glass spreaders or sterile single-use plastic spreaders.
8. 50-mL conical tubes.
9. Microbiological incubator.
10. Centrifuge capable of spinning 50-mL tubes at 3000–5000 × *g*.
11. Vortex mixer.
12. Vitek Densichek device (Bio Merieux) to determine yeast cell concentrations.

2.3 Determination of Fungal Burden

1. 1.8-mL or 2-mL cryovials.
2. Bead-beater homogenizer.
3. Glass beads, acid-washed, 425–600 μm.
4. YPD agar plates, prepared as in **item 2** of Subheading 2.2.
5. Autoclavable glass spreaders or sterile single-use plastic spreaders.
6. Microbiological incubator.
7. Freezer (−80 °C).
8. Precision scale.

2.4 General Equipment and Consumables

1. 1-mL, 100-μL, and 10-μL pipettes.
2. Filter tips.
3. 1.5-mL and 2-mL microcentrifuge tubes.

3 Methods

3.1 Preparation of Antifungal Drug-Supplemented Fly Food Vials

1. Dilute antifungal stock suspensions to the desired concentration (e.g., 1 mg/mL) with PBS (*see* **Note 8**).
2. Sterilize a spatula by flaming with a Bunsen burner, and let it cool for 30 s.
3. Create grid-like abrasions on the surface of the fly food. Abrasions should be superficial, not exceeding 2–3 mm in depth.

4. Add 200 μL of antifungal drug suspension to the abraded fly food surface. Higher volumes are not recommended as flies may get stuck to the fly food surface and die.
5. Fill a 1-mL pipette tip with dry, inactive yeast particles, and slowly drop them on the damp food surface to soak up the antifungal suspension (*see* **Note 9**).
6. Repeat **steps 2** through **5** for each antifungal drug suspension and concentration. Remember to add a “placebo” for mock treatment containing PBS and yeast particles but no antifungal agent. At least five vials are needed for each drug and experiment in order to transfer flies to fresh vials every other day.
7. Let antifungal-supplemented vials sit at room temperature for 24 h in order to let the yeast particles dry. Thereafter, vials can be stored at 4 °C for up to 10 days.

3.2 Antifungal (Pre) treatment of Flies

1. Place groups of 20–30 female adult Tl-deficient flies in empty vials to starve them for 6–8 h (*see* **Note 10**).
2. After the starvation period, transfer flies to the antifungal-supplemented fly food vials, and allow them to feed for 24 h before infection with *C. auris* yeast cells (*see* Subheading 3.4). As an alternative to this antifungal prophylaxis setting, flies may be infected and then be transferred to antifungal-supplemented fly food vials immediately after infection.

3.3 Preparation of the C. auris *Inoculum Suspension*

1. Inoculate YPD agar plates with the required *C. auris* strains, e.g., by using 5–10 μL of a glycerol stock.
2. Incubate the plates in a microbiological incubator at 37 °C for at least 48 h.
3. Pick a single colony of each isolate, and inoculate 5 mL of liquid YPD medium.
4. Incubate the culture overnight at 35 °C and 200 rpm in a shaking incubator.
5. Centrifuge the overnight culture for 5 min at 5000 × *g*, and discard the supernatant.
6. Wash the yeast cell suspension with 50 mL of PBS, and repeat the centrifugation step.
7. Resuspend the yeast cells in 5 mL of fresh, sterile PBS, and vortex the suspension thoroughly.
8. Measure the turbidity of the suspension with a Vitek Densichek instrument (Bio Merieux). Alternatively, yeast cells may be counted manually using a light microscope and hemocytometer (*see* **Note 11**).
9. Dilute the *C. auris* suspension to a 1.0-McFarland equivalent, resulting in a concentration of approximately 1×10^8 yeast cells/mL.

3.4 Infection of Flies with C. auris Yeast Cells

1. Prepare fly food vials containing 20–30 female adult (≥10-day-old) flies for each experimental cohort. Consider all relevant controls. Each experiment should be accompanied by a mock-infected control, i.e., pricking with a needle dipped into sterile PBS.
2. Anesthetize flies in one vial (experimental group) at a time by injecting CO_2 into the fly food vial through the plug (Fig. 4a) (*see* **Note 12**). Flies will fall asleep within 3–5 s.
3. Carefully transfer flies onto a fly pad that is connected to a CO_2 line, and space flies out with a brush (Fig. 4b).
4. Sterilize the injection needle (insect pin) by flaming with a Bunsen burner.
5. Let the needle cool for at least 30 s while vortexing the inoculum suspension (*see* **Note 13**).
6. Dip the insect pin into the yeast cell suspension.
7. Infect a fly with the fungal inoculum by inserting the insect pin midway into the dorsolateral aspect of the thorax. The correct anatomic location of the injection site is shown in Fig. 5.
8. Re-inoculate the insect pin after pricking each fly by dipping it into the yeast cell suspension.
9. Once all flies of an experimental cohort have been pricked, carefully transfer the infected flies with a brush into a new, appropriately labeled fly food vial placed on its side (Fig. 4c) (*see* **Note 14**).
10. Repeat **steps 2** through **9** for each experimental cohort. It is crucial to re-sterilize the needle by flaming when switching between different strains, inoculum concentrations, or fungal species.
11. We recommend performing at least three independent replicate experiments on different days in order to facilitate a sufficiently powered survival analysis (Tables 1 and 2) (*see* **Note 15**) and to determine inter-replicate variability. To avoid an impact of circadian variables on fly susceptibility (*see* **Note 16**), we recommend performing each replicate experiment on the same time of the day (e.g., always in the early afternoon).

3.5 Post-Infection Monitoring

1. Assess flies regularly during recovery from CO_2 anesthesia. Flies dying within the first 3 h after infection (usually <5%) have not died from candidiasis but from the injection injury or procedure/anesthesia-related stress. Therefore, these flies are removed and excluded from survival analysis.
2. Thereafter, incubate flies at 29 °C to maximize the susceptibility to microbial challenge (*see* **Note 17**). The 12-hourly

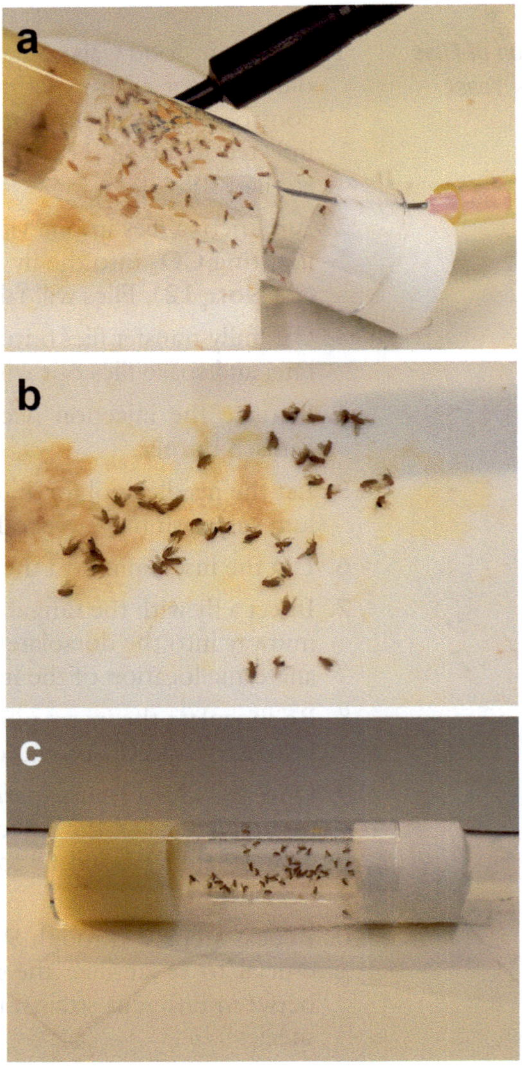

Fig. 4 Anesthesia of flies. (**a**) CO_2 is injected into the vial to anesthetize the flies. (**b**) The anesthetized flies are spread out on the CO_2 fly pad for infection. (**c**) Flies are returned to fresh fly food vials placed on their side until recovery from anesthesia

alternating light-dark cycle needs to be maintained throughout the post-infection monitoring period.

3. Count and remove dead flies daily. Consider preserving them for fungal burden monitoring (*see* Subheading 3.6). Alternatively, dispose of dead flies in a fly morgue filled with 70% ethanol.
4. Transfer flies to new vials every other day during the post-infection monitoring period (*see* **Note 18**).

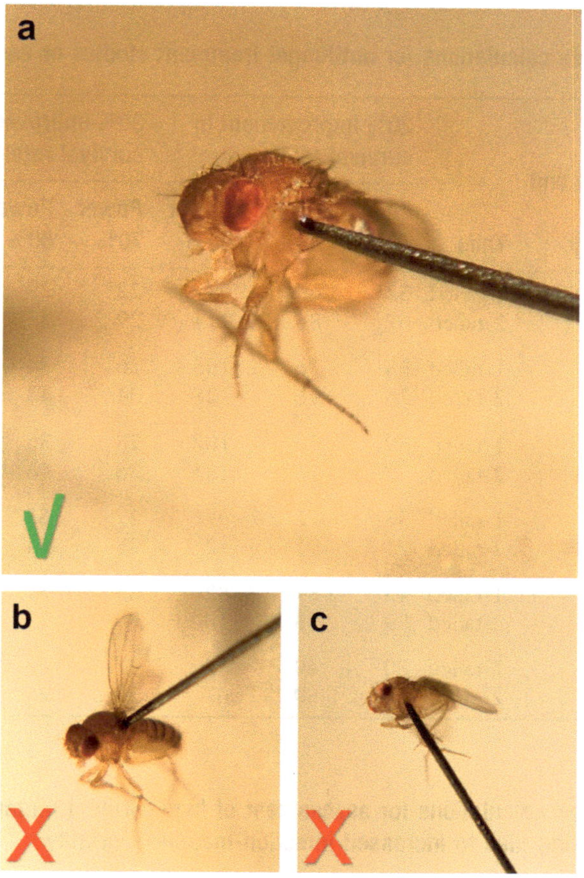

Fig. 5 Anatomic location of the injection site. (**a**) Correct location of the injection site at the dorsolateral thorax. (**b**) Incorrect location, which is too close to the wing hinge. (**c**) Incorrect location at the ventrolateral thorax. Injections into locations (**b**) and (**c**) will result in excess procedure-related mortality

5. Terminate the survival monitoring period between day 6 and 10 post-infection (*see* **Note 19**) by anesthetizing flies with CO_2 and disposing them of in the fly morgue. Alternatively, euthanize and preserve flies for fungal burden determination by freezing at −80 °C (*see* Subheading 3.6).

3.6 Determination of Fungal Burden

1. Collect dead flies daily, and store them in 1.5-mL microcentrifuge tubes at −80 °C until further analysis. Avoid contamination of the samples with fly food.
2. At the end of the survival monitoring period (e.g., on day 7 post-infection), euthanize and preserve the surviving flies by freezing at −80 °C until further analysis.
3. Thaw and combine all flies that belong to the same experimental cohort (*see* **Note 20**) in a bead-beating tube that has been weighed on a precision scale before.

Table 1
Cohort size calculations for antifungal treatment studies or assessment of hypovirulent mutants

Baseline end point mortality	Tails	20% improvement in survival rates			30% improvement in survival rates			40% improvement in survival rates		
		Power 70%	Power 80%	Power 90%	Power 70%	Power 80%	Power 90%	Power 70%	Power 80%	Power 90%
40%	1-tailed	52	68	93	22	29	40	12	16	22
	2-tailed	67	86	114	29	37	49	16	20	27
50%	1-tailed	58	76	105	26	34	47	15	20	27
	2-tailed	76	96	128	34	43	58	19	25	33
60%	1-tailed	59	77	107	28	36	50	16	21	29
	2-tailed	77	98	131	36	46	61	21	27	36
70%	1-tailed	55	72	99	27	35	48	16	21	30
	2-tailed	71	91	121	35	44	59	21	27	36
80%	1-tailed	45	59	82	23	31	42	15	20	27
	2-tailed	59	75	100	30	39	52	20	25	33
90%	1-tailed	30	40	55	17	23	31	12	16	22
	2-tailed	40	50	67	23	29	38	16	20	27

Table 2
Cohort size calculations for assessment of hypervirulent mutants or exogenous risk factors predisposing flies to increased infection-induced mortality

Baseline end point mortality	Tails	20% increase in mortality			30% increase in mortality			40% increase in mortality		
		Power 70%	Power 80%	Power 90%	Power 70%	Power 80%	Power 90%	Power 70%	Power 80%	Power 90%
40%	1-tailed	59	77	107	27	35	48	15	20	27
	2-tailed	77	98	131	35	44	59	20	25	33
50%	1-tailed	55	72	99	23	31	42	12	16	22
	2-tailed	71	91	121	30	39	52	16	20	27
60%	1-tailed	45	59	82	17	23	31	5	6	8
	2-tailed	59	75	100	23	29	38	6	8	10
70%	1-tailed	30	40	55	5	7	9	Not applicable		
	2-tailed	40	50	67	7	8	11			
80%	1-tailed	6	7	10	Not applicable			Not applicable		
	2-tailed	7	9	12						

4. Weigh the bead-beating tube again after addition of the flies in order to determine the net weight of the fly sample.
5. Add 1 mL of sterile, cold PBS (4 °C).

6. Carefully homogenize the fly tissue with a bead beater (*see* **Notes 21** and **22**).
7. Mix 10 μL of the homogenate with 990 μL of sterile, cold PBS. Thoroughly vortex the mixture.
8. Plate 10-μL and 100-μL aliquots of the diluted homogenate on YPD agar plates by using single-use plastic spreaders or sterilized glass spreaders.
9. Incubate plates in a microbiological incubator for 24–48 h at 37 °C.
10. Count *C. auris* colonies, and use the following formula to determine the amount of colony-forming units (CFUs) per gram tissue (*see* **Note 23**):

$$\text{CFUs/g tissue} = \frac{\text{number of colonies} \times 10^5}{\text{net tissue weight (g) in the homogenate} \times \text{number of } \mu\text{L of the diluted homogenate plated}}$$

4 Notes

1. Although the fly model is less complex than mammalian models and most institutions do not require formal training for manipulation of invertebrates, we recommend the following practice for inexperienced investigators prior to engaging in actual infection experiments: (1) Prepare a pathogen suspension (e.g., a 10^8/mL *Candida auris* yeast cell suspension), dip the needle/insect pin into the suspension, and "infect" a sterile 1-mL aliquot of PBS in a microcentrifuge tube or cell culture plate by superficially touching the liquid once with the needle. At least ten such aliquots should be "infected" and subsequently plated on YAG agar plates to determine the number of colony-forming units inoculated. A coefficient of variation of less than 30% should be achieved. (2) Prick sets of 25–30 anesthetized flies with a needle dipped into sterile PBS, let them recover, and monitor them for 7 days. A survival rate of >95% would be expected. Excess mortality would indicate that flies are pricked to deeply or at an unsuitable anatomic location (Fig. 5). (3) In the third step, prick flies with a pathogen that usually causes near-universal mortality (e.g., a 10^8/mL *Staphylococcus aureus* suspension) [24]. Seven-day survival rates >30% may indicate that the injection is performed too shallowly.
2. All experiments described in this chapter use $Tl^{r632}/Tl^{I\text{-}RXA}$ mutant flies, which display an impaired phagocytic response and antimicrobial peptide release [16]. At many academic institutions, such fly stocks can be obtained from a local *Drosophila* core facility, genetic resource core facility, or a similar core resource. Alternatively, we have previously published a detailed step-by-step protocol for generation of $Tl^{r632}/Tl^{I\text{-}RXA}$ mutant

flies by crossing thermosensitive allele of *Toll* (*Tl*^{r632}) flies with null allele of *Toll* (*Tl*$^{I\text{-}RXA}$) flies [16].

3. We recommend pairing five female and two male flies for optimal breeding results. During breeding, flies should be maintained at 25 °C as this temperature increases the yield of emerging adults. It is necessary to add distilled water to the vials every 3–5 days to maintain the optimal humidity and prevent the food from drying out. Under these optimal conditions, it takes approximately 8–12 days from the time the female flies lay their eggs in the fly food until the adults emerge.
4. We recommend the use of female flies since they are larger and, in our experience, less susceptible to injection injuries than male flies [10].
5. We use and recommend Fly Food Type M from LabExpress, Ann Arbor, Michigan, USA.
6. Antifungal drugs should be prepared as high-concentration stock solutions in DMSO, e.g., 40 mg/mL, in order to facilitate high antifungal drug concentrations or combination of drugs in the fly food while limiting DMSO toxicity.
7. Laboratories in the United States can request isolates from the Centers for Disease Control and Prevention (CDC) Antimicrobial Resistance Isolate Bank (AR-bank) free of charge. AR-bank isolates are validated by sequencing, and susceptibility breakpoints for several common antifungal agents are published on the AR-bank website [32]. The available panel contains isolates with varying degrees of azole susceptibility and even a multidrug-resistant isolate (AR-0384). Of note, the use of these CDC AR-bank isolates will facilitate immediate comparison of survival and fungal burden data in *D. melanogaster* with our published results [11]. Alternatively, *C. auris* strains are available for purchase from the Westerdijk Fungal Biodiversity Institute (former "Centraalbureau voor Schimmelcultures," CBS) or the American Type Culture Collection (ATCC).
8. We recommend a concentration of 1 mg/mL as a starting point [11, 15] and fourfold higher or lower concentrations for dose–effect studies.
9. Add enough yeast particles to fully absorb the antifungal drug suspension in order to prevent flies from dying by sticking to the damp surface. On the other hand, do not add more yeast particles than needed to soak up the antifungal drug suspension. Otherwise, the particles at the top (that are preferably eaten by the flies) will not contain sufficient drug quantities; thus, drug exposure of flies may be suboptimal.
10. Starvation for 6–8 h encourages increased ingestion of the antifungal drug-supplemented food. However, flies should

not be starved for longer periods as many flies will die, e.g., starvation for 24 h will result in the deaths of 50–75% of flies [10].

11. Manual counting can be imprecise since some *C. auris* strains have an aggregative phenotype. In order to increase the precision of manual counting, the inoculum solution may be passed through a 40-μm cell strainer to remove large yeast cell aggregates.

12. Caution: Prolonged exposure to CO_2 can cause headache, drowsiness, confusion, and, eventually, lethal CO_2 intoxication [33]. Therefore, ensure that the area for fly manipulations is well-ventilated and that the CO_2 line is properly closed after completion of the infection procedure. In planning large-scale experiments that require the use of the CO_2 fly pad for extended periods, consider splitting the experiment into multiple short sessions with a brief ventilation break every 30–45 min.

13. Vortex the yeast cell suspension every 4–5 min in order to ensure consistency of the inoculum and to prevent aggregation.

14. Anesthetized infected flies should be returned to food vials placed on their side until the flies recover from anesthesia. This usually only takes a few minutes and prevents flies from sticking to the food and dying.

15. Tables 1 and 2 provide an overview of suggested fly numbers per replicate depending on the baseline mortality at the end of the post-infection monitoring period, the expected change in survival rates (e.g., due to an investigational treatment), and desired power (1-β error). All numbers are based on a pre-set α error (*p*-value threshold for significance) of 0.05. Minimum numbers of flies per cohort for 1- and 2-tailed analysis, respectively, are provided. Of note, these numbers refer to pairwise comparisons. Multi-group comparisons may require considerably higher numbers of animals per cohort, depending on the type of multiple comparison compensation applied. Nonetheless, our recommended aggregate cohort size of 75 flies per condition (3 independent replicates with 25 flies per condition and replicate) would provide sufficient power for most interventions resulting in >20–25% changes in survival rates.

16. While systematic studies of the many interrelated experimental variables of infection experiments in mini-host models remain to be performed, we have observed that fruit flies tend to be more susceptible to fungal infection in the morning than in the afternoon (own unpublished data). This effect appears to be more pronounced in younger flies and may contribute to

considerable inter-replicate variability if independent experiments are performed at different times of the day.

17. Susceptibility to infection is maximal at 29 °C due to the temperature-sensitive loss of function (Tl^{r632}) allele.
18. For drug protection assays, antifungal drugs need to be added to each new set of fly food vials in order to maintain adequate drug exposure throughout the post-infection monitoring period.
19. Using our standard inoculum of 1×10^8/mL, a median survival time of 3–4 days and 7-day mortality rates of ≥75% are expected in untreated *C. auris*-infected Tl-deficient flies (Fig. 1).
20. We recommend a minimum of 20 flies per condition and replicate. With lower numbers, determination of the net tissue weight will be imprecise (e.g., due to minor remainders of fly food or moisture). Furthermore, higher numbers of flies per replicate help even out inter-individual differences in fungal burden and will thus allow for improved reproducibility of fungal burden quantification. At least three to four independent replicates from independent experiments are recommended.
21. We recommend two 15- to 20-s pulses at a maximum speed. However, optimal settings may need to be optimized depending on the specific properties (e.g., amplitude) of the devise used. Avoid bead beating for extended periods. Otherwise, fungal viability may be impacted, and the efficacy of *C. auris* colony recovery may diminish.
22. If a bead beater is not available, a vortex adapter or rubber foam bead tube holder can facilitate bead beating with common laboratory vortex mixers. When using a vortex adaptor instead of a bead beater, the number of 20-s pulses may need to be increased due to the smaller amplitude and force kinetics of a vortex mixer compared to most bead beaters.
23. The factor of 10^5 in the numerator takes into account (a) the 100-fold dilution step after bead beating (10^2) and (b) the unit conversion from μL of homogenate (plated) to the 1 mL of diluted homogenate prepared (10^3). The proposed simplified formula is derived from the following published formula [11]:

$$\text{CFUs per g fly tissue} = \frac{\text{number of colonies per mL plated} \times \text{dilution factor}}{\text{tissue (g)/mL original homogenate}}$$

Acknowledgment

DPK acknowledges the Robert C. Hickey Chair in Clinical Care endowment.

References

1. Conti HR, Huppler AR, Whibley N et al (2014) Animal models for candidiasis. Curr Protoc Immunol 105:19.6.1–19.617
2. Segal E, Frenkel M (2018) Experimental in vivo models of candidiasis. J Fungi (Basel) 4(1):E21
3. Muñoz JF, Gade L, Chow NA et al (2018) Genomic insights into multidrug-resistance, mating and virulence in *Candida auris* and related emerging species. Nat Commun 9(1): 5346
4. Wall G, Chaturvedi AK, Wormley FL Jr et al (2018) Screening a repurposing library for inhibitors of multidrug-resistant *Candida auris* identifies ebselen as a repositionable candidate for antifungal drug development. Antimicrob Agents Chemother 62(10):e01084-18
5. O'Brien B, Chaturvedi S, Chaturvedi V (2020) In vitro evaluation of antifungal drug combinations against multidrug-resistant *Candida auris* isolates from New York outbreak. Antimicrob Agents Chemother 64(4):e02195-19
6. Wurster S, Kumaresan PR, Albert ND et al (2019) Live monitoring and analysis of fungal growth, viability, and mycelial morphology using the IncuCyte NeuroTrack processing module. mBio 10(3):e00673-19
7. Chamilos G, Lionakis MS, Lewis RE et al (2006) *Drosophila melanogaster* as a facile model for large-scale studies of virulence mechanisms and antifungal drug efficacy in *Candida* species. J Infect Dis 193(7): 1014–1022
8. Chamilos G, Lewis RE, Hu J et al (2008) *Drosophila melanogaster* as a model host to dissect the immunopathogenesis of zygomycosis. Proc Natl Acad Sci U S A 105(27): 9367–9372
9. Hamilos G, Samonis G, Kontoyiannis DP (2012) Recent advances in the use of *Drosophila melanogaster* as a model to study immunopathogenesis of medically important filamentous fungi. Int J Microbiol 2012: 583792
10. Lionakis MS, Kontoyiannis DP (2012) *Drosophila melanogaster* as a model organism for invasive aspergillosis. Methods Mol Biol 845: 455–468
11. Wurster S, Bandi A, Beyda ND et al (2019) *Drosophila melanogaster* as a model to study virulence and azole treatment of the emerging pathogen *Candida auris*. J Antimicrob Chemother 74(7):1904–1910
12. Du H, Bing J, Hu T et al (2020) *Candida auris*: epidemiology, biology, antifungal resistance, and virulence. PLoS Pathog 16(10): e1008921
13. Cortegiani A, Misseri G, Fasciana T et al (2018) Epidemiology, clinical characteristics, resistance, and treatment of infections by *Candida auris*. J Intensive Care 6:69
14. Troha K, Buchon N (2019) Methods for the study of innate immunity in *Drosophila melanogaster*. Wiley Interdiscip Rev Dev Biol 8(5): e344
15. Lionakis MS, Lewis RE, May GS et al (2005) *Toll*-deficient *Drosophila* flies as a fast, high-throughput model for the study of antifungal drug efficacy against invasive aspergillosis and *Aspergillus* virulence. J Infect Dis 191(7): 1188–1195
16. Lionakis MS, Kontoyiannis DP (2010) The growing promise of *Toll*-deficient *Drosophila melanogaster* as a model for studying *Aspergillus* pathogenesis and treatment. Virulence 1(6):488–499
17. Rowley AF, Powell A (2007) Invertebrate immune systems specific, quasi-specific, or nonspecific? J Immunol 179(11):7209–7214
18. Govind S (2008) Innate immunity in *Drosophila*: pathogens and pathways. Insect Sci 15(1): 29–43
19. Kounatidis I, Ligoxygakis P (2012) *Drosophila* as a model system to unravel the layers of innate immunity to infection. Open Biol 2(5):120075
20. Lamoth F, Kontoyiannis DP (2018) The *Candida auris* alert: facts and perspectives. J Infect Dis 217(4):516–520
21. Rossato L, Colombo AL (2018) *Candida auris*: what have we learned about its mechanisms of pathogenicity? Front Microbiol 9:3081
22. Enkler L, Richer D, Marchand AL et al (2016) Genome engineering in the yeast pathogen *Candida glabrata* using the CRISPR-Cas9 system. Sci Rep 6:35766

23. Davis MM, Alvarez FJ, Ryman K et al (2011) Wild-type *Drosophila melanogaster* as a model host to analyze nitrogen source dependent virulence of *Candida albicans*. PLoS One 6(11): e27434
24. Ben-Ami R, Watson CC, Lewis RE et al (2013) *Drosophila melanogaster* as a model to explore the effects of methicillin-resistant *Staphylococcus aureus* strain type on virulence and response to linezolid treatment. Microb Pathog 55: 16–20
25. Christofi T, Apidianakis Y (2013) *Drosophila* immune priming against *Pseudomonas aeruginosa* is short-lasting and depends on cellular and humoral immunity. F1000Res 2:76
26. de Cássia Orlandi Sardi J, Silva DR, Soares Mendes-Giannini MJ et al (2018) *Candida auris*: epidemiology, risk factors, virulence, resistance, and therapeutic options. Microb Pathog 125:116–121
27. Osei Sekyere J (2018) *Candida auris*: a systematic review and meta-analysis of current updates on an emerging multidrug-resistant pathogen. Microbiologyopen 7(4):e00578
28. Shirazi F, Farmakiotis D, Yan Y et al (2014) Diet modification and metformin have a beneficial effect in a fly model of obesity and mucormycosis. PLoS One 9(9):e108635
29. Heys C, Lizé A, Blow F et al (2018) The effect of gut microbiota elimination in *Drosophila melanogaster*: a how-to guide for host-microbiota studies. Ecol Evol 8(8):4150–4161
30. Xin H, Mohiuddin F, Tran J et al (2019) Experimental mouse models of disseminated *Candida auris* infection. mSphere 4(5): e00339-19
31. Ben-Ami R, Berman J, Novikov A et al (2017) Multidrug-Resistant *Candida haemulonii* and *C. auris*, Tel Aviv, Israel. Emerg Infect Dis 23(1):195–203
32. CDC Antimicrobial Resistance Isolate Bank. https://www.cdc.gov/drugresistance/resistance-bank/index.html. Last accessed April 28, 2021
33. Amparan AA, Djoufack-Momo SM, Grunden B et al (2014) Exposure of research personnel to carbon dioxide during euthanasia procedures. J Am Assoc Lab Anim Sci 53(4): 376–380

Chapter 25

An Immunocompromised Mouse Model of *Candida auris* Systemic Infection

Janet Herrada, Kyle Roberts, Ahmed Gamal, Lisa Long, and Mahmoud A. Ghannoum

Abstract

With the recent emergence of multidrug-resistant *Candida auris*, there is an urgent need for new antifungal compounds with novel pharmacodynamic and pharmacokinetic properties that can treat systemic fungal infections caused by this emerging yeast. Historically, testing the efficacy of treatment for disseminated candidiasis was accomplished using a diverse array of in vivo animal models, including mice which offer an advantage both in their similarities to humans and their lower cost of maintenance. However, in order to create effective in vivo models for testing new antifungal compounds designed to treat systemic infections, it is important that these models also mimic several of the relevant predisposing conditions that can lead to disseminated candidiasis. Here, we describe an immunocompromised mouse model of hematogenously disseminated *C. auris* infection, which may have utility to test the efficacy of candidate antifungal compounds.

Key words *Candida auris*, Mouse model, Animal model, Disseminated candidiasis

1 Introduction

Candida auris is an emerging multidrug-resistant pathogenic yeast that has the potential to be a major global health threat due to several of its unique virulent traits and antifungal susceptibility patterns. The latter is the most concerning of these traits, with some isolates exhibiting multidrug resistance and others characterized as pan-resistant to every class of clinically available antifungals [1]. Additionally, *C. auris* is capable of surviving for extended periods on a variety of both wet and dry surfaces found in a hospital setting, demonstrating the potential for nosocomial transmission of the organism [2]. In this regard, the risk posed by *C. auris* in healthcare settings was confirmed in a recent report by the CDC

Janet Herrada and Kyle Roberts contributed equally with all other contributors.

Alexander Lorenz (ed.), *Candida auris: Methods and Protocols*, Methods in Molecular Biology, vol. 2517,
https://doi.org/10.1007/978-1-0716-2417-3_25,

in July 2021, which described the first evidence of nosocomial transmission of pan-resistant *C. auris* in the United States [3]. *C. auris* preferentially colonizes the skin, where it can develop into differentiated biofilms on the intact dermis and remain viable after desiccation on shed skin cells for upwards of 7 days [4]. Several factors put a patient at a greater risk for developing a systemic *Candida* infection, such as having a compromised immune system and the usage of indwelling medical devices including central line catheters [5]. Once inside the body, *C. auris* is capable of invading the bloodstream, which can lead to systemic candidiasis with a high mortality rate that ranges from 30% to 60% [6].

Since systemic candidiasis can affect multiple organ systems, new antifungal therapies with novel pharmacokinetic and pharmacodynamic properties need to be studied that may offer better treatment options than the available armamentarium for *C. auris* bloodstream infections [7, 8]. In order to test the effectiveness of these compounds, a suitable in vivo model should be developed that can mimic predisposing factors that make a person susceptible to a systemic *Candida* infection. Previous in vivo disseminated candidiasis models have used a variety of species including mice, rats, guinea pigs, and rabbits [9]. However, mice are the most commonly used species in systemic infection models due to their lower phylogenetic status, lower cost, and the reliability and reproducibility of the results they provide. Due to these factors, mouse models for *Candida albicans* blood stream infections have been widely used [10, 11] and also serve as suitable model for disseminated *Candida auris* disease. Since experimental evidence shows that *C. auris* is less virulent than *C. albicans* [10, 12], it is necessary to immunosuppress mice prior to challenging them with a highly concentrated yeast inoculum in order to establish a robust infection. Over the last 3 years, our team established such a model and tested a number of experimental and clinically available antifungals [7, 8, 13]. This chapter describes the method used to prepare the said model and evaluate in vivo treatment efficacy.

2 Materials

2.1 C. auris *Inoculum Preparation*

1. *C. auris* strain of choice (*see* **Note 1**).
2. Disposable hemocytometers.
3. Sterile 50-mL polypropylene conical tubes.
4. Serological pipette.
5. Adjustable volume micropipettes.
6. Microscope (white light).
7. Vortex mixer.
8. Centrifuge.

9. Manual hand tally counter.
10. Orbital platform shaker.
11. General-purpose incubator.
12. Sterile phosphate-buffered saline (PBS) pH 7.2 (1×): 210-mg/L potassium phosphate monobasic, 9.0-g/L sodium chloride, and 726-mg/L sodium phosphate dibasic (*see* **Note 2**).
13. 0.85% normal saline (NaCl) solution in water.
14. Sabouraud Dextrose Broth (SDB): 20.0-g/L dextrose, 5.0-g/L pancreatic digest of casein, and 5.0-g/L peptic digest of animal tissue (*see* **Note 3**).
15. Potato dextrose agar (PDA) with antibiotics: 4.0-g/L potato starch, 20.0-g/L dextrose, 0.125-g/L chloramphenicol, 0.03-g/L gentamicin (sulfate salt), and 15.0-g/L agar (*see* **Note 4**).
16. Stir bar.
17. Aluminum foil.
18. 2-mL microcentrifuge tubes.
19. Sterile Petri dishes.
20. Granulated agar (solidifying agent).
21. Distilled water.

2.2 Mice Immunosuppression and Inoculation

1. Female BALB/c mice with a body weight of approximately 20 g.
2. Cyclophosphamide monohydrate.
3. 1-mL syringe with 27-G × ½ (0.4 mm × 13 mm) needle for intraperitoneal (IP) administration.
4. 20-μL capillary tubes for white blood cell counting (*see* **Note 5**).
5. Mouse restrainer (*see* **Note 6**).
6. Insulin syringes with needle: 28 G × ½ for intravenous (IV) administration.
7. Wintergreen oil (essential oil with vasodilating properties) (*see* **Note 7**).
8. Water-circulating blanket (Stryker T/Pump Temperature Therapy System) (*see* **Note 7**).
9. Vortex mixer.
10. Adjustable volume pipettes.

2.3 Treatment Administration

1. Drug to be administered.
2. 1-mL syringe attached to an 18-G × 1.2″ disposable plastic animal feeding for oral (PO) treatment.

3. Insulin syringes with needle: 28 G × ½.
4. 1-mL syringe with a 27-G × ½ needle.
5. Mouse restrainer for intravenously administration of treatments.

2.4 Euthanasia and Sample Collection

1. CO_2 tank and euthanasia chamber.
2. Sterile forceps.
3. Sterile scissors.
4. 2-mL microcentrifuge tubes.
5. 5-mm stainless steel beads.
6. PBS (*see* Subheading 2.1, **step 12**).
7. Tissue homogenizer.
8. L-shaped sterile cell spreader.
9. PDA plates (*see* Subheading 2.1, **step 15**).

2.5 Survival Analysis and Determination of Tissue Fungal Burden

1. Excel spreadsheet to record daily animal mortalities in each treatment group. Data is expressed as percent survival in a line graph.
2. Excel spreadsheet to record and calculate tissue fungal burden (*see* **Note 8**).
3. Statistical analysis software (*see* **Note 8**).

3 Methods

3.1 Inoculum Preparation

1. Subculture *C. auris* strain of choice on PDA plates, streaking for isolation so that single colonies can be obtained.
2. Invert the subcultured plates and incubate at 37 °C for 48 h.
3. Subculture the *C. auris* isolate again by transferring four single colonies into a 50-mL conical tube containing 30 mL of SDB each, replace the lid but leave it loose to allow for air flow. Leave a fifth 50-mL conical tube with 30-mL SDB un-inoculated as a sterility control. Incubate overnight at 37 °C on a shaker set to 180 rpm.
4. Next, remove all tubes from the incubator, and inspect the sterility control tube which should not have any growth present. If growth is present in the sterility control, then the inoculum is contaminated, and a fresh batch of broth will have to be made and the procedure repeated. If the sterility control tube contains no growth, then it can be used to counterbalance the centrifuge if needed or discarded. Centrifuge all the tubes at 1450 × *g* for 5 min to obtain a pellet.

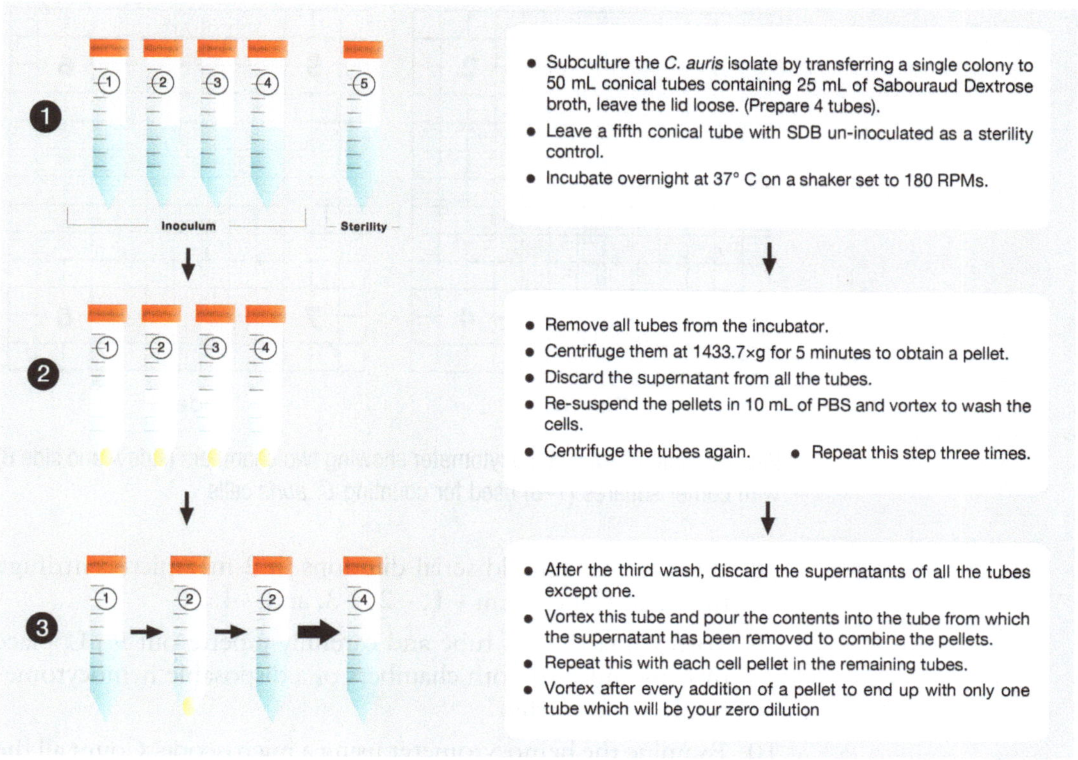

Fig. 1 Step-by-step procedure for culturing, washing, and preparing *C. auris* inoculum

5. Remove the supernatant from all the tubes, and resuspend the pellets in 10 mL of PBS using a vortex mixer. Then centrifuge the tubes again (as in **step 4**). Repeat this step for a total of three times to wash the cells.
6. After the third wash, discard the supernatants of all the tubes except one. Resuspend culture by vortex-mixing this tube, and pour the contents into the tube from which the supernatant has been removed to combine the pellets. Repeat this with each pellet in the remaining tubes. Vortex-mix after every addition of a pellet to end up with only one tube which will be your zero dilution (Fig. 1).
7. Centrifuge the zero dilution at 1450 × *g* for 5 min, discard the supernatant, and add an appropriate volume of PBS to suspend your pellet. To calculate this volume, use the following formula:

   \# animals × the volume of inoculum per animal (0.1 mL) × 1.5 (which will add 50% extra volume for contingencies).

   For example, if you are challenging 40 mice then the volume of inoculum is:

   40 × (0.1 mL/mouse) = 4 mL × 1.5 = 6.0 mL total volume needed. Resuspend using a vortex mixer.

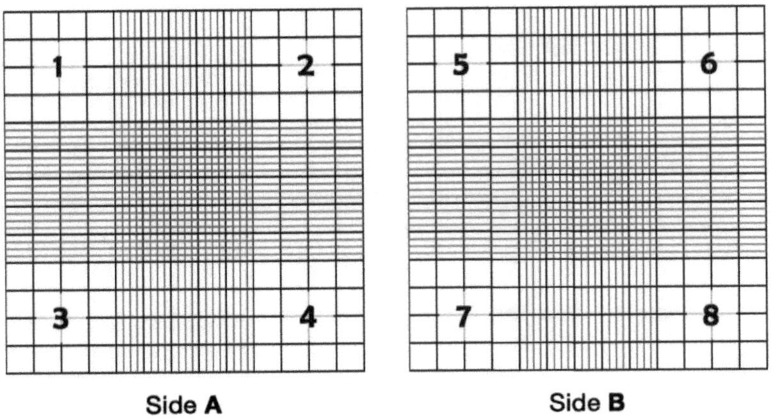

Fig. 2 Schematic of a hemocytometer showing two chambers (side A and side B) with corner squares (1–8) used for counting *C. auris* cells

8. Prepare four tenfold serial dilutions in 2-mL microcentrifuge tubes, and label them −1, −2, −3, and −4.
9. Vortex-mix the −4 tube and carefully pipette out 9 μL, place this amount into both chambers of a disposable hemocytometer. Do not overfill.
10. Examine the hemocytometer using a microscope. Count all the yeast cells in the eight corners of each side of the hemocytometer (Fig. 2, squares 1–8). Cells that overlap the outside boundary lines are not counted.
11. Average the counts of the squares. Based on the volume of the chambers of the hemocytometer, this represents the average count × 10^4 cells/mL (*see* **Note 9**).

3.2 Mice Immunosuppression, Inoculation, and Treatment Administration

1. Allow mice to acclimate for a minimum of 5 days prior to use.
2. Immunosuppression is induced by two intraperitoneal injection of cyclophosphamide: 200-mg/kg cyclophosphamide, 3 days prior to challenge, and 150-mg/kg cyclophosphamide 1 day after the challenge.
3. On the day of the challenge, collect 20-μL blood sample from the tail vein from one mouse from each group for a white blood cell count to verify immunosuppression [7].
4. On the day of the challenge, prepare a 3 × 10^8 cells/mL (3 × 10^7 cells in 100 μL per mouse) *C. auris* inoculum as explained in Subheading 3.1.
5. Place the mouse in a restrainer device (Fig. 3).
6. Locate the lateral tail vein of the mouse, and use a drop of wintergreen oil to dab the tail of the mouse (Fig. 3).
7. Inoculate each mouse with 100 μL of 3 × 10^7 *C. auris* yeast cells via the lateral tail vein [7].

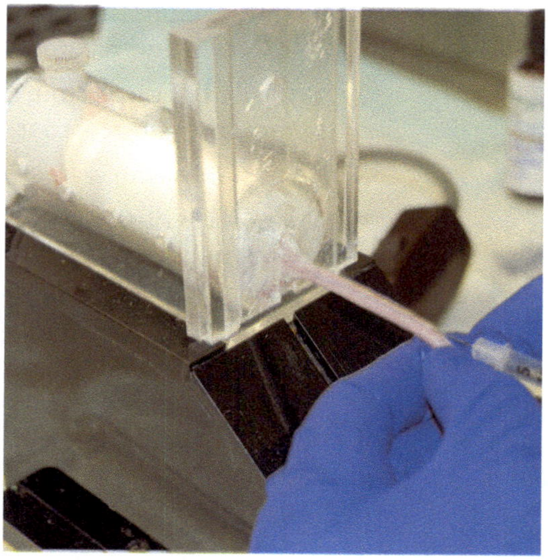

Fig. 3 Mouse in restrainer device with a view of lateral tail vein for intravenous injection

8. Fifteen mice per group should be used. Ten for survival and five for determining tissue fungal burden. Monitor animal survival, and record observations daily for 14 days (*see* **Note 10**).
9. Start mice treatments 2 h post challenge and continue for 7 days. Treatment dose regimen and route will depend on the pharmacokinetic/pharmacodynamics of the compound to be evaluated. IV or IP administrations are given in a volume of 100 μL per mouse and PO administration given in a volume of 200 μL per mouse.

3.3 Euthanasia and Sample Collection

1. Twenty-four hours following the last treatment administration, five mice per group are euthanized by CO_2 asphyxiation and cervical dislocation as a second method of euthanasia. Kidneys and brain are removed aseptically.
2. Place the collected organs in 2-mL microcentrifuge tubes, and measure and record the sample weight.
3. Add one steel bead to each tube containing the collected sample and 1-mL sterile 0.85% normal saline, and homogenize in tissue homogenizer (*see* **Note 11**).
4. Prepare four tenfold dilutions from each homogenized sample.
5. Plate 100 μL of each dilution in the middle of a PDA plate. Use an L-shaped spreader to disperse the sample on the entire surface of the plate.
6. Invert and incubate the plates at 37 °C for 48 h to allow the growth of fungal colonies.
7. Count the colony-forming units (CFUs), and express the tissue fungal burden as Log_{10} CFUs/g of tissue.

3.4 Survival Analysis and Determination of Tissue Fungal Burden

3.4.1 Survival Analysis

1. Monitor designated survival animals for a period of 14–28 days, and record it daily in a spreadsheet.
2. At the end of the period, calculate the percentage of survival in each group, and record it.
3. Plot a survival curve.
4. The efficacy of the test compounds is assessed by comparing survival rates between treatment and untreated control group.
5. Analyze the survival data using a Kaplan–Meier test (IBM SPSS version 27 or comparable program). A *P*-value of <0.05 is considered significant.

3.4.2 Determination of Tissue Fungal Burden

1. Record the number of CFUs in the prepared spreadsheet, and calculate the log CFUs/g of tissue; this represents the tissue fungal burden in the collected sample.
2. Calculate the average and mean in each sample.
3. Plot a graph showing the average log CFU/g ± standard deviation (SD).
4. Use statistical analysis software to evaluate the efficacy of the test compounds by comparing the tissue fungal burden between different treatment groups and the untreated control group and/or vehicle-treated group. The differences in mean CFUs in the kidneys and brain are compared to the untreated controls and other treatment groups using a one-way ANOVA with a post hoc Bonferroni. A *P*-value of <0.05 is considered statistically significant.

4 Notes

1. We used clinical isolate MRL #35368 to evaluate this protocol. This strain was chosen based on preliminary studies demonstrating that this strain causes disseminated infection in mice [7].
2. We use commercially available phosphate buffered saline pH 7.2 (1X) from Gibco (Life Technologies Corporation, Grand Island, NY, USA).
3. We use commercially available SDB from OXOID (Basingstoke, Hampshire, England). Fill a graduated polypropylene pitcher or a glass beaker with 1 L of distilled water. Weigh 30-g SDB powder and transfer to the pitcher, place a stir bar inside the pitcher, and mix well using a stir plate. The mix can be transferred to a glass flask, and heat to boiling to dissolve the medium completely. If this method is used, care should be taken as not to overboil and spill the mix. Cover the top of the pitcher with aluminum foil, and autoclave at 121 °C for 40 min. Allow to cool and store at 4–8 °C.

4. We use commercially available PDA from Difco (Becton, Dickinson and Company, Sparks, MD, USA). Fill a graduated polypropylene pitcher or a glass beaker with 1 L of distilled water. Weigh 39 g of PDA powder, 1 g of granulated agar, 0.125 g of chloramphenicol, and 0.03 g of gentamicin and transfer to the pitcher, place a stir bar inside the pitcher, and mix well using a stir plate. Cover the top of the pitcher with aluminum foil, and autoclave at 121 °C for 40 min. Pour the media into sterile Petri plates. The media needs to be stirred before plates are poured, if not some of the plates may not harden correctly.
5. We use Leuko-TIC kit with 20-μL capillary tubes for white blood cell counting. Collect a blood sample (20 μL collected with Leuko-TIC end-to-end capillary tube) from the tail vein from one mouse from each group, and place in Leuko-TIC tubes for a white blood count.
6. Inoculation in a disseminated model is performed by intravenous injection. We use a mouse tail illuminator restrainer consisting in a plexiglass restraining tube and sliding door to immobilize the mouse and allow the use of both hands to perform the tail vein injection without any sudden movements from the mouse. Select a device of appropriate size for the animal to be restrained. Monitor the breathing rate and color of the ears and nose while performing the procedure. If signs of difficulty breathing or change in color of ears and nose are observed, the mouse should be released immediately from the restrainer.
7. To facilitate the tail vein injection procedure, dilate the vein by placing a water-circulating blanket under the cage for 15–30 min before placing the mouse in the restrainer. Additionally, dab the tail with wintergreen oil (essential oil with vasodilation properties) prior to injection.
8. We use an Excel spreadsheet to enter CFU data, and calculate the average Log_{10} CFU ± standard deviation by treatment group (*see* Table 1). The efficacy of a treatment is evaluated using an ANOVA with a post hoc Bonferroni (IBM SPSS version 27 or comparable program). A *P*-value of <0.05 is considered significant.
9. We perform all animal inoculations in a biosafety cabinet. Follow your institutional and national guidelines while handling *C. auris*. For the inoculum preparation, follow this step-by-step example:
   - If the total count in the squares marked 1–8 = 348
   - Average count: 348/8 = 48 × 10^4 cells/mL.
   - Number of animals = 40

Table 1
Efficacy of treatment evaluated as a reduction in $\log_{10}$ CFU compared with all other groups

Treatment group	Mouse	Sample wt. (g)	CFU count/plate	Dilution factor	Total CFUs[a]	CFUs/g[b]	Log CFU/g[c]
Treatment 1	1						
	2						
	3						
	4						
	5						
Treatment 2	6						
	7						
	8						
	9						
	10						
Untreated control	11						
	12						
	13						
	14						
	15						

[a]Total CFU = CFU count × dilution × dilution factor
[b]Total CFU/sample wt. (g)
[c]Log (CFU/g)

- Volume of inoculum per animal = 0.1 mL (100 μL).
- Concentration of yeast cells per animal = 3×10^7 *C. auris* cells

Step 1. *Volume of inoculum needed*:

(# animals × the volume of inoculum per animal) × (1.5) which will add 50% extra volume for contingencies.

40 × (0.1 mL/mouse) = 4 mL × 1.5 = 6.0 mL total volume needed.

Step 2. *Number of yeast cells/mL*:

(Concentration of inoculum/100 μL) × 10:

(3×10^7 cells/100 μL) × 10 = 3×10^8 cells/mL

Step 3. *Total number of yeast cells needed*:

(Concentration of cells/mL) × (volume of inoculum needed):

$(3 \times 10^8 \text{ cells/mL}) \times (6.0 \text{ mL}) = 18 \times 10^8$ cells total

Step 4. *Count of yeast cells from dilution −4 in the hemocytometer*:

Average count: $348/8 = 48$

Thus, the count in −4 dilution is 48×10^4 *yeast cells*/mL.

The count in −3 dilution is 48×10^5 *yeast cells*/mL.

The count in −2 dilution is 48×10^6 *yeast cells*/mL.

The count in −1 dilution is 48×10^7 *yeast cells*/mL.

The count in 0 dilution is 48×10^8 *yeast cells*/mL (working dilution).

Step 5. *Volume needed from working dilution*:

(Total number of *yeast cells* needed)/(*yeast cells* count in working dilution):

$(18 \times 10^8 \text{ cells})/(48 \times 10^8 \text{ cells/mL}) = 0.375$ mL (375 μL)

Step 6. *Prepare inoculum*:

Amount of PBS needed:

(Total volume of inoculum) − (volume from working dilution):

$6 - 0.375 \text{ mL} = 5.625$ mL PBS

To complete the preparation of the inoculum, pipette 375 μL from the working dilution into a 15-mL conical tube, and add 5.625 mL of PBS.

10. Duration for which *C. auris*-infected mice are monitored for survival is dependent on the study being conducted. For example, in a study by Torres et al., survival of the control mice that had been infected with *C. auris* was monitored for 104 days post-infection; this was in order to detect whether the organism would be completely eliminated from the body [14]. Furthermore, we have observed that in most of our studies by day 14, the mortality rate of the infected mice does not increase. Thus, longer observation of survival may not be warranted.
11. Tubes containing the tissue sample, saline, and bead are placed in a TissueLyser adapter set and shaken at a very high speed, originating the tissue samples to be disrupted and homogenized by the beating and grinding effect of the beads. We employ a TissueLyser (QIAGEN) set for 5 min at 30 Hz. Other types of homogenizers need to be operated according the manufacturer's instructions.

References

1. Lockhart SR et al (2017) Simultaneous emergence of multidrug-resistant *Candida auris* on 3 continents confirmed by whole-genome sequencing and epidemiological analyses. Clin Infect Dis 64(2):134–140. https://doi.org/10.1093/cid/ciw691
2. Piedrahita CT, Cadnum JL, Jencson AL, Shaikh AA, Ghannoum MA, Donskey CJ

(2017) Environmental surfaces in healthcare facilities are a potential source for transmission of *Candida auris* and other *Candida* species. Infect Control Hosp Epidemiol 38(9): 1107–1109. https://doi.org/10.1017/ice.2017.127

3. Lyman M, Forsberg K, Reuben J et al (2021) *Notes from the field:* transmission of pan-resistant and echinocandin-resistant *Candida auris* in health care facilities—Texas and the District of Columbia, January-April 2021. Morb Mortal Wkly Rep 70(29):1022–1023. https://doi.org/10.15585/mmwr.mm7029a2
4. Horton MV et al (2020) *Candida auris* forms high-burden biofilms in skin niche conditions and on porcine skin. mSphere 5(1):e00910-19. https://doi.org/10.1128/mSphere.00910-19
5. Kean R, Brown J, Gulmez D, Ware A, Ramage G (2020) *Candida auris*: a decade of understanding of an enigmatic pathogenic yeast. J Fungi (Basel) 6(1):30. https://doi.org/10.3390/jof6010030
6. Keighley C, Garnham K, Harch AJ, Robertson M, Chaw K, Teng JC, Chen SCA (2021) *Candida auris*: diagnostic challenges and emerging opportunities for the clinical microbiology laboratory. Curr Fungal Infect Rep:1–11. https://doi.org/10.1007/s12281-021-00420-y
7. Hager CL, Larkin EL, Long LA, Ghannoum MA (2018) Evaluation of the efficacy of rezafungin, a novel echinocandin, in the treatment of disseminated *Candida auris* infection using an immunocompromised mouse model. J Antimicrob Chemother 73(8):2085–2088. https://doi.org/10.1093/jac/dky153
8. Hager CL, Larkin EL, Long L, Abidi FZ, Shaw KJ, Ghannoum MA (2018) *In vitro* and *in vivo* evaluation of the antifungal activity of APX001A/APX001 against *Candida auris*. Antimicrob Agents Chemother 62(3): e02319-17. https://doi.org/10.1128/AAC.02319-17
9. Hager CL, Long L, Imamura Y, Ghannoum MA (2019) Experimental animal models of invasive fungal infections. In: Ghannoum MA, Perfect JR (eds) Antifungal therapy, 2nd edn. CRC Press, Boca Raton
10. Xin H, Mohiuddin F, Tran J, Adams A, Eberle K (2019) Experimental mouse models of disseminated *Candida auris* infection. mSphere 4(5):e00339-19. https://doi.org/10.1128/mSphere.00339-19
11. Szabo EK, MacCallum DM (2011) The contribution of mouse models to our understanding of systemic candidiasis. FEMS Microbiol Lett 320(1):1–8. https://doi.org/10.1111/j.1574-6968.2011.02262.x
12. Wang X et al (2018) The first isolate of *Candida auris* in China: clinical and biological aspects. Emerg Microbes Infect 7(1):93. https://doi.org/10.1038/s41426-018-0095-0
13. Herrada J, Gamal A, Long L, Sanchez SP, McCormick TS, Ghannoum MA (2021) *In vitro* and *in vivo* antifungal activity of AmBisome compared to conventional amphotericin B and fluconazole against Candida auris. Antimicrob Agents Chemother 65(6):e00306-21. https://doi.org/10.1128/AAC.00306-21
14. Torres SR, Pichowicz A, Torres-Velez F, Song R, Singh N, Lasek-Nesselquist E, De Jesus M (2020) Impact of *Candida auris* infection in a neutropenic murine model. Antimicrob Agents Chemother 64(3):e01625-19. https://doi.org/10.1128/AAC.01625-19. PMID: 31818824; PMCID: PMC7038241

Chapter 26

Mouse Gastrointestinal Colonization Model for *Candida auris*

Stylianos C. Simantirakis and Donna M. MacCallum

Abstract

Animal colonization and infection models are frequently used to investigate host–pathogen interactions and disease progression. Here, we describe an effective model to investigate the ability of the newly emerged fungal pathogen *Candida auris* to persistently colonize the gut of immunocompetent mice. In our model, mice are inoculated by gavage and are subsequently monitored for colonization by determining daily fungal stool burdens. At the end of the experiment, mice are culled, and their spleen, liver, kidneys, lungs, heart, and caecum harvested to determine the fungal burdens in order to investigate colonization and potentially dissemination of *C. auris* to other host organs.

Key words *Candida auris*, Mouse model, Gut colonization, Infection model

1 Introduction

Animal infection models are an important tool for understanding host–pathogen interactions and elucidating the mechanisms used by pathogens to invade a host and ultimately cause infection. Mimicking such interactions in a live host (in vivo) provides better insight of how a pathogen colonizes the host or how a disease progresses and disseminates compared to in vitro or ex vivo models using host tissues or cells.

So far, *C. auris* is being characterized mainly as a skin colonizer, as it has been isolated from various patient body sites [1]. However, the main infections caused by this fungus are systemic infections, or infections of the respiratory, urinary, and gastrointestinal tracts [2], particularly in seriously ill or immunocompromised patients, including patients that are hospitalized in intensive care units (ICUs) due to underlying conditions such as diabetes [3], transplant recipients [4], patients who have undergone invasive surgery [5], and even patients with COVID-19 [6], and infections have been associated with the use of central venous catheters (CVCs)

Alexander Lorenz (ed.), *Candida auris: Methods and Protocols*, Methods in Molecular Biology, vol. 2517,
https://doi.org/10.1007/978-1-0716-2417-3_26,

[1, 7, 8]. Because of this, the murine *C. auris* infection models developed to date have either focused on systemic infection [3, 9] or skin colonization [10].

However, in clinical studies [1, 10, 11, 12], the rectum has been identified as the main colonization sites in patients with *C. auris* fungemia, and *C. auris* has been isolated from patient stool samples [5, 11], indicating potential gut colonization by this fungus. *C. auris* gut colonization and its contribution to pathogenicity have not yet been fully explored, but it has been shown for another fungal pathogen, *Candida albicans*, that it disseminates from the gut into the host, initiating systemic infections, only if there is mucosal damage to the gastrointestinal barrier [13]. One published study suggests that *C. auris* is capable of colonizing the gastrointestinal tract, but this has only been investigated in an immunocompromised mouse model [14]. It is, therefore, crucial to fully investigate how effectively *C. auris* can colonize the host gut and if such colonization predisposes to systemic infection.

In this protocol, we address the lack of a *C. auris* gastrointestinal colonization and dissemination model in immunocompetent mice by modifying a murine GI tract colonization model published previously for *Candida albicans* [13]. We detail a protocol that can be adapted for investigating colonization by different *C. auris* strains or to investigate various parameters, such as the effect of antibiotics or immune system competency. We demonstrate that *C. auris* does colonize the mouse gut, potentially acting as a reservoir for dissemination and causing infections in the host.

2 Materials

1. Mice: Six-to-eight-week-old (approximately 20–25 g) immunocompetent female mice (strain C57BL/6) (*see* **Note 1**).
2. *C. auris* strains (*see* **Note 2**).
3. Duran bottles (250 or 500 mL).
4. Benchtop autoclave.
5. Square polystyrene Petri dishes (120 mm × 120 mm).
6. Round polystyrene Petri dishes (100 × 15 mm).
7. YPD medium/agar: 1% (w/v) yeast extract, 2% (w/v) mycological peptone, 2% (w/v) glucose, 2% (w/v) agar (if required for plates). Add powders in to a Duran bottle, then add the appropriate volume of water. and autoclave at 121 °C at 1.08 bar for 28 min to sterilize. Agar was cooled to ~40 °C before addition of antibiotics, if required. Round Petri dishes require 20 mL of agar, while the square ones require 30 mL.

8. Water-soluble gentamicin (20 mg/mL) and 20-mg/mL chloramphenicol. Antibiotics were suspended in dH_2O, filter-sterilized, and stored at 4 °C for a few weeks or at −20 °C for a prolonged period. Add to YPD agar to give a final concentration of 50 mg/L for each antibiotic.
9. Individually ventilated cages (IVCs) to protect mice from exposure to environmental factors.
10. Mouse drinking fluid A: For 1 L, 2% (w/v) sucrose, 200-mL prepared PenStrep (10,000 units/mL penicillin and 10-mg/mL streptomycin), 128-mg/L fluconazole in sterile dH_2O.
11. Mouse drinking fluid B: For 1 L, 2% (w/v) sucrose, 200-mL prepared PenStrep (10,000 units/mL penicillin and 10-mg/mL streptomycin) in sterile dH_2O.
12. 1× sterile phosphate-buffered saline: 137-mM NaCl, 2.7-mM KCl, 10-mM Na_2HPO_4, and 1.8-mM KH_2PO_4 (this is commercially available).
13. Benchtop microcentrifuge.
14. Disposable microcentrifuge tubes (1.5 mL).
15. Disposable oral gavage tubes (FTP-18-30 polypropylene plastic feeding tubes).
16. Syringes (1 mL).
17. Dissecting scissors.
18. Dissecting scalpels.
19. Micropipettes (10, 20, 200, 1000 μL).
20. Pipette tips (20, 200, 1000 μL).
21. Disposable inoculating sterile loops.
22. Neubauer hemocytometer.
23. Tissue homogenizer.
24. Electronic analytical and precision balances.
25. Light microscope.

3 Methods

3.1 Inoculum Preparation

1. A single colony from a *C. auris* growth plate (Fig. 1) is inoculated into 5 mL of YPD medium, put in a shaking incubator at 30 °C at 200 rpm, and allow to grow for 16 h.
2. On the day of inoculation, 1 mL of *C. auris* culture is harvested by centrifugation at 2500 × *g* for 5 min in a microcentrifuge.
3. Without disturbing the pellet, remove the medium, and add 1 mL of phosphate-buffered saline (PBS), then resuspend the cells.

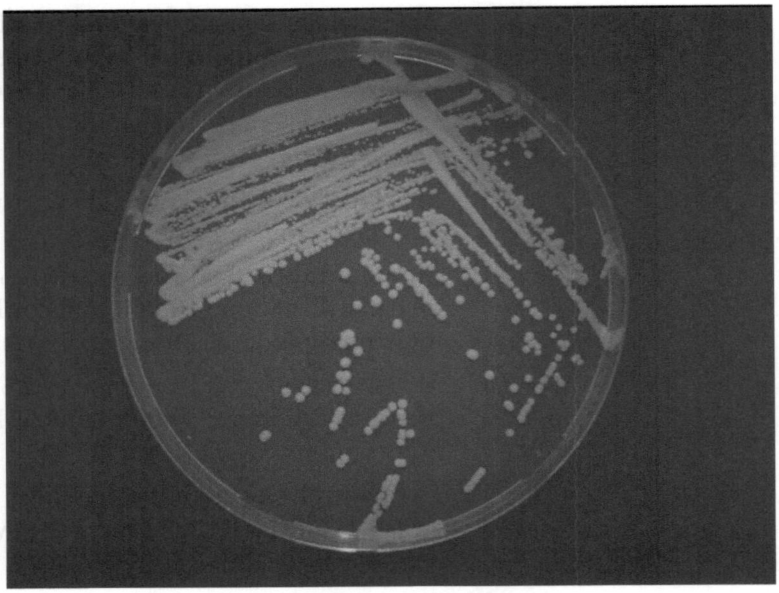

Fig. 1 *C. auris* colonies on a YPD agar growth plate

4. Re-centrifuge the new cell suspension at 2500 × g for 5 min.
5. Remove the supernatant, and resuspend the cells in 1 mL of fresh PBS.
6. Make a 1:100 dilution of the cells, and mount 15 μL of the diluted cell suspension on a hemocytometer slide.
7. Mount the hemocytometer on a light microscope, and determine the *C. auris* cell concentration by counting 5 of the 25 large squares in the center of the hemocytometer.
8. Calculate the *C. auris* colony-forming units (CFU)/mL in the undiluted suspension using the following formula:

$$\text{CFU/mL} = (\text{total cells counted}/5) \times 25 \times 10^{n} \times 10^{4}$$

where $n = \log_{10}(\text{dilution})$, which in this case where the dilution is 1 in 100, $n = 2$.

9. After determining the CFU/mL, dilute the cell suspension appropriately to achieve the desired cell concentration for inoculation.
10. Plate each inoculum produced this way on YPD agar to confirm the inoculum cell concentration. To do this, make five tenfold serial dilutions of each inoculum in a 96-well plate by pipetting 25 μL of the undiluted inoculum into 225 μL of PBS (dilution 1), mix by pipetting, change the pipette tip, then take 25 μL of dilution 1 and add it to 225 μL PBS (dilution 2), mix and change the pipette tip. Continue in the same pattern until five dilutions are made (*see* **Note 3**). Pipette 25 μL of each dilution on to the surface of dry YPD agar in a square Petri

dish, and, without spreading it, allow it to dry. Then, turn the plate upside down, and place in a 30 °C incubator.

11. Once colonies are visible in ~48 h, count the colonies, and calculate the CFU/mL for each inoculum to confirm that the hemocytometer counting was accurate.

3.2 Mouse Model

In the UK, mouse handling and procedures are carried out under Establishment, Project and Personal Licenses issued by the UK Home Office, following ethical approval of the research. Guidelines for the use of experimental animals in research may vary depending on the institution or country where the experiment is conducted (*see* **Note 4**).

1. Mice obtained from commercial suppliers should arrive at least 7 days prior to the experiment start date to allow animals to recover from travel and to acclimatize to their new environment.
2. Five days prior to inoculation with *C. auris*, randomly allocate mice into groups of 6 (*see* **Note 5**), assign each group to an IVC (*see* **Note 6**), and individually mark each mouse (*see* **Note 7**).
3. Change the water to drinking fluid A. This reduces the natural mouse gut microflora. Record the initial weight of each mouse.
4. Three days before inoculation, swap the drinking fluid to drinking fluid B.
5. Record the weight of each mouse.
6. On inoculation day (day 0), record the weight of each mouse.
7. Carefully place a mouse onto a cage lid and gently scruff it. The mouse should be immobilized within the researchers grip and lying flat. Insert the gavage tube into the mouse mouth, and pass over the tongue and into the esophagus. There should not be any resistance as the gavage tube is inserted and as the inoculum is dispensed. Gently gavage 100 μL of the inoculum directly into the mouse stomach. Place the mouse onto the floor of its home cage. Be careful that the mouse does not bite its tongue in discomfort or the tube scratches the back of the mouse's throat (*see* **Note 8**).
8. Mice should be placed in an individual plastic box, weighed on an analytical balance, and monitored each day after inoculation for adverse effects, weight loss, and any evidence of infection symptoms. Mice in this model are not expected to become ill, but mice which show signs of infection (such as decreased water/food consumption, self-isolation from other mice, raised/ruffled fur, decreased activity, twitching, tremor, and narrow or shut eyes) and/or which lose more than 20% of their initial body weight should be culled.

9. Stool samples should be collected every day, after weighing and monitoring. Allow the mouse to settle in its individual plastic box from **step** 7, and make sure it stays inside it for as long as necessary to collect the required stool samples. Do not pressure the mouse in any way, but only gently place it back inside should it attempt to leave the container. For each mouse, collect sufficient stool (0.8–1.3 g) to obtain accurate fungal burden counts. Collect stool samples for several days (at least 7) until stool fungal burdens remain relatively constant and mice have not developed any signs of infection.
10. Boxes used to collect stool samples are sterilized with 70% (v/v) ethanol before and after being used for each individual mouse to avoid cross-contamination.
11. During the experimental time period, change drinking fluid B every 3 days to ensure that the fluid is fresh and uncontaminated and that the mice receive fresh antibiotics.

3.3 End of Experiment: Culling and Dissecting

On the final day of the experiment, mice are humanely culled by a schedule 1 method, e.g., cervical dislocation or inhalation euthanasia using carbon dioxide (CO_2). Cervical dislocation, when performed correctly, causes lower amount of stress to animals compared to CO_2 inhalation and, therefore, is the method used in this protocol [15]. Death is confirmed prior to dissecting the mice.

1. Prior to culling, weigh mice and collect stool samples as described in Subheading 3.2.
2. Place an individual mouse in a cage where other mice cannot witness or hear culling.
3. Gently pick up one mouse and transfer it to the culling area. Ideally this area should be located as far away from the home cage as possible.
4. Scruff the mouse gently to immobilize it, and, in one rapid movement, using an unsharpened long metallic object, apply pressure on the neck and move it toward the base of the skull while pulling on the tail until the spinal cord breaks, essentially severing the link between the brain and spinal cord.
5. Death should be confirmed prior to commencing dissection.
6. Once all mice have been euthanized, position each one on its back, and spray the belly with 70% (v/v) ethanol to create a sterile environment.
7. Using a pair of scissors that has been soaked in 70% (v/v) ethanol, cut through the skin and abdominal wall, and open the body cavity all the way through the diaphragm and ribs in order to open up the chest.

8. Locate the following organs: spleen, liver, kidneys, lungs, heart, and caecum.
9. Remove the organs mentioned above with minimal damage, and place them carefully in a 1.5-mL microcentrifuge tube. In the case of the caecum, use a 7-mL bijou tube instead.
10. Using scissors that have been soaked in 70% (v/v) ethanol, cut through the skull with extreme care, and remove the brain with minimal damage. Place in a 1.5-mL microcentrifuge tube.
11. Dispose the mouse bodies in a double-bagged biohazard bag, and autoclave prior to disposal.
12. Take the organs to the laboratory for analysis.
13. Organs can be stored in 500-μL PBS at 4 °C for a short period of time (up to 48 h), but it is preferred to process them on the day of collection or the next day at the latest.

3.4 Stool and Organ CFU Determination (See Note 9)

Cut each organ in half, saving one half in formalin for histology and one half for fungal burden determination.

To determine fungal burdens:

1. Weigh each organ or stool sample using an analytical balance, and place into a 1.5-mL microcentrifuge tube (caecum is placed into a 7-mL bijou tube).
2. Add 500 μL of PBS into each microcentrifuge tube or 7-mL bijou containing samples (organs or stools).
3. Using a homogenizer, homogenize each sample until a smooth homogenate is obtained.
4. Create 5 or 6 tenfold serial dilutions for each sample in PBS (*see* **Note 3**).
5. Pipette 25 μL of each dilution of a sample onto a dry YPD agar containing chloramphenicol and gentamicin (*see* Subheading 3.1, **step 2**) in a 120 mm × 120 mm square Petri dish. Each dilution forms a single spot on the agar surface.

   Drop each dilution of the same sample next to each other in descending order, allowing enough space so that the spots will not touch or overlap (Fig. 2).

   Allow the spots to dry for about 1 h, turn the Petri dishes upside down, and place them in a 30 °C static incubator. Incubate for 48 h.
6. After incubation, count the number of colonies for each dilution. Use the lowest dilution that has between 20 and 100 colonies. Try to count the colonies in the dilution directly before the chosen dilution in order to confirm that it contains ten times more colonies, therefore confirming accuracy of the dilutions.
7. To calculate the CFU/mL for each sample, use the following formula:

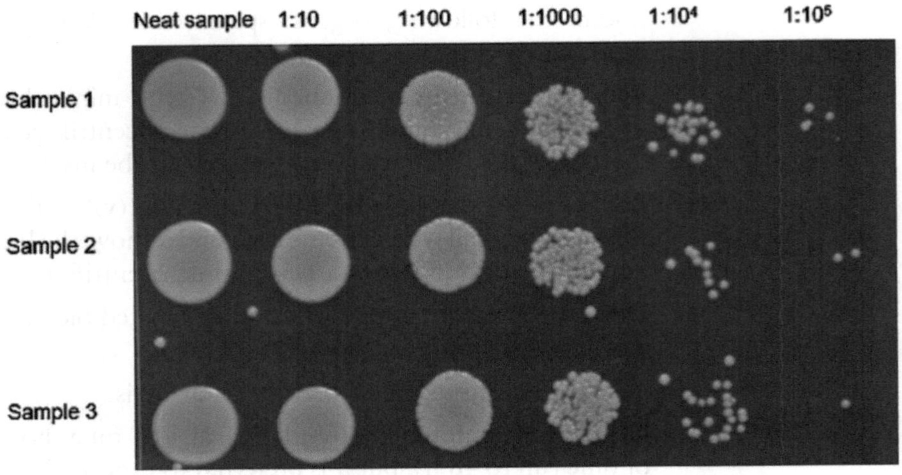

Fig. 2 Serial dilutions of homogenized samples. Samples (25 μL) of each dilution are plated onto the surface of a 120 mm × 120 mm square Petri dish containing 50 mg/L of gentamicin and chloramphenicol. Sufficient space is allowed between each sample row and each dilution column for the colonies to remain distinct and to eliminate the danger of the spots overlapping

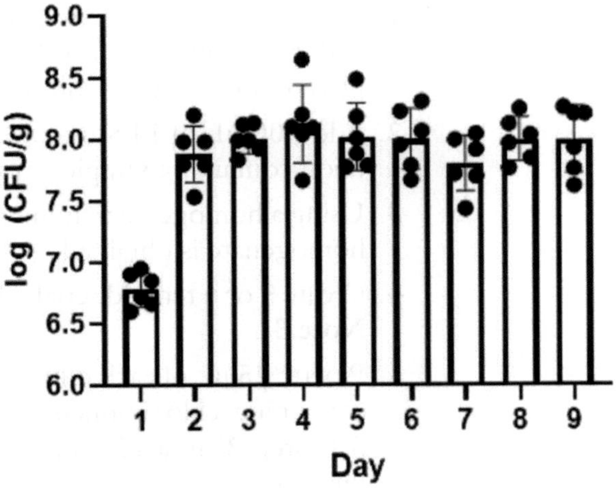

Fig. 3 *C. auris* mouse stool burdens. *C. auris* is persistently detected in mouse stools from day 1 post-inoculation. Mice ($n = 6$) were inoculated with 10^7 CFU of *C. auris* and have stool fungal burdens of 5×10^6–10^7 of CFU/g at 24 h after gavage, a number that increases to ~10^8 CFU/g of stool by 48 h post-inoculation and remains steady until the end of the experiment. The same six mice were sampled on each day of the experiment

$$\text{CFU/mL} = \text{number of colonies} \times 40 \times 10^n$$

where $n =$ dilution number, e.g., 4.

8. To subsequently convert CFU/mL into CFU/g, divide the CFU/mL by the weight (g)/mL (based upon the weight of the organ or stool homogenized in 0.5 mL PBS).
9. Using a statistical program of choice (IBM SPSS or GraphPad Prism is recommended), plot a graph of $\log_{10}$ (CFU/g) against time (Fig. 3).

3.5 Statistical Analysis

Depending on the number of experimental groups and the normality of the data, either unpaired t-test or the non-parametric equivalent Mann–Whitney U test was used to compare two groups or a one-way ANOVA or its non-parametric equivalent Kruskal–Wallis test was used out to compare more than two groups. We recommend the use of the latest versions of either IBM SPSS statistical program or GraphPad Prism program for the statistical analysis of data.

3.6 Reporting Animal Experiments

Experiments using animals should be published adhering to animal research reporting standards, for example, the ARRIVE reporting guidelines [16]. This ensures that there is an accurate description of the experimental design and improves reproducibility of the research. These guidelines are recommended by a wide range of biomedical scientific journals.

4 Notes

1. The choice of mouse strain and sex is not random. Black mice (C57BL/6) show higher tolerance to changes of their drinking fluid compared to white mice (BALB/c), thus making it easier to be certain that the mice will continue drinking after the water has been swapped with water-containing antibiotics and sucrose. Furthermore, female mice are used as domination fighting can occur between male mice. Fighting can result in open wounds on the mouse skin which can impact on both animal well-being and experimental accuracy.
2. Virulence studies typically compare isogenic strains. Therefore, we recommend comparing *C. auris* strains from the same clade as they are highly similar [17], and strains from an outbreak tend to be clonal. However, it has been shown in *C. albicans* that strains from the same clade do not necessarily show similar virulence or disease progression within the host [18].
3. We recommend using a 96-well plate to make dilutions. Add 225 μL of sterile PBS in a number of wells equal to the number of dilutions that need to be performed. Perform one in ten dilutions by pipetting 25 μL of the sample that is to be diluted into the first well (D1) containing 225-μL PBS and mix. Subsequently, pipette 25 μL from D1 into the second well (D2). Proceed likewise until all the dilutions have been completed. For maximum accuracy, it is recommended to mix each dilution by pipetting up and down three or four times before transferring 25 μL into the next well.
4. We recommend following the guidelines for the use of experimental animals in research offered by the National Centre for the Replacement, Refinement and Reduction of Animals in

Research (NC3Rs) [19]. NC3Rs guidelines aim to reduce stress experienced by animals undergoing experimental procedures. An example of this is using a plastic tube to pick up mice to move them to the next procedure area [20]. It is also recommended that this plastic tube is part of the mouse cage environment as this will encourage them to enter the tube as they will be familiar with it.

5. Determining the appropriate sample size (i.e., number of animals per group) is a crucial part of this protocol as it can both save time and determine the minimal number of animals required. For optimal experimental design, we recommend either consulting with a statistician to determine the sample size required for a correctly powered experiment or following the group size determination guidelines published [21]. NC3Rs Experimental Design Assistant (EDA) can also help with determining an appropriate group size [22].
6. Numbering individual mice is crucial for accurate monitoring of mice. To minimize stress for the animals, we recommend marking their tails using a surgical marker. Mice tend to clean the mark, so the ink might need to be renewed weekly.
7. Blinding is important to eliminate experimental bias in the study. We recommend that the person gavaging the mice is blinded to the identity of the inocula and/or a second person handles the mice and collects and analyzes samples. After the experiment has been completed, the blinding should be lifted.
8. Gavaging can be performed with either metal reusable gavage needles or disposable plastic ones. We recommend using disposable needles to minimize the risk of injuring the mouse esophagus. Avoiding such injury ensures both that the mouse suffers the minimum amount of stress and that *C. auris* will not find a way to enter the bloodstream through an open wound, compromising the study. Furthermore, the disposable tube is transparent allowing us to confirm that the inoculum has been successfully administered [23].
9. All samples, stools, or organs should be homogenized and plated on the day they are collected or the next day at the latest. While remaining in the tubes, even if they are in PBS and at 4 °C, there is a risk of cells dying, therefore altering the study results.

Acknowledgments

We thank the University of Aberdeen Medical Research Facility staff, as well as Dr. Ambre F. Chapuis, for their assistance with the mouse experiments. We also thank the A. G. Leventis Foundation for supporting Mr. Stylianos C. Simantirakis during his PhD studies. Dr. Donna MacCallum's research was supported by BBSRC (BB/P02050X/1), European Commission (847507 HDM-FUN), and NC3Rs (NC/S001557/1 and NC/N002482/1).

References

1. Vallabhaneni S, Kallen A, Tsay S et al (2016) Investigation of the first seven reported cases of Candida auris, a globally emerging invasive, multidrug-resistant fungus – United States, May 2013-August 2016. MMWR Morb Mortal Wkly Rep 65:1234–1237
2. PHE (2017) *Candida auris*. https://www.gov.uk/government/collections/candida-auris. Accessed 26 March 2021
3. Ben-Ami R, Berman J, Novikov A et al (2017) Multidrug-Resistant *Candida haemulonii* and *C. auris*, Tel Aviv, Israel. Emerg Infect Dis 23: 195–203
4. Azar MM, Turbett SE, Fishman JA (2017) Donor- derived transmission of *Candida auris* during lung transplantation. Clin Infect Dis 65:1040–1042
5. Schelenz S, Hagen F, Rhodes JL et al (2016) First hospital outbreak of the globally emerging *Candida auris* in a European hospital. Antimicrob Resist Infect Control 5:35
6. Christopher Prestel M, Erica Anderson M, Kaitlin Forsberg M et al (2021) *Candida auris* outbreak in a COVID-19 specialty care unit—Florida, July–August 2020. MMWR Morb Mortal Wkly Rep 70:56–57
7. Chowdhary A, Sharma C, Duggal S et al (2013) New clonal strain of *Candida auris*, Delhi, India. Emerg Infect Dis 19:1670–1673
8. Lee WG, Shin JH, Uh Y et al (2011) First three reported cases of nosocomial fungemia caused by *Candida auris*. J Clin Microbiol 49: 3139–3142
9. Fakhim H, Vaezi A, Dannaoui E et al (2018) Comparative virulence of *Candida auris* with *Candida haemulonii*, *Candida glabrata* and *Candida albicans* in a murine model. Mycoses 61:377–382
10. Huang X, Hurabielle C, Drummond RA et al (2021) Murine model of colonization with fungal pathogen *Candida auris* to explore skin tropism, host risk factors and therapeutic strategies. Cell Host Microbe 29:210–221
11. Armstrong PA, Rivera SM, Escandon P et al (2019) Hospital-associated multicenter outbreak of emerging fungus *Candida auris*, Colombia, 2016. Emerg Infect Dis 25: 1339–1346
12. Escandon P, Chow NA, Caceres DH et al (2019) Molecular epidemiology of *Candida auris* in Colombia reveals a highly related, countrywide colonization with regional patterns in amphotericin B resistance. Clin Infect Dis 68:15–21
13. Koh AY, Kohler JR, Coggshall KT et al (2008) Mucosal damage and neutropenia are required for *Candida albicans* dissemination. PLoS Pathog 4:e35
14. Abe M, Katano H, Nagi M et al (2020) Potency of gastrointestinal colonization and virulence of *Candida auris* in a murine endogenous candidiasis. PLoS One 15:e0243223
15. NC3Rs (2017) Euthanasia. https://nc3rs.org.uk/euthanasia. Accessed 26 April 2021
16. Percie du Sert N, Hurst V, Ahluwalia A et al (2020) The ARRIVE guidelines 2.0: updated guidelines for reporting animal research. PLoS Biol 18(7):e3000410
17. Lockhart SR, Etienne KA, Vallabhaneni S et al (2017) Simultaneous emergence of multidrug-resistant *Candida auris* on 3 continents confirmed by whole-genome sequencing and epidemiological analyses. Clin Infect Dis 64: 134–140
18. MacCallum DM, Castillo L, Nather K et al (2009) Property differences among the four major *Candida albicans* strain clades. Eukaryot Cell 8:373–387
19. NC3Rs/BBSRC/Defra/MRC/NERC/Royal Society/Wellcome Trust (2019) Responsibility in the use of animals in bioscience research: expectations of the major research councils and charitable funding bodies, 3rd edn. NC3Rs, London. https://nc3rs.org.uk/responsibility-use-animals-bioscience-research. Accessed 15 April 2021
20. NC3Rs (2010) Recommended techniques for handling lab mice to reduce anxiety. https://www.nc3rs.org.uk/news/recommended-techniques-handling-lab-mice-reduce-anxiety. Accessed 26 April 2021

21. Charan J, Kantharia ND (2013) How to calculate sample size in animal studies? J Pharmacol Pharmacother 4:303–306
22. NC3Rs The Experimental Design Assistant – EDA. https://www.nc3rs.org.uk/experimental-design-assistant-eda. Accessed 26 April 2021
23. Instech Labs (2020) Stainless steel vs. flexible plastic feeding needles for mice or rats. https://www.instechlabs.com/blog/stainless-steel-vs.-flexible-plastic-feeding-needles-for-mice-or-rats. Accessed 30 March 2021

Index

Alexander Lorenz (ed.), *Candida auris: Methods and Protocols*, Methods in Molecular Biology, vol. 2517, https://doi.org/10.1007/978-1-0716-2417-3,

MIX
Papier aus verantwortungsvollen Quellen
Paper from responsible sources
FSC® C105338

If you have any concerns about our products,
you can contact us on
ProductSafety@springernature.com

In case Publisher is established outside the EU,
the EU authorized representative is:
Springer Nature Customer Service Center GmbH
Europaplatz 3, 69115 Heidelberg, Germany

Printed by Libri Plureos GmbH
in Hamburg, Germany